# Productivity Enhancing Technologies for Horticultural Crops

*BY THE SAME AUTHOR*

- Diseases of Horticulture Crops: Nematode Problems and their Management ISBN: 978-81-7233-543-4
- Integrated Nematode Management in Horticultural Crops ISBN: 978-81-7233-523-6
- Organic Farming for Sustainable Horticulture: Principles and Practices ISBN: 978-81-7233-501-4
- Plant Protection in Horticulture (Vol. 1-3)
- Insect, Mite and Vertebrate Pests and their Management ISBN: 978-81-7233-615-8
- Fungal Diseases and their Management ISBN: 978-81-7233-616-5
- Bacterial and Viral Diseases and their Management ISBN: 978-81-7233-617-2

# Productivity Enhancing Technologies for Horticultural Crops

**P. Parvatha Reddy**

Former Director and Former Head
Division of Entomology & Nematology
Indian Institute of Horticultural Research
Bangalore-560 089

SCIENTIFIC
PUBLISHERS (INDIA)
www.scientificpub.com

*Print: 2021*

*Published by:*
Scientific Publishers (India)
5-A, New Pali Road, P.O. Box 91,
Jodhpur – 342 001, India

E-mail: info@scientificpub.com
Web: www.scientificpub.com



ISBN: 978-81-7233-672-1

Lasertype set : Rajesh Ojha

*Binding at:* Rajasthan Law Book Binding Works, Jodhpur
Printed in India

*Affectionately dedicated to*
*to the memory of Karmayogis*

*Smt. Tangemma N. Patil*
*Dr. Ninganagouda P. Patil*

*for their concern to the people*

# PREFACE

Horticulture in India is fast emerging as a major commercial venture, because of higher remuneration per unit area and the realization that consumption of fruits and vegetables is essential for health and nutrition. Horticultural crops form an important part of wholesome food containing carbohydrates, proteins, minerals, dietary fibers and vitamins in adequate amount. They produce more edible energy and protein per unit area and time than many other food crops. There is an urgent need to increase horticultural productivity in order to provide nutritional security to the fast growing population of the country (increasing at 1.6% annually) which is expected to reach 1331 million by 2020. India plans to increase the production of horticultural crops to 200 million tonnes by 2020 from current level of 125 million tonnes (NHB, 2008).

In India, prospects of increasing production of horticultural crops by increasing land under cultivation are limited. Hence, it is essential to increase productivity of horticultural crops in order to meet the future demand and fulfillment of country's commitment to the sustainable horticultural development with the objective of nutritional security at individual level, poverty alleviation and employment generation through horticulture sector.

Even though India is leading in productivity of some horticultural crops like mango, banana, cauliflower, pea, coconut, cashew nut, ginger, turmeric and black pepper; still there is scope to increase productivity in other horticultural crops compared to other countries.

The average productivity of most horticultural crops in India is low. There is a wide gap between yields obtained and potential yields with improved varieties and technologies. For many crops, yield realized are much less than 50% of the potential. The productivity of horticultural crops is very low on farmers' fields as compared to that of the co-ordinated trials. The productivity in co-ordinated trials is also higher than that of the developed countries. The yield gap between the co-ordinated trials and on-farm trials could be bridged by adopting improved HYV/hybrids and good horticultural practices, thereby achieving the production target of 200 million MT by 2020.

Programmes, therefore, need to be taken up to reduce the yield gap by improving productivity. The main emphasis should be on improving the productivity by use of disease-free, quality planting material of only released and recommended varieties/ hybrids, both in the public and private sector. High density planting by reduction in planting distance or by use of plant growth inhibitors and dwarfing rootstocks as recommended for crops like mango, citrus, apple, banana, pineapple and some temperate fruits will go a long way. Use of protected cultivation under controlled conditions for growing fruits like strawberry, vegetables such as cucumber, cabbage, capsicum, tomato and temperate vegetables in plain areas and high value cut flowers for domestic use and export need to be promoted. The cultivation of crops, which produce higher biomass/unit area/unit time like, banana, pineapple, papaya, potato, sweet potato, tapioca, elephant foot yam should be taken up in areas requiring poverty alleviation and nutritional security.

The information on productivity enhancing technologies in horticultural crops (fruits, vegetables, plantation, spice, tuber, ornamental, medicinal and aromatic crops) is very much scattered. There is no book at present which comprehensively and exclusively deals with the above aspects on horticultural crops. The present book is divided into two parts. The first part deals with the principles of productivity enhancing technologies covering aspects such as use of high yielding varieties/hybrids, high density planting, micro-irrigation, fertigation, integrated nutrient management, protected cultivation, bioregulators, biotechnological approaches, post-harvest management, integrated weed management, integrated pest management, integrated disease management and integrated nematode management. The second part of the book deals with crop-wise productivity enhancing technologies. The book is illustrated with excellent quality photographs enhancing the quality of publication. The book is written in lucid style, easy to understand language along with adoptable recommendations for enhancing the productivity.

In view of greater emphasis being given by the Government of India to horticulture by establishing horticulture mission and exclusive State Horticultural Universities, there is an urgent need for good text books to teach courses on different disciplines of horticulture. In this context, the present book is of immense help to all those concerned with enhancement of productivity in horticultural crops.

This book can serve as a practical guide to practicing farmers of horticultural crops. Further, it is a useful reference to policy makers, research and extension workers and students. The material can also be used for teaching post-graduate and under-graduate courses. Suggestions to improve the contents of the book are most welcome (E-mail: reddy_parvatha@yahoo.com).

The publisher, Scientific Publishers (India), Jodhpur, deserves commendation for their professional contribution.

June 5, 2010

**P. Parvatha Reddy**
34, UAS Layout,
Ist Main, 7th Cross
Sanjay Nagar,
Bangalore-560 094

# CONTENTS

*Section - I*

# PRINCIPLES OF PRODUCTIVITY ENHANCING TECHNOLOGIES

## Chapter 1

# INTRODUCTION

Horticultural crops form an important part of wholesome food containing carbohydrates, proteins, minerals, dietary fibers and vitamins in adequate amount. They produce more edible energy and protein per unit area and time than many other food crops. There is an urgent need to increase horticultural productivity in order to provide nutritional security to the fast growing population of the country (increasing at 1.6% annually) which is expected to reach 1331 million by 2020 (present population – 1170 million as on July 2009) (Table 1.1). In India, every year we add one Australia to our population needing additional 4 to 5 million MT of food grains. We have 17.5% of total population sustaining on only 2.4% of global land. It is anticipated that we may even surpass the population of China by 2020.

**Table 1.1.** Percentage increase in population since 1950

| Period | % increase in population |
|---|---|
| 1951-61 | 21.50 |
| 1961-71 | 24.80 |
| 1971-81 | 24.70 |
| 1981-91 | 23.90 |
| 1991-01 | 21.34 |

## 1.1. Growth performance of horticultural sector

The stock of agricultural land will diminish and as a consequence the arable land area/person which was 0.3 ha in 1981 will decline to 0.13 ha in 2050. Since the available land is likely to remain constant, there is a need to accelerate our productivity growth in horticulture to achieve nutritional security on a long term basis. Horticulture must become economically viable and ecologically sustainable to support crop productivity.

Among the horticultural crops, fruits recorded a three-fold increase in area and four-fold increase in production over the last five decades. India is the largest producer of mango, banana, cauliflower, pea, coconut, cashew nut, ginger, turmeric, and black pepper; and the second largest producer of fruits and vegetables.

In 1991-92, India produced 28.6 million tonnes of fruits from 2.87 million hectares of land and 58.74 million tonnes of fruits from 5.50 million hectares of land in 2005-06 recording the increase of 105.17 per cent in production and 91.97 per cent in area, respectively (Table 1.2). In case of vegetables in the same period the increase in area and production was observed to be 92.12 and 147.60 per cent, respectively (Table 1.3). In the reported period under major spices, the increase in area and production was recorded to be 6.98 and 83.73 per cent, respectively (Table 1.4).

**Table 1.2.** Area and production of fruit crops

| Year | Area ('000 Ha) | Production ('000 MT) | Productivity (MT/ha) | % increase over 1991-92 | | |
|---|---|---|---|---|---|---|
| | | | | Area | Production | Productivity |
| 1991-92 | 2870 | 28630 | 9.97 | - | - | - |
| 1998-99 | 5140 | 58530 | 11.17 | 79.09 | 104.44 | 12.03 |
| 2005-06 | 5509.6 | 58740.3 | 10.66 | 91.97 | 105.17 | 6.92 |

**Table 1.3.** Area and production of vegetable crops

| Year | Area ('000 Ha) | Production ('000 MT) | Productivity (MT/ha) | % increase over 1991-92 | | |
|---|---|---|---|---|---|---|
| | | | | Area | Production | Productivity |
| 1991-92 | 3729 | 44042 | 11.81 | - | - | - |
| 1998-99 | 5870 | 87530 | 14.91 | 57.41 | 98.74 | 26.24 |
| 2005-06 | 7164 | 109050 | 15.22 | 92.12 | 147.60 | 28.87 |

**Table 1.4.** Area and production of spice crops

| Year | Area ('000 Ha) | Production ('000 MT) | Productivity (MT/ha) | % increase over 1991-92 | | |
|---|---|---|---|---|---|---|
| | | | | Area | Production | Productivity |
| 1991-92 | 1618 | 1211 | 0.75 | - | - | - |
| 1998-99 | 1751 | 1884 | 1.07 | 8.22 | 55.57 | 29.33 |
| 2005-06 | 1731 | 2225 | 1.28 | 6.98 | 83.73 | 70.66 |

### 1.1.1. Growth in major fruit crops

A perusal of the Table 1.5 indicates a very disappointing picture with respect to growth rate of productivity in different perennial fruit crops. The yield of mango declined by 53.4 per cent, followed by citrus (46.4 per cent) and guava (18.8 per cent) from 1970 to 2006. There has been negative growth in the productivity of mango, citrus and guava fruit crops whereas the productivity of banana and papaya registered a positive growth rate of 2.94 and 1.80 per cent, respectively (Table 1.5).

**Table 1.5.** Growth rate in area, production and yield in major fruit crops in India during 1970-2006

| Growth Rate | Banana | Papaya | Citrus | Mango | Guava |
|---|---|---|---|---|---|
| Area (Per cent/annum) | 2.79 | 5.97 | 6.53 | 4.12 | 2.36 |
| Production (Per cent/annum) | 5.72 | 7.54 | 4.44 | 1.34 | 1.92 |
| Yield (Per cent/annum) | 2.94 | 1.80 | -2.12 | -2.77 | -0.44 |

The contribution of productivity is negative in all the perennial fruits, whereas in case of banana and papaya yield effect is positive but low. Thus, in case of all the fruit crops, the predominant factor responsible for increasing the production is area but not the productivity (Table 1.6).

**Table 1.6.** Area effect and yield effect on production growth of major fruit crops in India during 1970-71 to 2006-07

| Particulars | Banana | Papaya | Citrus | Mango | Guava |
|---|---|---|---|---|---|
| Area effect (Per cent) | 31.00 | 58.00 | 206.00 | 437.00 | 144.00 |
| Yield effect (Per cent) | 27.00 | 6.00 | -15.00 | -88.00 | -17.00 |

### 1.1.2. Growth in major vegetable crops

During the period 1970 to 2004, the increase in area under potato was observed approximately three times followed by onion (2.6 times). The production of potato and onion increased approximately five times and three times, respectively. The yield of potato increased 1.8 times followed by onion 1.13 times only. The growth of productivity of onion is very low (Table 1.7).

**Table 1.7.** Growth rate in area, production and yield in major vegetable crops in India during 1970-2006

| Particulars | Potato (1970-71 to 2004-05) | | | Onion (1978-79 to 2004-05) | | |
|---|---|---|---|---|---|---|
| | Area | Production | Yield | Area | Production | Yield |
| Growth Rate (Per cent/ annum) | 2.99 | 4.98 | 1.99 | 3.19 | 3.42 | 0.23 |

### 1.1.3. Growth in major spices

The area doubled in case of black pepper whereas in turmeric area increased 1.9 times followed by coriander (1.8 times). In chilli and cardamom, the area is almost same from 1970 to 2003. The area registered the maximum positive growth of 2.71 per cent in black pepper followed by turmeric (2.56 per cent). The production of coriander increased approximately 4 times followed by turmeric and black pepper (3.5 times), cardamom (3.3 times) and chilli (2.4 times). The production registered highest positive and significant growth rate of 6.01 per cent in turmeric followed by cardamom (4.47 per cent), black pepper (3.97 per cent), coriander (2.83 per cent) and chilli (2.54 per cent). The productivity increased 4 times in cardamom and more than 2 times in chilli and coriander followed by turmeric (1.9 times) and black pepper (1.4 times). In case of all spices the yield registered positive growth rate. It registered highest growth in cardamom (3.77 per cent) followed by turmeric (3.47 per cent), chilli (2.03 per cent), coriander (1.34 per cent) and black pepper (1.31 per cent) (Table 1.8).

**Table 1.8.** Growth rate in area, production and yield in major spices in India during 1970-2003

| Growth Rate | Turmeric | Chilli | Black pepper | Cardamom | Coriander |
|---|---|---|---|---|---|
| Area (Per cent/annum) | 2.56 | 0.49 | 2.71 | 0.61 | 0.93 |
| Production (Per cent/annum) | 6.01 | 2.54 | 3.97 | 4.47 | 2.83 |
| Yield (Per cent/annum) | 3.46 | 2.03 | 1.31 | 3.77 | 1.34 |

In case of turmeric the contribution of area and yield effect is almost equal. However, in case of chilli, cardamom and coriander yield effect is more than the area effect. In case of black pepper, area effect is more (52.95 per cent) than the yield effect (Table 1.9).

**Table 1.9.** Area effect and yield effect on production growth of major spices in India during 1970-71 to 2003

| Particulars | Turmeric | Chilli | Black pepper | Cardamom | Coriander |
|---|---|---|---|---|---|
| Area effect (Per cent) | 33.46 | 10.22 | 52.95 | 7.10 | 15.85 |
| Yield effect (Per cent) | 33.93 | 81.61 | 23.26 | 74.74 | 68.75 |

India plans to increase the production of horticultural crops to 200 million tonnes by 2020 (GOI, 2002) from current level of 125 million tonnes (NHB, 2008).

It is estimated that to fulfill the increased demand for providing the nutritional security, the production of high value horticultural crops in South Asia has to be increased by 142 per cent by 2020.

## 1.2. Gap in potential and realized yields

Even though India is leading in productivity of some horticultural crops like grapes, cassava, ginger and turmeric, still there is scope to increase productivity in other horticultural crops compared to other countries (Table 1.10).

**Table 1.10.** Productivity of major horticultural crops in India *vis-à-vis* other countries

| Horticultural crop | National (MT/ha) | World highest (MT/ha) | Country | Increase over India (%) |
|---|---|---|---|---|
| Banana | 34.00 | 52.54 | Costa Rica | 54.50 |
| Mango | 6.75 | 12.50 | Brazil | 85.00 |
| Citrus | 8.40 | 25.00 | Brazil | 297.60 |

| | | | | |
|---|---|---|---|---|
| Grape | 25.69 | 25.69 | India | --- |
| Apple | 7.68 | 84.01 | Austria | 993.88 |
| Pear | 5.59 | 283.19 | Austria | 4966.01 |
| Potato | 17.05 | 43.87 | Netherlands | 157.00 |
| Tomato | 17.08 | 26.69 | USA | 70.45 |
| Eggplant | 16.08 | 17.48 | Japan | 34.70 |
| Chilli | 9.18 | 14.40 | Spain | 44.5 |
| Onion | 13.41 | 51.19 | USA | 73.80 |
| Garlic | 4.93 | 17.71 | USA | 72.20 |
| Okra | 9.59 | 6.47 | Jordan | 17.78 |
| Peas | 9.14 | 8.35 | Lithuania | 20.00 |
| Coconut (copra) | 0.51 | 0.61 | Indonesia | 20.00 |
| Black pepper | 0.315 | 2.95 | Malaysia | 836.00 |
| Cardamom | 0.175 | 0.25 | Guatemala | 42.8 |
| Ginger | 3.47 | --- | India | --- |
| Turmeric | 3.92 | --- | India | --- |
| Cassava | 31.44 | 31.44 | India | --- |
| Sweet potato | 8.94 | 24.24 | Japan | 171.00 |

In India, prospects of increasing production of horticultural crops by increasing land under cultivation are limited. Hence, it is essential to increase productivity of horticultural crops in order to meet the future demand and fulfillment of country's commitment to the sustainable horticultural development with the objective of nutritional security at individual level, poverty alleviation and employment generation through horticulture sector.

## 1.3. Potential for enhancing productivity

### 1.3.1. Vegetable crops

The per capita availability of vegetables (210 g/caput/day) is still behind the recommended quantity (285 g/caput/day). To meet the per capita requirement of vegetables at the recommended level of 285 g/caput/day, the vegetable production has to be increased substantially to match the population growth rate of 1.8% per year. To provide the balanced diet to Indian population, the production level has to be increased to 200 million MT of vegetables by 2020 as against approximately 125 million MT of present vegetable production. Hybrid vegetable technology is one of the better options particularly at this juncture because of the fact that growing of vegetables is four times more remunerative than cereals. Hence in coming decade, hybrid vegetable technology has to go a long way in this country to meet challenges.

There is significant gap in the national average productivity of different vegetable crops as compared to the world average productivity as well as the potential productivity for the respective vegetable crops (Table 1.11). Thus there is a vast scope for improving the productivity of these crops at national level. The critical gap between the experimental yield and national average can be minimized by a strong extension service, training and demonstration. There is an urgent need to recast the vegetable production strategy so that vegetable production may get impetus. Further for enhancing the vegetable production in the country, our target should be to achieve 25 MT/ha productivity by 2020, which is presently around 16 MT/ha.

**Table 1.11.** Vegetable productivity scenario (MT/ha)

| Vegetable crop | India (Average) | World (Average) | Potential productivity | Maximum productivity |
|---|---|---|---|---|
| Tomato | 14.07 | 26.69 | 60-80 | 70.45 (USA) |
| Eggplant | 16.08 | 17.48 | 40-50 | 34.7(Japan) |
| Chilli | 9.18 | 14.4 | 30-40 | 44.5 (Spain) |
| Okra | 9.59 | 6.47 | 15-20 | 17.78 (Jordan) |
| Peas | 9.14 | 8.35 | 18-20 | 20 (Lithuania) |
| Melons | 20.48 | 20.95 | 30-40 | 45.83 (Cyprus) |
| Cucurbits | 9.72 | 12.97 | 25-30 | 41.33 (Israel) |
| Cucumber | 6.67 | 16.98 | 40-50 | 67.67 (Korea) |
| Watermelon | 12.75 | 27.13 | 30-40 | 40.96 (Spain) |
| Cabbage | 21.43 | 21.1 | 30-40 | 42.59 (Japan) |
| Cauliflower | 17.14 | 18.36 | 35-40 | 45.25 (New Zealand) |
| Onion | 10.38 | 17.53 | 40-50 | 60.33 (Korea) |
| Garlic | 4.17 | 12.37 | 15-20 | 23.23 (Egypt) |

### 1.3.2. Spice crops

Considerable increase in productivity has been reported from research stations especially in black pepper, turmeric and coriander (Table 1.12). The effective transfer of technology attempts should be made to identify the main production constraints and to workout appropriate and economic strategies to overcome the same (Peter, 1999).

**Table 1.12.** Potential for productivity increase in spice crops (kg/ha)

| Spice crop | National | Progressive farmer | Research station | Abroad | Scope for increase |
|---|---|---|---|---|---|
| Black pepper | 315 | 2000 | 2445 | 2925 (Malaysia) | 1500 |
| Cardamom | 154 | 1625 | 450 | 250 (Guatemala) | 1000 |
| Ginger | 3477 | 5500 | 8250 | --- | 2000 |
| Turmeric | 3912 | 6200 | 10700 | --- | 2000 |
| Coriander | 591 | --- | 1900 | 515 (Morocco) | 1000 |
| Cumin | 578 | --- | 2000 | --- | 500 |

| | | | | | |
|---|---|---|---|---|---|
| Clove | 400 | --- | 1100 | --- | 700 |
| Nutmeg | 600 | --- | 885 | --- | 250 |
| Cinnamon | 200 | --- | 400 | --- | 200 |
| Fennel | 1800 | --- | 2500-3000 | --- | 700 |
| Fenugreek | 600-800 | --- | 1500-2000 | --- | 700 |

## 1.4. Interventions needed for enhancing productivity

It has been observed that in most of the horticultural crops the productivity gap between national level and leading states is more than 100% as compared to demonstration plots or experimental station yields. Major focus on research should be devoted to productivity enhancement technologies like development of high yielding varieties/F1 hybrids, use of micro-irrigation, integrated nutrient management, fertigation, protected cultivation of ornamental and vegetable crops, use of biotechnological approaches, use of bioregulators, post-harvest management, integrated weed, pest, disease and nematode management. This calls for efficient transfer of technology and training of the stake holders both private and public coupled with providing appropriate inputs, technical knowledge and infrastructure facilities.

## 1.5. Nutritional security

In the world, about 815 million people suffer from hunger and malnutrition, mostly in the developing world. Nearly one-third of world's hungry people (233 million) are homed in India. In India, around 87.5% pregnant women and 56% pre-school children suffer from anemia.

The prosperity and advancements of a nation depend largely on achieving nutritional security to its citizens. By and large the entire world points at horticultural crops particularly fruits and vegetables, easily available, as the best means of providing nutritional security. Horticultural crops are the major, cheap and sustainable source of fibre, carotenoids, vitamin C, folate, potassium, iron, calcium, zinc and other vitamins, minerals and bioactive compounds which prevent degenerative processes. There is ample evidence that diets high in fruits and vegetables protect against cancer, coronary artery disease, cataract, diabetes through its antioxidants, micronutrients and phytochemical composition. Vegetables such as amaranthus, parsley, celery, carrot, spinach, water melon and tomato are rich in carotenoid contents (precursor of vitamin A). Green leafy vegetables are the abundant source of mineral, vitamin and fibre contents.

Hence, it is essential to increase productivity of horticultural crops in order to meet the future demand and fulfillment of country's commitment to the sustainable horticultural development with the objective of nutritional security at individual level, poverty alleviation and employment generation through horticulture sector.

Chapter 2

# HIGH YIELDING VARIETIES AND F1 HYBRIDS

During the last six decades (since 1950), over 1500 improved varieties/ hybrids of about 130 horticultural crops have been developed. As a result, productivity of banana and potato has gone up three times each and cassava two times. In the last four years itself, India has recorded 8% increase in potato production. Regular-bearing mango hybrids, export quality grapes, multiple-disease resistant vegetable hybrids, high-value spices and tuber crops of industrial uses have been developed. Tomato varieties resistant to leaf curl virus, bacterial and Fusarium wilts have been developed. Improved varieties/hybrids have revolutionized the horticulture sector. For instance, high-yielding varieties of sweet potato viz. Gauri, Sankar and Sree Bhadra have brought a revolution by minimizing malnutrition, improving nutritional security in the Kandhamal district of Orissa state.

The per capita availability of vegetables has increased from 76 g in 1951 to 256 g in 2001 as against the minimum dietary requirement of 285 g/day/ person recommended by Indian Council of Medical Research (ICMR) for a balanced diet. The vegetable production target for 2020 has been fixed at 200 million MT.

## 2.1. High yielding varieties

Vegetable production is still dominated by the locally available genotypes or inferior land races all over the country. It is estimated that under eggplant approximately 32.2% area, cauliflower 46.71%, chilli 60%, gourds 77.56%, melons 70%, okra 14.62% and tomato 18.49% area are under unidentified local varieties. Thus there exists a need to replace the local cultivars with the improved high yielding and disease resistant varieties.

As a result of multi-disciplinary, multi-location testing of new research materials during the last two decades, 119 improved varieties in 16 major vegetable crops have been identified and recommended for cultivation in various agro-climatic regions of the country (Table 2.1).

**Table 2.1.** Improved varieties of vegetable crops

| Vegetable crops | Varieties released |
|---|---|
| **Solanaceous crops** | |
| Brinjal | Pusa Anupama, Pant Samrat, RU-IC, ARU-2C, PH-4, Pusa Kranti, Pusa Azad Kranti, H-7, Purple Cluster, Punjab K-202-9, NDB 25, Pusa Barasti, Pusa Purple Long, Jamuni Gole Baingan, Arka Navneet, BB-7, Pant Rituraj, BWR-12, H-8, Pusa Hybrid-6, T-3, Aruna, Arka Kusumakar, Annamalai, Azad B-1, CO-1, Gujarat Brinjal-6, Junagadh Oblong, Kalianpur Type 3, MDU-1, Punjab Neelam, Sada Bahar Baingan, Punjab Moti, BH-1, BH-2 |
| Tomato | HS-101, Pusa Early Dwarf, BT-1,CO 3, KS-2, Lanate, Pusa Gaurav, S-12, Bonita, Punjab Chhauhhara, Punjab Kesri, Sel-7, Arka Vikas, Pant Bahar, Pant T-3, Arka Saurabh, Pusa Ruby, Sel-120, Sioux, Pant T-2, Azad T-2, Kalianpur Angoorlata, Kalianpur-1, NTDR-1, PKM-1, Punjab Tropic, PNR-7, TH 2312, TH 802. |
| Chillies | Chanchal, CO-1, CO-2, DH-76-6, Gujarat Chilli, K-1, MDH-1, Pant C-1, Sindhur, CH-1, Andhra Jyoti, J-218, K-2, LDC-206, Bhagyalakshmi, Musalwadi, Pusa Jwala, Sel-IX-235. |
| **Cole crops** | |
| Cabbage | Pusa Mukta. |
| Cauliflower | Early Kunwari, Pusa Early Synthetic, Pusa Deepali, 235-S, Pusa Synthetic, Improved Japanese, Pant Shubhra, Pusa Shubra, Pusa Snowball-1, Snowball-16, Pusa Snowball-2, Pusa Snowball-K-1. |
| **Leguminous crops** | |
| Cowpea | Pusa Komal. |
| Dolichos | Deepaliwal. |
| French Bean | VL-Boni, Arka Komal, Pant Anupma. |
| Pea | Arkel, Jawahar Matar-4, Early December, PM-2, Mattar Agata-6, JP-4, Jawahar Matar-1, Bonneville, Lincoln, VL-3, P-88, PRS-4, Pant Uphar. |
| **Cucurbitaceous crops** | |
| Muskmelon | Hara Madhu, Arka Jeet, Arka Rajahans, Pusa Sharbati, Punjab Madhu, M-3, Pusa Sunheri, Madhuras. |
| Pumpkin | Arka Chandan, Ambali, Arka Suryamukhi, Pusa Vishwas. |
| Watermelon | Arka Manik, Sugar Baby, Arka Jyoti, Durgapura Meetha. |
| **Bulb crops** | |
| Garlic | Jamuna Safed, Agri-Found White. |
| Onion | Agrifound Dark Red, Arka Kalyan, Arka Niketan, N-2-4-1, N-257-9-1, Pusa Madhavi, Punjab Selection, Punjab Red Round, Pusa Ratnar, Punjab Naroya, Pusa Red, VL-3, ALR, Pusa White Flat, Pusa White Round, PB-48. |
| **Root crops** | |
| Carrot | Pusa Yamdagni. |
| **Malvaceous crops** | |
| Orka | P-7, Parbhani Kranti, Arka Abhay, Arka Anamika. |

## 2.2. F1 Hybrids

F1 hybrids are well known for increasing vegetable production due to their high yield potential, earliness, quality and resistance attributes. Heterosis breeding in vegetable crops in India has received serious attention only in recent years (Table 2.2). The first F1 hybrid of tomato (Karnataka Hybrid) and capsicum (Bharat) were released for commercial cultivation in 1973 by a private seed company M/s Indo-American Hybrid Seeds followed by 28 other hybrids in nine vegetable crops.

**Table 2.2.** Heterosis for % increase in yield of vegetable crops

| Vegetable crop | Heterosis for yield (%) | Inbreeding depression |
|---|---|---|
| Asparagus | 64-140 | Severe |
| Cabbage | 12-60 | High |
| Broccoli & cauliflower | 40-90 | High |
| Carrot | 25-40 | High |
| Onion | Up to 80 | High |
| Cucumber | More than 100 | High |
| Tomato | 15-50 | Not high |
| Brinjal | Up to 75 | High |
| Chillies | 10-60 | High |

F1 hybrids have been developed in tomato, brinjal, cabbage, cauliflower, capsicum, chilli, cucumber, carrot, muskmelon and watermelon (Table 2.3). Presently, Karnataka, Maharashtra, Gujarat, Andhra Pradesh, Uttar Pradesh and Madhya Pradesh are leading producers of hybrid tomato. The entire cabbage production in southern parts of Maharashtra and West Bengal is under F1 hybrids.

**Table 2.3.** List of F1 hybrids in different vegetable crops.

| Vegetable crops | F1 Hybrids |
|---|---|
| Tomato (63) | Pusa Divya, Pusa Hybrid 1, Pusa Hybrid 2, Pusa Hybrid 4, Arka Vishal, Arka Varadan, Arka Abhijit, Arka Shreshta, Karnataka Hybrid, Vaishali, Rupali, Mangala, Naveen, Apoorva, Abhimanyu, Mruthyunjaya-2, Asha, MTH 1, MTH 2, MTH 3, MTH 4, MTH 6, MTH 15, MTH 16, Pant Hybrid 2, Pant Hybrid 10, KT 1, KT 2, KT 3, KT 4, HG 9, HG 12, HG 18, Sonali, Gulmohar, Sadabahar, Avinash, IAHS 88-1, IAHS 88-2, IAHS 88-3, Rajani, Rashmi, Sheetal, Coth 1, NS 815, NS 812, NS 1094, NS 816, NS 536, NS 535, NS 2535, # 5130, # 4130, # 1237, Utsav, Amogh, Swaraksha, DRD 813, ARTH-3, ARTH-4, NA 601, NA 501, BSS 20. |
| Brinjal (26) | Arka Navaneeth, Vijay, Pusa Anmol, Pusa Kranti, Pusa Ankur, Suphal, Azad Hybrid, Pusa Hybrid 5, Pusa Hybrid 6, Pusa Hybrid 9, MHB 1, MHB 2, MHB 9, MHB 10, MHB 20 (Kalpatharu), MHB-39, Sumex, NDBH 1, NDBH 2, NDBH 6, |

| | |
|---|---|
| | Phule Hybrid 2, Phule Hybrid 13, Phule Hybrid 14, ABH-1, ABH-2, ARBH-201. |
| Chilli (35) | Krishna, Festival, Kiran, INDM 5, Agni, Bijli, Punjab F1 Hybrid, Tejaswini, Sujata, Surya, Tej, NS 101, NS 1101, NS 1701, NS 555, NS 999, HOE 888, HOE 808, NPH 70, NPH 120, Sankranti, Soldier, Vardhan, ARCH 001, ARCH 006, ARCH 236, BSS 138, BSS 141, BSS 273, Wonder Hot, Guntur Hope, Kodian Hot, US 635, US 515. |
| Capsicum (9) | Pusa Deepti, Bharath, Early Bounty, Green Gold, IAHS P-2, KT- F1 Hybrid, NS 436, NS 625, Sel-2. |
| Muskmelon (14) | Pusa Rasraj, Punjab Hybrid, Durgapura Hybrid, Swarna, Sona Madhubala, MH-10, MHC-5, MHC-6, MHC-2, MMCS 5, MMCS 6, M3, M4, NS 7455, Abhijit. |
| Watermelon (37) | Arka Jyothi, Pusa Bedan (seedless), Amrut, Madhu, Milan, MHW 4, MHW 5, MHW 6, MHW 11, MHW 15, Nath 101, Nath 102, Santrupti, Century No. 2, Mohini, NS 295, NS 246, NS 1004, NS 321, NS 2332, NS 251, NS 252, NS 253, NS 271, NS 272, NS 281, NS 262, NS 263, NS 291, NS 261, NS 221, NS 211, NS 1032, Suruchi, Kushboo, Madhubala. |
| Cucumber (5) | Pusa Samyog, Priya, Superior Slicer, Liberty, Aman. |
| Bottle gourd (15) | Pusa Megadoot, Pusa Manjari, Pusa Naveen, Pusa Hybrid-3, NS 422, NS 421, Kaveri, Gutka, Harit, Warad, New Swathi, CBH-4, # 204, # 1720, MGH-1. |
| Bitter gourd (10) | Pusa Hybrid-1, NS 433, NS 431, NS 432, MBTH-101, MBTH-102, # 49, # 711, Vivek, Tijarti. |
| Ridge gourd (8) | Sureka, Rohini, Gaurav, NS 3, NS 401, NS 402, NS 403, Sanjeevini. |
| Sponge gourd (5) | Harita, Utsav, MSGH-1, NS 441, NS 442. |
| Snake gourd (1) | MDU 1. |
| Summer squash (1) | Pusa Alankar. |
| Pumpkin (3) | Pusa Hybrid-1, MBTH 101, MBTH 102. |
| Okra (23) | Adhunik, Panchali, Nath Shobha, Supriya, Sun Gro 35, # 1, # 7, # 8, # 10, # 11, CO 2, CO 3, Varsha, Vishal, Vijay, NS 501, NS 502, NS 503, IIVR-10, DVR-1, DVR-2, DVR-3, US-7109. |
| Cabbage (32) | Pusa Synthetic, Pusa Mukta, Ganga, Kaveri, Yamuna, Green Express, Hari Rani Gol, Green Boy, Green Gold, Gloria, NI 40, Ganesh Gol, NS 22, NS 25, NS 27, Nath Laxmi 401, Bejo Sheetal 32, # 8, Nath 501, Nath Indu, Quisto, BSS-16, BSS-50, BSS-32, KGAT-1, H-44, H-64, NL-138, Maharani, Kranti, Sel 507-4, Sel-528, Sel 507-22. |
| Cauliflower (9) | Pusa Synthetic, Pusa Early Synthetic, Himani, Swathi, Pusa Hybrid 2, NS 60, NS 60N, NS 84, NS 90. |
| Broccoli (1) | NS 50. |
| Gherkin (1) | Calypso. |
| Turnip (2) | Pusa Swamima, Pusa Kanchan. |
| Onion (2) | Arka Kirtiman, Arka Lalima. |
| Carrot (1) | NS 10. |

Growing of hybrid vegetables has resulted in increased yields, better transportation capacity and good shelf life. Today hybrid acceptance in tomato, cabbage, capsicum, watermelon, chilli, cauliflower, okra, brinjal and gherkins is very high.

In vegetables, several F1 hybrids have been developed and the productivity has increased from 6.5 MT/ha to 16.0 MT/ha. In tomato, OP cvs. yield about 30 to 40 MT/ha, while F1 hybrids yield 70 to 80 MT/ha.

Hybrid varieties are very remunerative in tomato, cabbage, cauliflower, capsicum, okra and chillies. But in low value crops like brinjal, gourds, melons and cucumber they are only moderately economical.

Area under cabbage and tomato has increased rapidly as a result of increased use of F1 hybrids. In tomato and cabbage, adoption of hybrid technology tops the list covering an area slightly above 30% in each. There has been steady increase in hybrid coverage in brinjal, okra and chillies. Imported hybrids are exclusively grown (more than 90%) in cabbage. With respect to other vegetables like chilli, brinjal, gourds, melons, cucumber and okra, most of the hybrids grown are indigenous.

Looking back at the performance and adoption of vegetable hybrids during the past 25 years, there is no doubt that productivity has shown an upward trend especially in tomato and cabbage. What is now required is cost-effective technology to promote widespread adoption of vegetable hybrids. Resistance breeding should be integrated with hybrid technology to boost productivity of vegetables.

However, it is estimated that presently only about 10% area of vegetable is under the hybrids of which tomato covers 36%, cabbage 30%, brinjal 18%, okra 7%, melons and gourds each 5%, cauliflower 2% and chilli 1% (Table 2.4). It is expected that in coming five years period there will be more than 25% area, and after ten years hybrids will occupy more than 50% area, which will substantially contribute in enhancing the vegetable production of the country. Hybrids resistant to multiple diseases and pests are being developed in tomato, brinjal, chilli, capsicum, okra, muskmelon, watermelon, cucumber, bitter gourd, carrot, radish, etc.

**Table 2.4.** Estimated share of hybrid, OP and local or unidentified varieties in total cropped area in India (1997-98).

| Crop | Total area (ha) | Area under F1 hybrids (ha) | Share of area (%) | Share of OP varieties (%) | Share of unidentified varieties (%) |
|---|---|---|---|---|---|
| Brinjal | 479095 | 85300 | 17.80 | 50.00 | 32.2 |
| Cabbage | 242140 | 76000 | 31.39 | 68.61 | 0.0 |
| Cauliflower | 304156 | 10000 | 3.29 | 56.00 | 46.71 |
| Chilli | 573529 | 14000 | 2.24 | 37.56 | 60.00 |
| Gourds | 409270 | 10000 | 2.44 | 20.00 | 77.56 |
| Melons | 169350 | 6800 | 4.02 | 25.98 | 70.00 |
| Okra | 371665 | 20000 | 5.38 | 80.00 | 14.62 |
| Tomato | 485520 | 153000 | 31.51 | 50.00 | 18.44 |

## 2.3. Resistant varieties

Research on breeding for disease/pest resistance has resulted in the release of several varieties. 'Pusa Sawani' variety of okra developed as resistant to yellow-vein-mosaic virus is the first example of successful disease resistance breeding in vegetable crops in India. Okra hybrids of Indo-American Hybrid Seeds, Bangalore such as Varsha, Vijay and Vishal are resistant to YVMV. In case of indigenous hybrids of brinjal, gourds, melons and cucumber, more area can be covered if they carry additional attributes like disease/pest resistance or any special quality attribute.

Several varieties have also been released which are resistant to abiotic stresses such as heat and moisture. A list of varieties of different crops resistant to abiotic stresses released so far is given in Table 2.5.

**Table 2.5.** Vegetable crops varieties resistant to major abiotic stresses.

| Vegetable crop | Abiotic stress | Resistant varieties |
|---|---|---|
| Tomato | Moisture | Arka Meghali |
| | Cold set | Pusa Sheetal |
| | Hot set | Pusa Hybrid-1 |
| | Hot & cold set | Pusa Sadabahar |
| Chilli | Moisture | Arka Lohit |
| Field bean | Moisture | Arka Jay, Arka Vijay, Konkan Bushan |
| Cowpea | Moisture | Arka Garima |
| French bean | Heat tolerant | Arka Komal |
| Cluster bean | Moisture | Pusa Nav Bahar, Pusa Sada Bahar |
| Lima bean | Moisture | IIHR Sel-1, IIHR Sel-4 |
| Round melon | Heat tolerant | Arka Tinda, Punjab Tinda |
| Long melon | Heat tolerant | Arka Sheetal, Punjab Long melon |
| Bottle gourd | Heat tolerant | Pusa Summer Prolific Long |
| Bitter gourd | Heat tolerant | Pusa Do Mousami, Kalyan Sona |
| | Warm humid climate | Arka Harit, Coimbatore Long, Konkan Tara, Priya, CO-1, MC-84, MDU-3 |
| Cabbage | High temp. tolerant | Pusa Ageti |
| Cauliflower | Curd devp. in May in lower hills | Pusa Himajyoti |
| Turnip | Hot & humid climate | Pusa Sweta |
| Radish | High temp. tolerant | Pusa Chetki |
| | Low temp. tolerant | Pusa Himani |
| Carrot | Temperate type, bolting & seed setting under high temp. | Pusa Meghali |
| Palak | Not bolting in plains | Pusa Harit |
| Turmeric | Tolerant to draught | CO-1, BSR-1 |

## 2.4. Nutrition rich varieties

More than 800 million people worldwide are thought to be malnourished. In the developing world, around 192 million children and about 2000 million others experience malnutrition, particularly the pregnant women leading to infant mortality. British, American and Australian researchers are all working on development of a range of fruit and vegetable varieties that the industry claims could counter malnutrition and illness.

Some of the varieties of horticultural crops which are rich in nutrients are listed in Table 2.6.

**Table 2.6.** Nutrient rich horticultural crop varieties.

| Horticultural crop | Variety | Nutrient/s | Developed at |
|---|---|---|---|
| Amla | BSR 1 | Ascorbic acid – 610 mg/100g | ARS, TNAU, Bhavanisagar |
| Fig | YCD1 TIMLA | Vitamin C – 500 mg/100g, ß-carotene, lycopene | HRS, TNAU, Yercaud |
| Pomegranate | CO 1 | Vitamin C – 16 mg/100g | TNAU, Coimbatore |
| | YCD 1 | Ascorbic acid of 15.5 mg/100g | HRS, TNAU, Yercaud |
| Guava | TRY (G) 1 | Ascorbic acid -180.8 mg/100 g | AC&RI, TNAU, Trichy |
| Papaya | CO 2 | Vitamin C – 50.8 mg/100g | TNAU, Coimbatore |
| | CO 7 | Carotenoids – 7.95 mg/100g | TNAU, Coimbatore |
| Tomato | Pusa Uphar, Pusa Hybrid-2 | Vitamin C | IARI, New Delhi |
| | CO 1 | Ascorbic acid – 17.50 mg/100g | TNAU, Coimbatore |
| | CO 2 | Ascorbic acid -19.5 mg/100g | TNAU, Coimbatore |
| | CO 3 | Vitamins (25 ml/100 ml) | TNAU, Coimbatore |
| | PKM 1 | Vitamin C – 23.7 mg/100g fruit pulp | HC&RI, TNAU, Periyakulam |
| | Azad T 2 | Lycopene (3.68 mg/g), Vitamin C (15 mg/100 g) | CSAUAT, Kanpur |
| Brinjal | COBH 1 | Ascorbic acid (16.65 mg/100 g) | TNAU, Coimbatore |
| | MDU 1 | Protein -1.5%, ascorbic acid – 14.0 mg/100g, carbohydrate – 4.3% | TNAU, Coimbatore |
| | PKM 1 | Ascorbic acid 14 mg/100g pulp | HC&RI, TNAU, Periyakulam |
| | PLR 1 | Ascorbic acid – 2 mg/100 g | TNAU, Coimbatore |

| | | | |
|---|---|---|---|
| Chilli | Gujarat veg. chilli 121 | Ascorbic acid | GAU, Anand |
| | PKM 1 | Ascorbic acid – 306.0, 269.8 and 67.0 mg/100 g of green, ripe and dry pods, respectively | HC&RI, TNAU, Periyakulam |
| Okra | CO 1 | Crude fibre -14.06%, starch – 14.80%, protein – 18.08% | TNAU, Coimbatore |
| | CO 2 | 12.6% crude fibre, 15.2% starch, 19.2% protein, 2.84% fat and 50.5mg of ascorbic acid in 100 g pulp | TNAU, Coimbatore |
| Onion | CO 3 | 0.53 % sulphur | TNAU, Coimbatore |
| Broccoli | Palam Vichitra | Vitamins, minerals | HPKV, Palampur |
| French bean | Pusa Parvati | Protein | IARI, New Delhi |
| Pea | Arkel | Protein | IARI, New Delhi |
| | Ooty 1 | Protein (19.94%), ascorbic acid (18.2 mg/100 g) | HRS, TNAU, Ooty |
| Cowpea | Sel 263 | Crude protein (3.57%) | PAU, Ludhiana |
| Field bean | Pusa Sem-2 | Protein | IARI, New Delhi |
| | CO 1 | 4.69% protein, 0.12% vitamin C, 0.2% Fe, 0.39% calcium, 0.14% phosphorous, 1.33% crude fibre/100g edible pods | TNAU, Coimbatore |
| Broad bean | Pusa Sumeet | Protein | IARI, New Delhi |
| Muskmelon | Arka Jeet | Vitamin C | IIHR, Bangalore |
| | Durgapura Madhu | Vitamin C | ARS, Durgapura |
| Ridge gourd | CO 1 | 3.5 g carbohydrates, 0.9 g proteins, 0.2 g fat, 0.4 g fibre, 42.0 mg calcium, 37 mg phosphorous, 37 mg potassium, 1.7 mg iron, 4.2 mg sodium, 12 mg sulphur, 5 mg ascorbic acid/100 g edible fruit | TNAU, Coimbatore |
| Bottle gourd | Pusa Samridhi | Vitamin C-8.07 mg, calcium-19.35 mg, phosphorus-20.39 mg, iron-1.35 mg/100 g | IARI, New Delhi |
| | Pusa Santushti | Vitamin C-7.12 mg, calcium-19.45 mg, phosphorus-18.15 mg, iron-1.10 mg/100 g | IARI, New Delhi |
| Bitter gourd | Arka Harit | Vitamin C, iron | IIHR, Bangalore |

| | | | |
|---|---|---|---|
| | Priya | Vitamin C, iron | KAU, Thrissur |
| | Pusa Hybrid-1 | Vitamin C, iron, calcium | IARI, New Delhi |
| Snake gourd | MDU 1 | Vitamin-A (44.4 mg/100g) | TNAU, Coimbatore |
| Pumpkin | Pusa Hybrid-1, Pusa Vikas | Vitamin A | IARI, New Delhi |
| | Arka Chandan | Vitamin A | IIHR, Bangalore |
| | CO 2 | 9.2% starch, ascorbic acid (10 mg/100 g) | TNAU, Coimbatore |
| Carrot | Pusa Meghali | ß-carotene (Vitamin A) (11571 IU/100 g) | IARI, New Delhi |
| | Pusa Kesar | ß-carotene (Vitamin A) (7753 IU/100 g) | IARI, New Delhi |
| | Arka Suraj | Carotene (11.27 mg/100 g) | IIHR, Bangalore |
| | Hisar Gairic | Carotene (96.2 mg/100 g) | CCSHAU, Hisar |
| | Ooty 1 | ß -carotene | HRS, TNAU, Ooty |
| Beet root | Ooty 1 | Protein – 1.52%, carbohydrate – 10.25%, Vitamin C – 6 mg/100 g | HRS, TNAU, Ooty |
| Radish | Palam Hriday | Vitamin C | HPKV, Palampur |
| | CO 1 | Vitamin C-10.9 mg/100g, 1.20% crude fibre | TNAU, Coimbatore |
| Amaranth | Pusa Kiran, Pusa Kirti, Pusa Lal Chauli | Vitamin A & C, iron, calcium | IARI, New Delhi |
| | CO 2 | Iron - 19.0 mg, calcium – 20.0 mg/100g | TNAU, Coimbatore |
| | CO 3 | 25.2 mg Vitamin C, 1.74% crude fibre, 0.8% iron and 2.48% Ca/100g | TNAU, Coimbatore |
| | CO 4 (grain type) | Protein -15.95%, lysine – 7.5 mg/100g, phenylalanine – 5 mg/100g, leucine – 1.2 mg, isoleucine – 1.8 mg/100g | TNAU, Coimbatore |
| Spinach beet (Palak) | Pusa Jyoti, Pusa Bharati | Vitamin A & C, iron, calcium | IARI, New Delhi |
| | Ooty 1 | Vitamin A | HRS, TNAU, Ooty |
| Chenopodium | Pusa Bathua-1 | Vitamin A, iron, calcium | IARI, New Delhi |
| Sweet potato | Pusa Sunehari | Carotene (Vitamin A) | IARI, New Delhi |
| | Sree Kanaka | Carotene (8.8-10.0 mg/100 g) | CTCRI, Trivandrum |
| | CO 1 | Ascorbic acid – 20 mg/100g | TNAU, Coimbatore |

| | | | |
|---|---|---|---|
| | CO 2 | Starch – 29.5%, carotenoids -3.2 mg/100g | TNAU, Coimbatore |
| | CO 3 | Ascorbic acid (21.1 mg/100g) | TNAU, Coimbatore |
| Cassava | Sree Visakham | Carotene (466 IU/100 g) | CTCRI, Trivandrum |
| Colocasia | CO 1 | Starch – 22.5%, protein – 2.4% | TNAU, Coimbatore |
| Greater yam | CO 1 | Protein – 2.5%, carbohydrate –28% | TNAU, Coimbatore |
| Coleus | CO 1 | Starch – 21.5% | TNAU, Coimbatore |

Chapter 3

# HIGH DENSITY PLANTING

Accommodation of the maximum possible number of plants per unit area to get the maximum possible productivity and profit per unit of the tree volume without impairing the soil fertility status is called High Density Planting (HDP). Basically, the availability of a dwarf plant is the first and foremost pre-requisite for establishing any high density orchard. HDP is one of the methods to enhance productivity per unit area both in short duration and perennial horticultural crops. In perennial, it is more useful since it permits efficient use of land and resources, realizing higher yields, net economic returns per unit area, easy canopy management suited for farm mechanization and cultural operations, efficient spray and weed control, improvement in fruit quality and easy and efficient harvest of high quality produce, etc.

Planting of fruit trees rather at a closer spacing than the recommended one using certain special techniques with the sole objective of obtaining maximum productivity per unit area without sacrificing quality is often referred as high density planting. This technique was first established in apple in Europe during sixties and now majority of the apple orchards in Europe, America, Australia and New Zealand are grown under this system. In this system, four planting densities are recognized for apples viz., low HDP (< 250 trees/ha), moderate HDP (250-500 trees/ha), high HDP (500 to 1250 trees/ha) and ultra high HDP (>1250 trees/ha). Recently, super high density planting system has been also established in apple orchards with a plant population of 20,000 trees per ha. In some orchards, still closer, planting of apple trees is followed (say 70,000 trees/ha) which is often referred as 'meadow orchards'.

High density planting is referred as semi-intensive system accommodating 500-1,000 trees/ha, intensive system accommodating 1,000 to 10,000 trees/ha employing specialized training systems and super-intensive system with 20,000 to 1,00,000 per ha.

### 3.1. Methods to achieve HDP

There are five important methods to achieve HDP namely

- Use of dwarf scion varieties.
- Adopting dwarfing rootstocks and inter-stocks.
- Efficient training and pruning.
- Use of plant growth regulating chemicals.
- Suitable crop management practices.

There are several fruit crops where desired success has been achieved using HDP. These crops include apple, peach, plum, sweet cherry and pear among temperate fruits, and banana, pineapple and papaya among tropical fruits. In India, HDP technology has been successfully demonstrated in banana, pineapple, papaya and recently in mango, guava and citrus where two to three times increase in yields are realized. In plantation crops like cashew, coconut and areca nut, such dense plantings can best be adopted using multi-crop species cropping systems. These systems need popularization. Some of the major concerns in adoption of HDP is that being an intensive system, requires high capital to establish though it is more productive and profitable, if followed scientifically. Non-availability of adequate planting material of dwarf varieties in different horticultural crops, and non-availability of clonal rootstocks in some fruit and plantation crops are the major constraints in promotion of HDP. However, taking into consideration the need for doubling the production and productivity, it is important that such intensive systems be adopted in India to replace the traditional planting densities and methods of fruit growing. The area expansion programmes under HDP need to be promoted among farmers. Accordingly, the scheme for HDP promotion under National Horticulture Mission should be continued with more budget allocation in targeted crops. There is also need to promote HDP systems along with micro-irrigation and fertigation methods standardized for different crops and agro-climatic conditions. The mono-species HDP is costly and technology-intensive, hence should be exploited for commercial sole fruit production. For small and marginal farmers, multi-species HDP approach should be standardized. In plantation crops, multiple cropping systems like coconut and areca nut with pepper; tree spices, tuber crops, banana and pineapple should be made popular in south India.

High density planting in fruit growing was first established in apple in Europe in the early sixties. Presently, the continued decline in the availability of cultivable land, rising energy and land costs together with the mounting demand for horticultural produce, have given thrust to adoption of HDP in horticultural crops. Furthermore, it is of main concern to the growers with small land holdings. High density planting is more efficient since it is precocious, easily manageable, has higher yield potential with better quality fruits and higher returns/unit area.

### 3.2. Advantages of HDP

The advantages of HDP are:

- Early cropping and higher yields for a long time; the average yield in apple is about 5.0 MT/ha under normal system of planting and it is about 140.0 MT/ha under HDP.
- Reduced labour costs.
- Improved fruit quality.

Characteristics of HDP are:

- The trees of HDP should have maximum number of fruiting branches and minimum number of structural branches.
- The trees are generally trained with a central leader surrounded by nearly horizontal fruiting branches.
- These branches should be so arranged and pruned in such a way that each branch casts a minimum amount of shade on other branches.
- The height should be one and half its diameter at the base. A key to successful HDP depends upon the control of tree size.

This is achieved by

- Use of size controlling rootstocks. In apple, dwarfing rootstocks and intermediate stocks like MM 106, MM 109, and MM 111 are used to control the size of the plant. In pears, Quince A, Adam and Quince-C are commonly used as dwarfing rootstocks.
- Use of spur type scions – In temperate fruit crops like apple, the cultivars can be classified into a spur type or non-spur type. The spur types which have restricted annual growth are alone suitable for HDP.
- Training and pruning methods to induce dwarfness – Under Indian conditions, apple trees trained under spindle bush, dwarf pyramid, cordon systems are found to contain the growth of the trees appreciably for HDP systems.
- Mechanical device and use of chemicals to control size – Growth regulators such as diaminozide, ethephon, chlormaquat and paclobutrazal are extensively used to reduce shoot growth by 30 to 50%. This results in increased flowering in the subsequent years and may be useful in encouraging earlier commercial fruit production in strongly vegetative fruitful young trees. Besides chemical manipulation, mechanical devices employing the use of spreaders and tying down the branches to make them grow from near horizontal to an angle of 45° from the main stem are also some of the standard practices to control tree size.

## 3.3. Planting system

### 3.3.1. Planting season

The season of planting varies with different fruits and local conditions. There are two seasons of planting in vogue in India.

- Monsoon (June – August) and
- Spring (February – March)

Monsoon season is considered to be the best for planting evergreen fruit trees like citrus, mango, sapota and guava. If the trees are planted early in the rainy season they soon establish themselves and grow vigorously. Deciduous trees may be planted during the dormant period without shock. Care should be taken that planting is done before the growth starts, otherwise trees suffer severely and will be in poor condition to withstand the next hot weather.

### 3.3.2. Planting methods

The success of HDP depends upon the right choice of planting system. Generally, rectangular planting with single, double and three row plantings are followed. In single row planting, the distance within the row is close, whereas the distance between the row is wide (4 x 2m). In double row planting, a wider spacing is given after every two rows (4 + 2 x 2m) whereas in three row planting, a wider spacing is given after every three rows (4 + 2 x 2 x 2m). In meadow orchard system, a bed of 10 to 15 rows is closely planted (30 x 45 cm) and separated by alleys of 2.5m width between beds. This system is also called bed system.

The total number of trees per hectare for various important horticultural crops under (a) square (b) hexagonal and (c) triangular system of planting is given Table 3.1.

**Table 3.1.** Plant population of various horticultural crops under different planting systems

| Crop | Planting distance (m) | No. of trees per hectare | | |
|---|---|---|---|---|
| | | Square system | Hexagonal system | Triangular system |
| Mango | 10 x 10 | 100 | 115 | 89 |
| Sapota | 8 x 8 | 156 | 118 | 139 |
| Clove | 6 x 6 | 277 | 320 | 248 |
| Acid lime | 5 x 5 | 400 | 461 | 357 |
| Coconut | 7.5 x 7.5 | 177 | 205 | 159 |

It may be seen that hexagonal system of planting accommodates 15% more number of plants while triangular system accommodates 11% lesser number of plants. The calculation of the number of trees per hectare when planted under square or rectangular system is very easy, and is obtained by dividing the total area 'by the area occupied by each tree (a x a in square system or l x b in rectangular system). The theoretical and the actual number of possible trees which can be planted in an orchard depends upon the shape of the field. In practice, in large fields, the percentage difference between the theoretical number and the actual number possible will be less.

After locating the positions of the orchard trees, it is important that the trees are planted exactly where the stakes stood. It can be easily done with the help of a planting board. The planting board is usually of 15m long, 10 cm wide and 2.5 cm thick with a central notch and one hole on either end, the central notch and the two holes (one on either end) are in a straight line. The planting board is placed in such a way that the stake (tree marker) fits into the central notch. Two small stakes are inserted one in each end hole. The planting board along with the tree marker is then lifted straight up without disturbing the end stakes. A pit of about 1 m cube or of the desired dimensions at the position of the tree marker is then dug.

The pits are allowed to wither for few weeks before planting in some cases. The pits are then filled with top soil already mixed with red earth and well rotten farm yard manure. Irrigation is then applied to enable the contents of the pits to settle down properly. In the event of depressions taking place as a result of irrigation, more soil should be added to the pits to fill them to the level of the land. The pits are then ready for planting.

Transplanting: The trees should be planted approximately where the original pegs were placed. This is achieved by replacing the planting board in position with the help of the guide pegs and the stem of the trees is brought to the central notch with the help of a hand hoe.

One of the most common mistakes is that of planting the trees too deep. The plants should be planted in such a way that the bud union remains slightly above the ground level. The trees in the field should be planted as deep as they stood in the nursery. The trees are irrigated soon after planting. This consolidates the soil and helps the roots to establish contact with it and to secure a supply of water quickly. A small basin may be made around the tree for this purpose. Planting if taken up during the rains, this basin should be demolished within a day or two so that water will not collect around the tree. This is more dangerous on heavy than light soils.

Spring winds cause damage to the growing plants by giving a constant shaking. To prevent this, plants should be staked when planted. Some young plants are subject to considerable injury from sunburn particularly if they have been trained to single stem with no branches for 45 cm or more from the ground. Such trunk can be protected by wrapping them with paper or other material or by painting them with white wash. The latter is probably best, as most materials wrapped around the trunk would be subject to termite attack.

Recommended planting densities of some fruit crops is presented in Table 3.2.

**Table 3.2.** Recommended planting densities of some fruit crops

| Fruit crop | Variety | Spacing | System of planting | No. of trees/ha |
|---|---|---|---|---|
| Mango | Amrapali | 2.5 x 2.5 m | Triangular system | 1600 |
| | Dashehari | 3.0 x 2.5 m | Rectangular system | 1333 |
| Citrus (mandarin) | Kinnow | 1.8 x 1.8 m | Square system using Troyer citrange rootstock | 3025 |
| | | 2.4 x 2.4 m | Square system using Karna Khatta rootstock | 1780 |
| | | 3.0 x 3.0 m | Square system using Soh Sarkar rootstock | 1111 |
| Acid lime | Kagzi lime | 4.5 x 4.5 m | Square system | 484 |
| Banana | Cavendish group | 1.2 x 1.2 m | Square system using Karna Khatta rootstock | 6944 |
| | | 1 x 1 x 2 m | Paired row system | 6666 |
| Pineapple | Kew | 25 x 60 x 78 cm | Double row planting system | 63000 |
| Apple | Spur type | 4 x 4 m | Square system using MM 111, MM 109 rootstocks | 625 |
| | | 3 x 3 m | Square system using MM 106, MM 7 rootstocks | 1111 |
| | Non-spur type | 5 x 5 m | Square system using MM 106, MM 119 rootstocks | 400 |
| | | 2 x 2 m | Square system using MM 9 rootstock | 2500 |
| Guava | Aneuploids | 3 x 3 m | Square system | 1111 |
| | Lucknow-49 | 6 x 2 m | Rectangular system | 833 |
| Papaya | Pusa Nanha | 1.25 x 1.25 m | Rectangular system | 6400 |

The planting density and yield potential of some fruit crops under conventional and high density planting are presented in Table 3.3.

**Table 3.3.** Planting density and yield potential of fruit crops in conventional and high density planting

| Crop | Spacing (m)/Plant density | | Fruit yield (MT/ha) | |
|---|---|---|---|---|
| | Conventional | HDP | Conventional | HDP |
| Apple | 5-8 x 5-8 (400 to 277) | 3.0 x 7.5 (2222) | 30-50 | 68.75 |
| Apricot | 6.0 x 6.0 (277) | 4.5 x 1.5 (1481) | 13-22 | 64.40 |
| Banana | 2.0 x 2.0 (2500) | 1.5 x 1.5 (6944) | 35-36 | 120 |

| | | | | |
|---|---|---|---|---|
| Sweet orange | 8.0 x 8.0 (156) | 3.2 x 2.5 (1250) | 10-15 | 135 |
| Mandarin | 6.0 x 6.0 (277) | 2.0 x 2.0 (2500) | 10-15 | 72.50 |
| Guava | 6.0 x 6.0 (277) | 2.0 x 2.0 (2500) | 15.70 | 58.30 |
| Litchi | 10.0 x 10.0 (100) | Double hedge row planting (222) | 9-10 | 20.30 |
| Mango | 10.0 x 10.0 (100) | 9.0 x 2.0 (555.5) | 8-10 | 44 |
| Papaya | 2.0 x 2.0 (2500) | 1.2 x 1.8 (4262) | 40-50 | 146 |
| Peach | 6.0 x 6.0 (277) | 3.0 x 3.0 (1111) | 7-10 | 34.65 |
| Pineapple | 15,000-20,000 | 43,000-63,750 | 15-20 | 60-70 |
| Sweet cherry | 8.0 x 5.0 (250) | 4.0 x 2.0 (1250) | 6.3-9.0 | 16 |

### 3.4. Constraints in adoption of HDP systems

- Non-availability of planting material in dwarf varieties in different horticultural crops.
- Non-availability of vegetatively propagated rootstocks in different fruit and plantation crops.
- Avoidance in use of inter- and mixed-cropping system.
- Overcrowding trees in mango HDP after 10 to 12 years?
- Harmful effects of chemicals like paclobutrazol causing barrenness in crops like grape etc.
- Non-availability of complete package for HDP and use of mechanization.

Chapter 4

# MICRO-IRRIGATION

Adequate, timely and assured availability of water is critical to horticulture for ensured yield. Water is getting depleted in terms of quantity as well as quality. Water management assumes paramount importance to reduce the wastage of water to increase the water use efficiency and ensure equitable distribution. Even though India has the largest irrigation system in the world, its water use efficiency is not more than 40%. If it continues, water crisis would lead to reduced production and productivity.

Currently, only 38% of cultivated area is irrigated and more efforts are needed to cover additional area to enhance productivity. Approaches have largely been for creation of water with less emphasis on enhancement of irrigation efficiency. To meet the nutritional security to the projected population in 2020, the additional requirement of fruits and vegetables have to come from the use of improved system of irrigation, which irrigates the plant rather than field and results in productive use of water.

## 4.1. Water losses under various irrigation systems

In traditional surface irrigation methods, the losses in water conveyance and application are large. These losses can be considerably reduced by adopting drip and sprinkler irrigation methods. Among all the irrigation methods, the drip irrigation is the most efficient and it can be practiced in a large variety of crops, especially in vegetables, orchard crops, flowers and plantation crops (Table 4.1).

**Table 4.1.** Water losses under various irrigation systems

| Method of irrigation | Water losses (%) |
|---|---|
| Surface | 30-45 |
| Gate pipe | 15-20 |
| Sprinkler irrigation | 5-10 |
| Drip irrigation | 3-6 |

Micro-irrigation system is an irrigation system with high frequency application of water in and around the root zone of plant system, which consists of a network of pipes along with a suitable emitting device. With micro-irrigation the yields of crops are better in addition to the saving of water.

Micro irrigation can be classified as drip irrigation and sprinkler irrigation. In drip irrigation, water is applied near the plant root through emitters or drippers, on or below the soil surface, at a low rate varying from 2-20 litres per hour. The soil moisture is kept at an optimum level with frequent irrigations. Drip irrigation results in a very high water application efficiency of about 90-95 per cent. A typical drip and sprinkler irrigation systems are shown in Figs. 4.1 and 4.2.

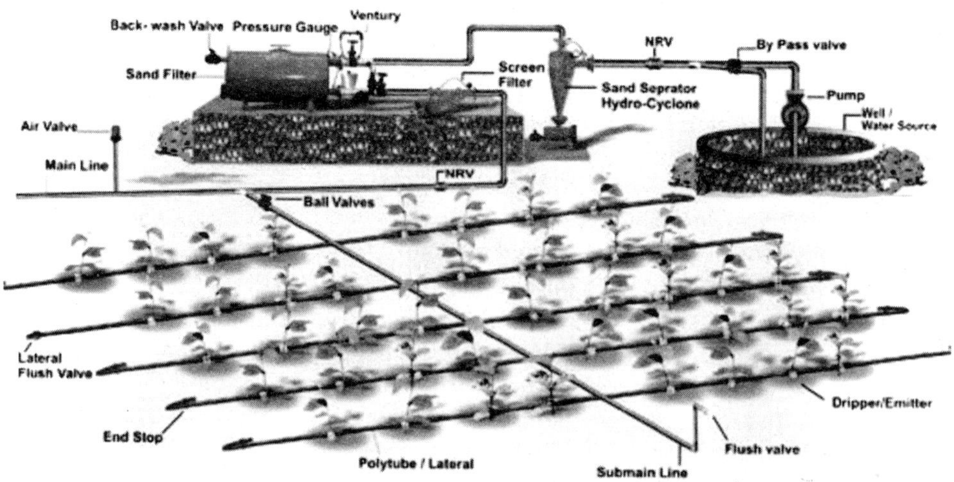


**Fig. 4.1.** A typical drip irrigation system

**Fig. 4.2.** Sprinkler irrigation system

## 4.2. Area under drip irrigation

In India, the irrigated area consists of about 38 per cent of the net sown area. There has been a tremendous growth in the area under drip irrigation during the last 15 years. At present, around 3.5 lakh ha area is under drip irrigation with the efforts of the Government of India, while it was only 40 ha in 1960. Maharashtra (94,000 ha), Karnataka (66,000 ha) and Tamil Nadu (55,000 ha) are some of the states where large areas have been brought under drip irrigation. Many crops are irrigated by the drip method in India with the tree crops occupying the maximum percentage of the total area under drip irrigation, followed by vine crops, vegetables, field crops, flowers and other crops (Table 4.2).

**Table 4.2.** Area under drip irrigation and production of horticultural crops (1999- 2000)

| Crop | All India | |
|---|---|---|
| | Area (Million ha) | Production (Million MT) |
| Fruits | 3.78 | 45.50 |
| Vegetables | 5.99 | 90.83 |
| Spices | 2.52 | 2.91 |
| Coconut | 1.78 | 8.42 |
| Cashew | 0.69 | 0.52 |
| Others | 2.78 | 1.75 |
| Total | 17.54 | 149.93 |

Drip irrigation system is suitable for all orchard and vegetable crops. This system has also been successfully employed for close growing crops including onion and okra. The National Committee on Plasticulture Applications in Horticulture (NCPAH), Ministry of Agriculture, Government of India (GOI) has estimated a total of 27 million hectares area in the country that has the potential of drip irrigation application.

There is a need to promote increased irrigation water use efficiency. Participatory action research programme initiated by the Ministry of Water Resources, Government of India during 2007-09 in over 2000 villages all over the country to assess the impact of water saving technologies has shown that yield and income can be increased by 50 to 100% in most crops by using water saving technologies.

Micro-irrigation comprising drip, mini sprinkler and sprinkler must become a pivotal element of integrated water use system with many agro-ecological, socio-economic and environmental advantages. Micro-irrigation could be a tool for effective management of resources which save water, fertilizers and also electricity.

There are automatic micro-processor based drip irrigation systems available, which ensure that the plants get right amount of water at right time. Soil moisture sensors are installed in the field, which activate the drip irrigation system when irrigation is required. Table 4.3. shows the advantages of different irrigation systems.

**Table 4.3.** Advantages of different drip irrigation systems

| Target | Simple drip system | Automated drip system |
|---|---|---|
| Water saving | 25-70% | Higher possible |
| Fertilizer saving | 30-40% | Higher possible |
| Human interface | Needed | Computer based |
| Maintenance | Higher | Lower |
| Uniformity in water application | 75-80% | More than 80% |
| Increase in yield | 20-40% | Still higher |
| Water application | Less precise | More precise |
| Precise irrigation scheduling | Not available | Available |

## 4.3. Advantages of micro-irrigation

The advantages of micro-irrigation are of two types – qualitative and quantitative. The qualitative advantages are production of quality horticultural produce and increase in water and fertilizer use efficiency. Excellent soil health could be maintained due to optimum soil moisture content, accelerated microbial activities and maximization of biomass decomposition converting into humus. The quantitative advantages are related to growth of micro-irrigation in terms of area coverage and overall increase in the productivity. Quantity of water or fertilizer saved due to imposition of drip system, the energy conservation and reduced labour cost are the other advantages. Data on increase in crop yields and percent increase over traditional surface irrigation methods by using micro-irrigation system are presented in Table 4.4. It shows that micro-irrigation could save water ranging from 14% to 84% with enhancement of yield from 13% and 514% under varied agro-climatic and crop conditions. The higher yield under micro-irrigation system is attributed to the fact that during crop growth favourable soil moisture (generally at ⅓ bar) is maintained and the required volume of soil mass is wetted through irrigation water, which provides optimum environment for root growth of the plant.

**Table 4.4.** Yield increase with drip irrigation system in horticultural crops

| Horticultural crop | Yield (MT/ha) | | % water saving | % increase in yield |
|---|---|---|---|---|
| | Surface irrigation | Drip irrigation | | |
| Beet root | 5.7 | 8.9 | 79.1 | 56.1 |
| Bitter gourd | 32.0 | 43.0 | 56.6 | 34.4 |
| Brinjal | 91.0 | 148.0 | 61.9 | 62.6 |
| Broccoli | 140.0 | 195.0 | 14.3 | 39.3 |
| Cauliflower | 171.0 | 274.0 | 33.3 | 60.2 |

| | | | | |
|---|---|---|---|---|
| Chilli | 42.3 | 60.9 | 61.7 | 44.0 |
| Cucumber | 155.0 | 225.0 | 55.6 | 45.2 |
| Okra | 100.0 | 113.1 | 84.0 | 13.1 |
| Onion | 284.0 | 342.0 | 50.0 | 20.4 |
| Potato | 172.0 | 291.0 | 54.2 | 69.2 |
| Radish | 10.5 | 11.9 | 76.1 | 13.3 |
| Sweet potato | 42.4 | 58.9 | 60.3 | 38.9 |
| Tomato | 61.8 | 88.7 | 78.5 | 43.5 |
| Banana | 575.0 | 875.0 | 45.0 | 52.2 |
| Grape | 264.0 | 325.0 | 47.2 | 23.1 |
| Papaya | 130.0 | 230.0 | 67.9 | 76.9 |
| Pomegranate | 34.0 | 67.0 | 23.8 | 97.0 |
| Water melon | 82.1 | 504.0 | 65.3 | 513.9 |

The details of crop yield and benefit-cost ratio for some of the horticultural crops covered under micro-irrigation are presented in Table 4.5.

**Table 4.5.** Benefits from micro-irrigation of horticultural crops

| Horticultural crop | Spacing (m) | Yield (MT/ha) | Benefit: cost ratio |
|---|---|---|---|
| Banana | 0.91 x 1.5 x 1.8 (pair) | 75 | 3.00 |
| Ber | 4.5 x 4.5 | 25 | 4.56 |
| Grape | 3.0 x 1.8 | 45 | 3.28 |
| Papaya | 1.81 x 1.81 | 60 | 4.09 |
| Tomato | 0.45 x 0.45 | 75 | 1.09 |

Results with experiments on micro-irrigation have categorically exhibited saving of 31 to 63% water with enhanced yield up to 23 to 88% under varied agro-climatic and crop conditions (Table 4.6).

**Table 4.6.** Water use efficiency with micro irrigation

| Crop | % increase in yield | % water saving | % increase in water use efficiency |
|---|---|---|---|
| Banana | 52 | 45 | 176 |
| Grape | 23 | 48 | 136 |
| Sweet lime | 50 | 61 | 289 |
| Pomegranate | 45 | 45 | 167 |
| Tomato | 50 | 31 | 119 |
| Chilli | 45 | 63 | 291 |
| Water melon | 88 | 36 | 195 |

The farmers are adopting the micro-irrigation system in high value horticultural crops like grapes, banana, citrus, mango, pomegranate, tomato, okra and chilli (Table 4.7).

**Table 4.7.** Yield increase with drip irrigation system in horticultural crops

| Horticultural crops | Yield (MT/ha) | | |
|---|---|---|---|
| | Conventional irrigation | Drip irrigation | % yield increase |
| Banana | 57.5 | 87.5 | 52 |
| Grape | 26.4 | 32.5 | 23 |
| Sweet lime | 100.0 | 150.0 | 50 |
| Pomegranate | 55.0 | 109.0 | 98 |
| Papaya | 13.4 | 23.5 | 75 |
| Tomato | 32.0 | 48.0 | 50 |
| Water melon | 24.0 | 45.0 | 88 |
| Okra | 15.3 | 17.7 | 16 |
| Chilli | 4.2 | 6.1 | 44 |
| Sweet potato | 4.2 | 5.9 | 39 |

The farmers of Andhra Pradesh, Karnataka, Madhya Pradesh, Gujarat, Kerala, Maharashtra, Tamil Nadu and Rajasthan states are well convinced that drip irrigation help them to get more yield with less inputs (fertilizers and pesticides) apart from saving of water (Table 4.8).

**Table 4.8.** Water used and yield for horticultural crops in drip and conventional methods

| Horticultural crops | Yield (MT/ha) | | | Water supplied (cm) | | |
|---|---|---|---|---|---|---|
| | Coven. | Drip | % increase | Coven. | Drip | % Water saving |
| Banana | 57.50 | 87.50 | 52 | 176.00 | 97.00 | 45 |
| Grape | 26.40 | 32.50 | 23 | 53.20 | 27.50 | 48 |
| Sweet orange | 10.00 | 15.00 | 50 | 166.00 | 64.00 | 61 |
| Pomegranate | 5.50 | 10.90 | 98 | 144.00 | 78.50 | 45 |
| Brinjal | 28.00 | 32.00 | 14 | 90.00 | 42.00 | 53 |
| Papaya | 13.40 | 23.48 | 75 | 228.00 | 73.30 | 68 |
| Chillies | 4.23 | 6.09 | 44 | 109.71 | 41.77 | 62 |
| Sweet potato | 4.24 | 5.89 | 39 | 63.14 | 25.30 | 60 |

Chapter 5

# FERTIGATION

Among the various factors that are responsible for increasing the productivity of horticultural crops, a regular and optimum supply of water and fertilizers at appropriate times is very crucial as it enables the crops to attain higher growth rate and give increased yields. As both water and fertilizers are costly inputs and are becoming scarce in recent years, there is an immediate need for adoption of techniques with which these inputs can be utilized more efficiently. Fertigation is one such technique of applying fertilizers to the crops along with water through micro irrigation systems (drip/sprinkler) on a continual basis in controlled manner so as to allow steady uptake of nutrients by the plants and to effect savings in the costly inputs including water and fertilizers. Salient features of the research conducted on fertigation of horticultural crops using different combinations of water soluble fertilizers in increasing the productivity are presented in Table 5.1 and 5.2.

**Table 5.1.** Saving in fertilizer and increase in horticultural crop yield under fertigation

| Horticultural crop | Saving in fertilizer (%) | Increase in yield (%) |
|---|---|---|
| Banana | 20 | 11 |
| Onion | 40 | 16 |
| Potato | 40 | 30 |
| Tomato | 40 | 33 |
| Okra | 40 | 18 |
| Broccoli | 40 | 10 |

## 5.1. Advantages of fertigation

The fertigation allows applying the nutrients exactly and uniformly only to the wetted root volume, where the active roots are concentrated. This remarkably increases the efficiency in the application of the fertilizer, which allows reducing the amount of applied fertilizer. This not only reduces the production costs but also lessens the potential of ground water pollution caused by the fertilizer leaching. Fertigation allows adapting the amount and concentration of the applied nutrients in order to meet the actual nutritional requirement of the crop throughout the growing season. In order to make a correct planning of the nutrients supply to the crop according to its physiological stage, we must know the optimal daily nutrient consumption rate during the growing cycle that results in maximum yield, quality and production. These functions are specific for each crop and climate, and were determined in different experiments for the main crops in Israel like tomatoes, cucumbers, melons, etc. (Table 5.3). The optimal curve of consumption of nutrients defines the minimal application rate of a certain nutrient that is required to maintain a constant nutrient concentration in the soil solution. These data constitute the base of the recommendations given by the Israeli Soil Extension Service for the farmers regarding the fertigation regime for the different crops.

**Table 5.2.** Response of horticultural crops to fertigation

| Treatment | Yield (kg/plant) | | |
|---|---|---|---|
| | Blue Grapes | Sapota | Cashew |
| 100% Normal fertilizers + Basin irrigation | 9.20 | 31.78 | --- |
| 100% Normal fertilizers + Micro irrigation | 12.75 | 31.34 | 2.59 |
| 80% water soluble fertilizers through micro irrigation | 20.25 | 75.66 | 2.87 |
| 100% water soluble fertilizers through micro irrigation | 22.16 | 48.82 | 3.23 |

**Table 5.3.** Consumption rate of nitrogen, phosphorus and potassium (kg $ha^{-1}$) of different vegetables grown under drip irrigation after emergence or planting.

| Particulars | Tomato greenhouse | | | Tomato industry | | | Eggplant | | | Broccoli | | | Melon | | |
|---|---|---|---|---|---|---|---|---|---|---|---|---|---|---|---|
| | N | P | K | N | P | K | N | P | K | N | P | K | N | P | K |
| Nutrients | 450 | 65 | 710 | 393 | 59 | 520 | 290 | 33 | 380 | 202 | 26 | 165 | 151 | 25 | 385 |
| Variety | F-144 | | | VFM82-1-2 | | | Black Oval | | | Woltam | | | Galia | | |
| Plants/ha | 23,000 | | | 50,000 | | | 12,500 | | | 33,000 | | | 25,000 | | |
| Soil | Sandy | | | Clay | | | Sandy | | | Loam | | | Sandy | | |
| Yield (MT/ha) | 195 | | | 160 | | | 51 | | | 13 | | | 56 | | |

There are three beneficial effects of fertigation:

### 5.1.1. Maximizing water and fertilizer use efficiency

- Delivers water and nutrients to the root zone.
- Increases the availability and uptake of nutrients.

- Decrease nutrient losses from leaching.
- Accurate and uniform application of water and nutrients.
- Maximum saving of water and nutrients.

### 5.1.2. Maximizing crop and soil productivity

- Higher yields and high quality produce.
- Steep-sloped and marginal lands can be made cultivable.

### 5.1.3. Minimizing production cost

- Economy in labour, water and nutrients.

### 5.1.4. Other advantages

- The saving of energy and labour.
- The flexibility of the moment of the application (nutrients can be applied to the soil when crop or soil conditions would otherwise prohibit entry into the field with conventional equipment).
- Convenient use of compound and ready-mix nutrient solutions containing also small concentrations of micronutrients which are otherwise very difficult to apply accurately to the soil.
- The supply of nutrients can be more carefully regulated and monitored. When fertigation is applied through the drip irrigation system, crop foliage can be kept dry thus avoiding leaf burn and delaying the development of plant pathogens.

Drip and micro irrigation have a characteristic not shared by other irrigation methods: fertigation is not optional, but is actually necessary. Fertigation provides the only good way to apply fertilizers physically to the crop root zone. On high value drip irrigated crops, such as lettuce, tomatoes, and peppers, the level of fertigation management for achieving high yields and crop qualities exceeds to what is found with other irrigation methods and crops.

**Table 5.4.** Efficiency of different nutrients and methods of application

| Element | Method of application | Efficiency (%) |
|---|---|---|
| N | Fertigation | 90-95 |
| | Surface Flow | 50-70 |
| | Broadcasting (Solid fertilizers) | 30-50 |
| P | Fertigation | 40-45 |
| | Surface Flow | 10-20 |
| K | Fertigation | 75-80 |
| | Surface Flow | < 50 |

## 5.2. Nutrient use efficiency

The crops do not efficiently utilize fertilizers applied under traditional methods of irrigation. The efficiency of different nutrients and methods of application is presented in Table 5.4.

## Chapter 6

# INTEGRATED NUTRIENT MANAGEMENT

The plant nutrition is concerned with the provision of plant with nutrients as well as nutrient uptake and their distribution in plant system. However, the practical side of plant nutrition is fertilizer and/or nutrient application. Fertilizer is one of the major inputs accounting for nearly one third of the cost of cultivation. Hence, efficient and judicious use of the fertilizers is imperative for obtaining more yields per unit area on a sustainable basis. The intensification of integrated nutrient management in crop production has become important, because it is directly related to the soil and water resources.

Integrated nutrient management refers to maintaining the soil fertility and plant nutrient supply to an optimum level for sustaining the desired crop productivity through optimization of the benefits from all possible sources of plant nutrients in an integrated manner. It is a holistic approach, to know what exactly is required by the plant for an optimum level of production, in what different forms these nutrients should be applied in soil and at what different timings in the best possible method, and how best these forms should be integrated to obtain the highest productive efficiency on the economically acceptable limits in an environment friendly manner.

The encouraging results obtained by conjunctive use of fertilizers, organic manure, green manure, biofertilizers and fertigation can provide leads to the future strategies for the rational use of fertilizers and manures for enhancing the productivity of horticultural crops without detrimental effect to the environment.

The nutrients have been classified into three groups: primary, secondary and micro-nutrients, which are determined by the quantity required by plants. Generally, carbon (C), hydrogen (H), oxygen (O), nitrogen (N), phosphorus (P) and potassium (K) are absorbed in large quantities and therefore, referred as primary nutrients. The nutrients required at medium level are calcium (Ca), magnesium (Mg) and sulphur (S) fall in the category of secondary nutrients. Iron (Fe), zinc (Zn), manganese (Mn), copper (Cu), boron

(B), molybdenum (Mo) and chlorine (Cl) are included in the micro-nutrient category.

### 6.1. Optimum leaf nutrient status

At optimum leaf nutrient status, the growth, yield and quality of the crop are satisfactory and there is no need to make any changes in the schedule of manures and fertilizers. Changes in nutrient concentration in the specified plant part do not increase or decrease growth or production. Leaf is the principal site of plant metabolism and changes in the nutrient supply are reflected in composition of the leaf. The concentrations of the nutrient in the leaf at a specific growth stage are related to the performance of the crop. The leaf nutrient levels required for optimum production in some important fruit crops have been presented in Table 6.1. Efforts in the nutritional management programme should be directed towards the maintenance of these levels in tree leaves before flowering.

**Table 6.1.** Optimum leaf nutrient status of some important fruit crops

| Fruit crop | Nutrients | | | | | | | | | | | Reference |
|---|---|---|---|---|---|---|---|---|---|---|---|---|
| | % | | | | | | ppm | | | | | |
| | N | P | K | Ca | Mg | S | B | Fe | Mn | Zn | Cu | |
| Mango | 1.23 | 0.66 | 0.54 | 1.71 | 0.91 | 0.12 | --- | 171 | 66 | 25 | 12 | Samra *et al.*, 1978 |
| Banana | 3.29 | 0.44 | 3.11 | 2.12 | 0.24 | --- | --- | --- | --- | --- | --- | Ramaswamy, 1971 |
| Citrus | 2.50 | 0.14 | 0.90 | 4.20 | 0.43 | 0.25 | --- | 90 | 112.5 | 62.5 | 10.5 | Reuther *et al.*, 1962 |
| Guava | 1.70 | 0.50 | 1.30 | 2.10 | 0.46 | --- | 10 | --- | --- | 77.5 | 75 | Singh, 1981 |
| Grape | 1.75 | 0.56 | 1.66 | 0.94 | --- | 0.20 | --- | --- | 125 | 92 | --- | Bhargava & Chadha, 1993 |
| Litchi | 1.47 | 1.03 | --- | --- | --- | --- | --- | --- | --- | --- | --- | Ghosh *et al.*, 1986 |
| Sapota | 1.66 | 0.08 | 0.80 | 0.83 | 0.48 | 0.07 | --- | 100 | 39.3 | 15.7 | 6.7 | Annapurna *et al.*, 1988 |
| Papaya | 1.66 | 0.50 | 5.21 | 1.81 | 0.67 | 0.38 | --- | --- | --- | --- | --- | Bhargava *et al.*, 1990 |
| Pineapple | 1.40 | 0.16 | 3.70 | --- | --- | --- | --- | --- | --- | --- | --- | Chadha & Reddy, 1993 |
| Pomegr-anate | 2.50 | 0.19 | 1.47 | --- | --- | --- | 96 | 199 | 196 | 55 | --- | Shende, 1977 Bhambal, 1987 |
| Ber | 2.50 | 0.25 | 1.82 | 1.21 | 0.45 | --- | --- | --- | --- | --- | --- | Khanduja & Garg, 1984 |

## 6.2. Manure and fertilizer recommendations

Appropriate manure and fertilizer recommendations provide plant nutrients needed to sustain maximum production and profitability while minimizing environmental impact of fertilizer use. The knowledge on nutrient removal, nutrient uptake patterns by crops and capacity and availability of nutrients in the soil reservoir helps to formulate correct manure and fertilizer recommendations.

To optimize the soil fertility level and to keep the plant nutrient status above the critical levels, a number of recommendations for the fruit crops are available region-wise. Based on the averages of these recommendations and the results available under AICRP on fruit crops, the following recommendations are being made for different fruit crops (Table 6.2). Besides these recommendations, application of zinc and iron has become common. Spray of boron has helped in amla and mango. Therefore, regular spraying of micronutrients is necessary to get an optimum production per unit area.

**Table 6.2.** Manure and fertilizer recommendations for some tropical and subtropical fruit crops

| Fruit crop | FYM (kg/plant) | Nutritional recommendation (g/plant/year) | | |
|---|---|---|---|---|
| | | N | $P_2O_5$ | $K_2O$ |
| Mango | 75 | 775 | 500 | 700 |
| Citrus (Acid lime) | 50 | 900 | 250 | 500 |
| Guava | 60 | 360 | 180 | 180 |
| Grape | 25 | 270 | 450 | 900 |
| Pomegranate | 30 | 600-700 | 200-250 | 200-250 |
| Ber | 50 | 500 | 200 | 300 |
| Sapota | 50 | 1000 | 500 | 500 |
| Amla | 50 | 1500 | 750 | 1000 |
| Date palm | 30 | 1000 | 500 | 500 |

## 6.3. Methods of application

### 6.3.1. Application in active root zone

For the effective utilization of applied nutrients, it is essential to know the active root zone, so that the nutrients may be placed around this zone, to be made available to the plant. The active root zone as well as place of application of fertilizers as summarized by Chundawat (1997) has been presented in Table 6.3.

**Table 6.3.** Active root zone of some important fruit crops and the placement of fertilizers

| Fruit crop | Area of the active root zone | Placement of fertilizer |
|---|---|---|
| Mango | 82 to 85% of active roots located within a radius of 300 cm with highest activity at 120 cm from tree trunk. | 1 m away from tree trunk under drip and mix thoroughly. |
| Citrus | Within a radial distance of 120 cm at depth of 15 to 30 cm. | 30 to 45 cm away from tree trunk under drip followed by light working. |
| Guava | At 30 cm distance from trunk at a depth of 30 to 60 cm. | Within drip at a depth of 15 to 20 cm. |
| Grape | In radius of 100 to 220 cm at a depth of 45 to 75 cm. | Deep placement. |
| Litchi | Shallow on surface. | Soil surface with light working. |
| Pineapple | Highly restricted. | 9 to 10 cm away in 10 to 15 cm deep strip around plants. |
| Ber | 50 to 75 cm deep under drip. | Under drip and mix in soil up to 30 cm. |
| Sapota | Under canopy within a depth of 0 to 30 cm. | Under canopy and mix in soil up to 15 cm depth. |
| Date palm | Up to a radius of 1.7 m from the bole. | Around tree up to a radius of 2.0 m. |

### 6.3.2. Foliar application

Foliar application is based on the principle that the nutrients are quickly absorbed by the leaves and transported to different parts of the plant to fulfill the functional requirement of nutrition. It is an ideal way evading the problems of nutrient availability. This method is highly helpful for correction of trace element deficiencies to restore disrupted nutrient supply and overcome stress factors limiting their availability. This method has been commercialized in a number of fruit crops like citrus, pineapple, guava, etc.

## 6.4. Biofertilizers

Of late, there is an increasing awareness in favour of adopting biological routes of soil fertility management for preventing soil degradation and for sustaining production. The atmosphere over a hectare of land consists of 80,000 tonnes of N. Though atmospheric N is present in sufficient quantity (80%), it is not available to plants since it exists in inert form. Biological nitrogen fixation (BNF) is the conversion of atmospheric N by living organisms into forms that plants can use. This process is carried out by a group of bacteria and algae which fix or convert the element of N into assimilable forms either in association with plants or free-living state. The conversion of inert N is facilitated by the enzyme nitrogenase, present in the

N fixing micro-organisms which are marked with 'nif' genes. They synthesize nitrogenase enzymes responsible for converting inert $N_2$ to plant usable $NH_3$.

Biofertilizers or microbial fertilizers or more appropriately 'microbial inoculants' are preparations containing live or latent cells of efficient strain of N fixing micro-organisms used for seed or soil application with the objective of increasing the numbers of such micro-organisms in soil or rhizosphere and consequently improve the extent of microbiologically fixed N for plant growth. They are used to either to fix N or to solubilize plant nutrients like phosphates or to otherwise stimulate plant growth through synthesis of growth promoting substances (PGPR) or to collect available P from remote places out of reach of plant root hairs by sending elongated filaments (AMF). Biofertilizers on application remain in soils, multiply and keep benefiting the growing crops.

Use of cost effective and eco-friendly biofertilizers with suitable integration of organic manures will restore the soil health and keep the soil productive and sustainable (Table 6.4). The beneficial microbes in the soil which are of greater significance to horticultural crops are the biological nitrogen fixers, phosphate solubilizers and the mycorrhizal fungi.

**Table 6.4.** Effect of bio-fertilizers on yield of some vegetable crops.

| Crop | Yield (q/ha) | | % increase over control |
|---|---|---|---|
| | Without biofertilizers | With biofertilizers | |
| Potato | 32.5 | 44.5 | 52.30 |
| Potato | 190.7 | 335.0 | 75.00 |
| Cabbage | 153.5 | 224.4 | 46.18 |
| Cabbage | 144.8 | 195.0 | 20.80 |
| Brinjal | 82.8 | 106.0 | 28.00 |

Biofertilizers improve soil physical properties and sustain soil fertility by providing aeration, biomass and nutrients. They not only make atmospheric N available but also solubilize and mobilize soil fixed P and improve P uptake in plants. They enhance plant growth due to release of hormones, vitamins, auxins, etc. They also help to proliferate beneficial micro-organisms which in turn suppress soil-borne pathogens.

### 6.4.1. Types of biofertilizers

There are 6 types of biofertilizers:

- Biological N-fixing micro-organisms.
- Phosphate solubilizing and mobilizing micro-organisms.
- Potash solubilizing micro-organisms.
- Sulphur mobilizing micro-organisms.
- Arbuscular mycorrhizal fungi.
- Growth promoting substance excreting micro-organisms.

### *(i) Biological nitrogen – fixing micro-organisms*

Biological N-fixing micro-organisms help in reduction of atmospheric $N_2$ to $NH_3$. Biological N-fixation contributes more than 175 million MT of N out of which legume N-fixation account for almost 40%.

The N-fixing organisms associated with horticultural crops are the *Rhizobium* spp. which lives in symbiotic association with roots of leguminous vegetables forming nodules and free living fixers-*Azotobacter* spp. and *Azospirillum* spp. which live in association with root system of crop plants. There are 2 types of rhizobia; (a) the slow growing *Bradyrhizobium* and (b) the fast growing *Rhizobium. Azospirillum* fix N from 10 to 40 kg per ha and saves N fertilizer inputs by 25 to 30%. *Azotobacter* inoculation saves N fertilizers by 10 to 20%.

#### *(a) Rhizobium and Bradyrhizobium*

They symbiotically fix N with leguminous plants increasing the amount of available N for uptake by plants. The quantum of N fixation ranges from 50-300 kg N/ha/crop under most optimum conditions (Table 6.5). An increase in yield of about 10 to 20% has been observed in pulses treated with *Rhizobium.*

**Table 6.5.** Quantity of N fixed by legume crop root nodules.

| Legume pulse crop | Nitrogen (kg/ha) |
|---|---|
| Cowpea | 80-85 |
| Bengal gram | 85-100 |
| Red gram | 168-200 |
| Soybean | 60-88 |
| Groundnut | 50-60 |
| Lucerne | 100-300 |
| Black gram/green gram | 50-55 |
| Clover | 100-150 |
| Cluster bean | 37-196 |
| Fenugreek | 44 |
| Lentil | 90-100 |
| Pea | 52-57 |

#### *(b) Azotobacter*

*Azotobacter* is a free living aerobic N-fixing bacterium (fixes about 10-25 kg N/ha/season) which is commonly found to be involved in close association with crops. 50% of N fertilizer application can be reduced through *Azotobacter* inoculation along with FYM. *A. chroococcum* is the dominant species in arable soils which rarely exceeds a population of $10^4$ to $10^5$/g of soil. *A. chroococcum, A. vinelandii, A. beijerinki* and *A. paspali* are known to form cysts to withstand adverse conditions. With the onset of favourable conditions, the cysts give rise to vegetative cells. These bacteria are also known to produce plant growth substances (IAA, gibberellins, Vitamin B, nicotinic acid, pantothenic acids, biotin and heteroauxin) which have beneficial effect on crop growth. It also acts as a crop protectant, since it releases antimetabolites which protect from root pathogens (*Fusarium, Alternaria* and *Trichoderma*). Vegetable crops in general (tomato, brinjal and cabbage) responded better to *Azotobacter* inoculation than other crops. It can also be used in fruit, plantation, spice and flower crops.

In case of pomegranate, the nitrogen use efficiency has been reported to be increased through the use of 250 g *Azotobacter* culture with 100 g N causing a net saving of 200 g N/plant (Wavhal *et al.*, 1985).

*(c) Azospirillum*

Azospirilla are a group of bacteria found in association with the root system of many crop plants. *Azospirillum* is known to fix nitrogen from 15 to 40 kg/ha. *Azospirillum* inoculation helps the plants in better vegetative growth and also in saving inputs of nitrogenous fertilizers by 25-30%. Four species of *Azospirillum* have been identified (*A. lipoferum, A. brasiliense, A. amazonense* and *A. irakense*). *A. lipoferum* and *A. brasiliense* are common in Indian soils. A reduction in the requirement of N by 25% to the crops receiving *Azospirillum* treatment is reported besides favourable effect on yield. The yield increase due to *Azospirillum* is mostly attributed to its capacity to produce hormones which establish early seedling vigour. *A. bangalorense* can fix 40 kg N/ha in the cortical cells of tomato cv. Pusa Ruby. *Azospirillum* biofertilizer can be used in fruit, vegetable and flower crops in which 11% yield increase can be obtained.

The inoculation of the nitrogen fixing bacteria, *Azospirillum* in combination with N fertilizer increased the banana fruit yield up to 13% in Poovan and resulted in saving of 280 g N/ha (Jeeva *et al.*, 1988).

*(d) Beijerinckia*

Its production is high in acidic soils. *B. indica* is a common species. *B. flumineuris, B. mobilis* and *B. derxii* are other species found in the tropical soils. It is generally present in the rhizosphere of plantation crops such as coconut, areca nut, cashew, cocoa and pepper.

### *(ii) Phosphate solubilizing and mobilizing micro-organisms*

It is known that Indian soils are very poor in phosphorus (P) content. Only about 25-30% of the applied P becomes available to the crop and the remaining part gets converted into insoluble/unavailable forms. It has been estimated that about 130 million tonnes of low grade P deposited (non-available) in our soil, but plant can not utilize this P directly. There are specific microorganisms which can soloubilize this P from non-available form to available form.

Several soil bacteria (*Pseudomonas, Bacillus*) and fungi (*Pencillium, Aspergillus*) possess the ability to bring insoluble phosphates in soil into soluble forms. Addition of these organisms (*Bacillus megaterium, B. arculans, B. subtilis, Pseudomonas striata, Aspergillus awamori, Pencillium digitatum*) to the soil increases the availability of phosphorus thereby saving the input of P. Bacteria and fungi solubilize bound phosphates in the soil by secreting organic acids such as formic, acetic, propionic, lactic, glycolic, fumaric and succinic acids which contribute about 10–25 kg $P_2O_5$/ha/season. These acids lower pH and bring about dissolution of bound form of phosphate. Some acids chelate with Ca and Fe resulting in effective solubilization and utilization of phosphate by crop plants.

The organic forms of P are phytin, phospholipids and nucleic acids which are added to the soil by way of decaying vegetation. Soil containing high organic matter are rich in organic forms of P. Large amount of P applied to various soils get fixed which is unavailable to the plants.

Microbial agents that regulate the availability of P in the rhizosphere have greater importance in organic farming. They help to convert the fixed forms of P– Al $(OH)_2H_2$ $PO_4$, Fe $(OH)_2$ $H_2$ $PO_4$, $Ca_{10}$ $(PO_4)_6$ $F_2$ into ionic available forms – H $PO_4^{2-}$, $H_2$ $PO_4$ that can be then taken up by plants. Indigenous and cheap source of P like rock phosphate can be more effectively used with these bio-inoculants. The most important mineral phosphate solubilizers are *Pseudomonas striata, P. fluorescens, Bacillus polymyxa* among bacteria, and *Aspergillus awamori, A. flavus, A. niger* and *Pencillium* spp. among fungi. In compost pits, the decaying plant materials release organic acids and other acids that can release the bound P in rock phosphate in the soil solution. Seed or soil inoculation with phosphate solubilizing organisms are known to increase the uptake of P by plants, increased growth and yield of beet root, cabbage and tomato to the extent of 10-20%. Field trials conducted have revealed that application of rock phosphate at 50 kg $P_2O_5$/ha to soil and inoculation with phosphate solubilizing organisms can give yield equivalent to those obtained with application of 50 kg as super phosphate.

Arbuscular Mycorrhizal Fungi (AMF) are also responsible for converting fixed P into available P. Through inoculation of efficient strains of AMF, 25 to 50% of P fertilizer can be saved.

### *(iii) Potash Mobilizing Micro-organisms*

It is known that Indian soils are moderate in potash (K) content. It is generally known that, directly and indirectly, K is a factor in the assimilation of $CO_2$ by plants. The plants well supplied with K contain more carbohydrates than plants deficient in this element.

Chandra *et al.* (1998) have isolated K mobilizing bacterium for the first time which is capable of mobilizing elementary or fixed K into a usable form to the plants. The mobilizing power is so high that it can save up to 50-60% of the cost of K fertilizer. The bacterium, *Frateuria aurantia* (Family: Pseudomonadaceae) was isolated from banana plant from Orissa soil. The bacterium has a solubilizing power of 90% within 22 days when the mineral source of K is in fixed form. The bacterium was tested on banana which increased the yield by 20%. It can be used as soil application for all types of crops at 2.5 kg/ha. It has to be mixed with 200-500 kg FYM in furrows before sowing. The bacterium can save up to 50-60% of the cost of chemical K fertilizer. *F. aurantia* has been reported to increase maximum level of K availability to the magnitude of 5-60 kg/ha.

### *(iv) Sulphur Mobilizing Micro-organisms*

Chandra *et al.* (1998) isolated the bacterium, *Acetobacter pasteurianus* which secretes the sulphur and makes it available to the plants in usable form. Sulphur present as insoluble sulphate form, are formed 30-35 cm deep in soil and are associated with oxides of Iron and Aluminium. *A. pasteurianus* helps in converting this non-usable form to usable form. The use of 625 g/ha of *A. pasteurianus* influenced the levels of sulphur in crops like vegetables, cabbage, turnip, onion, etc.

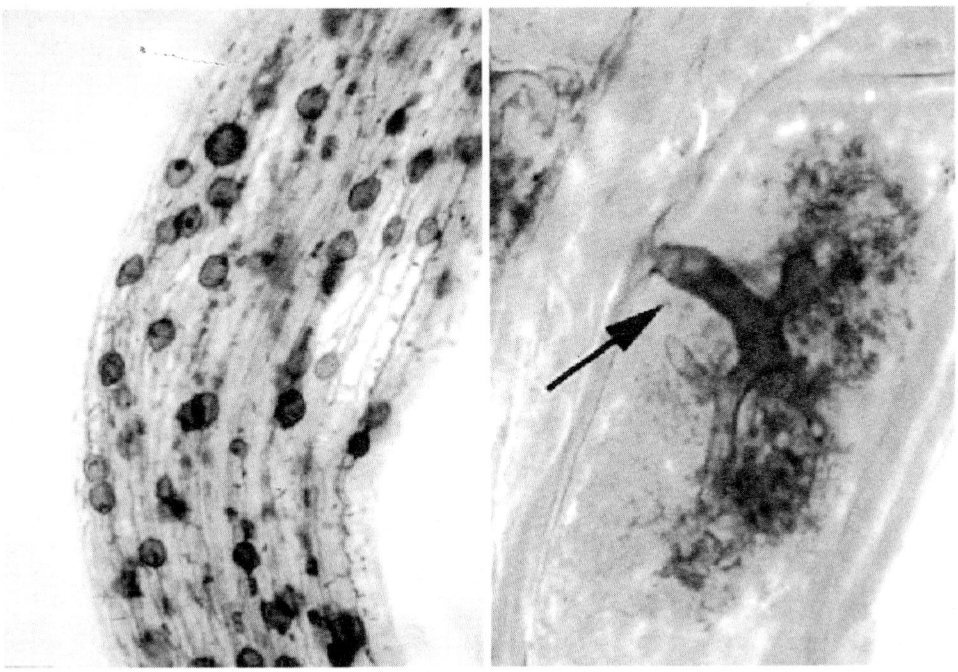

**Fig. 6.1.** Arbuscular mycorrhizal fungi. Left – Vesicles in root, Right – Arbuscle in a root cortical cell.

### *(v) Arbuscular mycorrhizal fungi (AMF)*

Arbuscular mycorrhizal fungi develop both intra and extra matrical hyphae that extends far areas away from the roots and increases the absorptive surface area of the P by mycorrhizal root system. AMF improve plant growth through better uptake of nutrients like P, Zn, Cu, etc. and make the root zone inamicable to root pathogens. They improve the activity of beneficial soil micro-organisms like N-fixers and P-solubilizers. AMF also help the plant to withstand water stress and transplant shock. They are obligate symbionts and can be used for transplanted crops like chilli, tomato, brinjal, potato, grapevine, apple, banana, tea, coffee, cocoa, rubber etc. AMF can be inoculated to nursery beds at 2 kg/m$^2$ and pre-colonized seedlings can be transplanted to harness the benefits of mycorrhization. AMF saves 25-50 kg P/ha in addition to enhancement of yield to an extent of 10-12%.

The most common mycorrhizal fungi belong to the genera *Glomus, Gigaspora, Acaulospora* and *Scelerocystis* (Family-Endogonaceae, Order-Mucorales, Class-Phycomycetes). They produce vesicles and arbuscules inside the root system (Fig. 6.1). Arbuscles are highly branched fungal hyphae while vesicles are the bulbous swellings of these hyphae. These AMF make more nutrients available to the host plant, improve soil texture, water holding capacity, disease resistance and help in better plant growth. Besides, AMF are also helpful in the biological control of root pathogens.

Seedling (tomato and chilli) inoculation with endomycorrhiza help better colonization, plant growth, nutrient uptake and increased crop yields, besides saving of P by 50%.

## Chapter 7

# BIOREGULATORS

Ever since the discovery that the plant hormones have the ability to influence several physiological processes in plants, attempts have been continuously made to manipulate the growth and development process in different horticultural crops to streamline the production. As the years passed, besides the natural compounds, a number of synthetic compounds were also found to have growth regulatory roles, which are now collectively known as bio-regulators. Many of the research accomplishments have led to the commercial applications.

The use of bioregulators to increase the vitality and therewith the productivity of horticultural crops is a distinct possibility. In several experiments the positive effect of different bioregulators used (Humate, Lactate, *Bacillus subtilis*) could be discovered if one of the growth factors was at a critical level. Humates are effective in nutrient and salt control as well as subtropical temperature. Lactates have a good effect if the pH-value is unstable. Microorganisms like *Bacillus subtilis* are effective under different stress conditions. They have an unspecific positive effect in particular if the nutritional status is suboptimal. However, horticultural crops with long vegetative or fruiting phase are often confronted with different stress factors. A combined application of bioregulators could be beneficial in such complex stress situations in order to improve the productivity of the horticultural crops.

Bioregulators have been used for many years to alter the behavior of fruit or fruit trees for the economic benefit of the fruit grower. Control of vegetative vigour, stimulation of flowering, regulation of crop load, reduction of fruit drop, and delay or stimulation of fruit maturity and ripening are important examples of processes in fruit and fruit trees that can be regulated with exogenous applications of bioregulators.

The use of plant bioregulators in ornamental crops is more prevalent than in edible crops. The most common types of regulators used are gibberellins (GAs, most often $GA_3$), ethylene and their antagonists. GAs are used to enhance elongation of many cut flowers and to promote bud break, thus producing more flowering shoots. They are commonly used to promote flowering of long day (LD) and cold requiring plants grown under marginal inductive conditions and of autonomous-flowering plants of the Araceae. GA also enhance post-harvest life of certain flowers and foliage plants. The main use of ethylene in ornamentals is to promote flower formation, especially in bromaliads and certain bulbous plants. When desired, ethylene also promotes lateral branching. The largest use of bioregulators in ornamental crops is probably that of GA antagonists - the growth retardants. They are commonly used in pot plant production to achieve more compact and attractive structure they promote flowering in certain woody ornamentals. Endogenous and external ethylene is the main factor reducing post-production life and promoting flower and foliage abscission of many ornamentals. Silver-thio-sulphate (STS) is the main ethylene antagonist used to protect plants from ethylene damage. Concern about the impact of STS on the environment has promoted a search for effective alternatives. Those presently available, however, are less effective than STS. An effective chemical may be a volatile ethylene binding-site inhibitor such as diazocyclopentadiene. Manipulation of cytosolic $Ca^{2+}$ level may become a novel means for controlling flowering, flower senescence and geotropic bending of flowers.

### 7.1. Rapid mass multiplication of the planting material

Accomplishment of a higher success in the production of true-to-type planting material in some fruits has become a mile-stone in fruit development in India. Some of the fruit species, which are either difficult-to-root or slow-to-root have been made to root easily and rapidly with the use of bioregulators. Indolebutyric acid (IBA) and Naphthalene acetic acid (NAA) have been especially found to be outstanding root promoting bioregulators.

Bioregulators have been used for successful air layering in sapota and gauva. Uthaiah *et al.* (1976) found that the application of a mixture of IBA and NAA in equal proportion of 15,000 ppm to the girdled portion of pre-etiolated and ringed shoots resulted in higher and hastened success of the root layers. Higher success in air layers of the juvenile guava shoots was obtained with the application of 3000 ppm each of IBA and NAA mixture (Rao and Sulikeri, 1997). NAA has synergistic effect on the rooting, when mixed with IBA formulation.

Bioregulators help in improving seed germination in sapota and *khirnee* rootstock. Farooki *et al.* (1971) obtained a better germination of the decoated sapota seeds in a solution containing 100 ppm ethrel and 50 ppm $GA_3$. Pre-soaking of the *khirnee* seeds in 400 ppm ethrel gave the highest germination (Pampanna, 1992).

### 7.2. Stimulation of lateral branching in young trees

Proper branch development is a critical feature in the care of young fruit trees. Good branch development begins in the nursery and continues in the orchard. Techniques for effective stimulation of branch development that avoid pruning could shorten the time needed to develop a fruitful canopy. Dr. Elfving's research discovered the lateral branching potential of a new candidate bioregulator, cyclanilide® (Fig. 7.1). This product effectively produces excellent lateral branch development in apple, pear and sweet cherry trees. It can be used in the nursery as well as the orchard.

Branched trees from the nursery, crop more quickly when planted in the orchard. Improved branching and more rapid canopy development could significantly reduce the time required for new orchards to amortize their establishment costs. This single fact would return more money earlier as profit to the grower, and that benefit would continue over the life of the orchard.

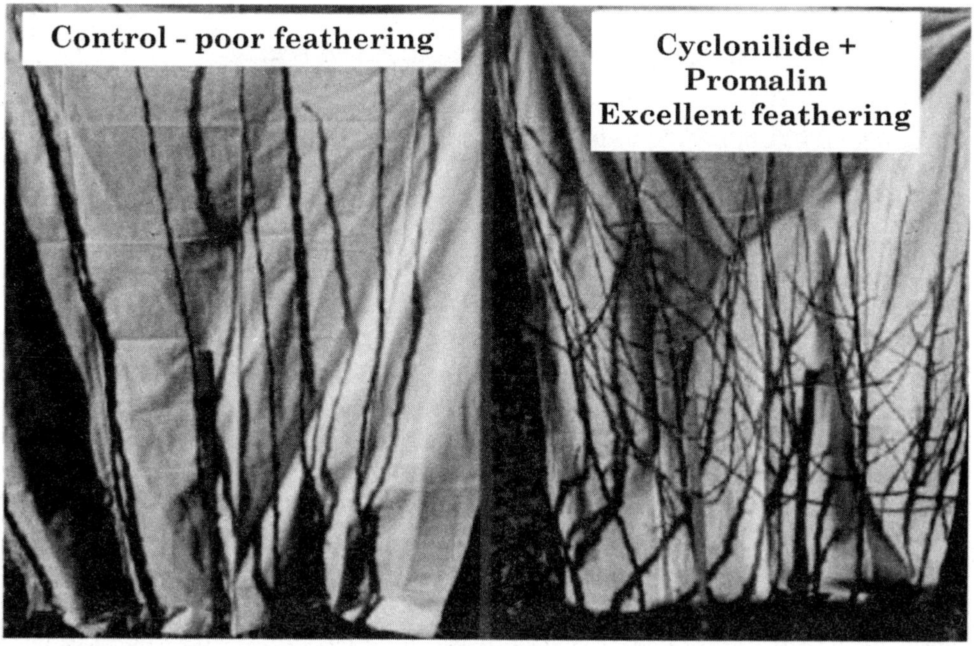


**Fig 7.1.** Stimulation of lateral branching in young trees with bioregulator cyclanilide

## 7.3. Induction of flowering

Bioregulators have been used to overcome the problem of alternate bearing in mango and apple. It has been reported that paclobutrazol (cultar) has good potential both as foliar spray and soil application in overcoming biennial bearing in mango. Soil application of paclobutrazol at 10 g mixed in 5 litres of water in the ring of 15 cm depth dug 60 cm away around the tree trunk in August resulted in appreciable flowering in the following expected "off" year in Alphanso (Rao *et al.*, 1997, Srihari and Rao, 1996).

Alternation in cropping from year to year has always been a problem in apple production. Chemical thinning techniques have been used for half a century to address this problem but with limited success. Removal of young fruitlets alone is not always sufficient. The use of gibberellic acid (GA) has been evaluated as a technique for reducing flower formation in the low crop year to see if the alternating cycle can be effectively interrupted with a combination of flowering control in the low crop year and good thinning practices in the high crop year. Consistent reductions have been obtained in flower formation in the 'Fuji' cultivar when GA is applied shortly after flowering.

Evening out cropping from year to year would greatly benefit the profitability of the tree fruit industry by diminishing the impact of reduced prices in heavy crop years and improving returns to more growers in what would have been light crop years. The use of bioregulators to reduce biennial bearing appears promising at this time, but much more work is necessary before it can be recommended for commercial practice.

## 7.4. Promoting the growth and development of fruits

Bioregulators help in improving the fruit size of banana and grapes. Spraying of 100 ppm $GA_3$ twice at 25 and 50 days after the shooting resulted in an increase of 32.5, 23.9 and 16.98% of length, girth and weight of the fingers, respectively in banana cv. Rajapuri (syn. Munavalli) which is popular in Belgaum district of Karnataka (Rao, 1977).

Application of $GA_3$ has caused multi-fold increase in berry size and consequently productivity has been raised from 6 to 8 MT/ha to 15 to 18 MT/ha in Thompson Seedless cv. of grapes.

Bioregulators have been used in improvement of colour and ripening in grapes. Rao and Shankarnarayanan (1982) could overcome the problem of uneven ripening in Gulabi cv. of grape by dipping the clusters twice at 5 days interval during the lag phase in 250 ppm ethrel containing 2% urea. Use of 50 ppm 2,4,5-TP was also found effective in this respect (Hanumashetti *et al.*, 1981).

Bioregulators help in controlling the berry drop in Anab-e-Shahi cv. of grape. Spraying of 100 ppm NAA or PCPA a week before harvest reduced the berry drop (Rao *et al.*, 1968). Mixing of 100 ppm BA along with 100 ppm NAA was much more effective. NAA helps in thickening of the pedicel attachment.

Chapter 8

# BIOTECHNOLOGICAL APPROACHES

## 8.1. Tissue culture

### 8.1.1. Tissue culture for micropropagation

This micropropagation technique offers not only means for mass propagation, but also plays an important role to conserve elite or rare plants that are threatened with extinction. Moreover, meristem tip culture offers the possibility of virus eradication. These technologies started about 20 to 25 years ago, especially for ornamental plants, and for a few other crops: potato, strawberry, bananas propagated clonally, it was very important to have a true to type propagation technique. African countries are producing potato, banana and cassava through *in vitro* micropropagation.

On the other hand, it is also possible to use tissue or cell culture to increase genetic variability. Undifferentiated cells obtained from callus, cells or protoplasts culture are produced and submitted to a selective pressure to improve and fix the somaclonal variants.

#### *(i) Production of true to type plant*

*(a) Meristem tip culture*

***Production of virus-free plants:*** In 1952 Morel and Martin were successful in regenerating a virus-free dahlia plant by the excision of some meristematic domes from virus infected shoots. Three years later, the same authors were able to eliminate virus A and Y from virus infected potato. Semal and Lepoivre (1992) reported that a virus-free sweet potato was producing 40 MT/ha in China by comparison of the 20 MT/ha produced before meristem culture.

Virus eradication is dependent on several parameters. But to take advantage of the non-uniform and imperfect virus distribution in the host plant body, the size of the excised meristem should be as small as possible. For Stone (1963), only tips between 0.2 and 0.5 mm most frequently produce virus free carnation plants. The explants smaller than 0.2 mm can not survive and those larger than 0.7 produce plants that still contain mottle virus.

Today, no definitive explanation can be given to understand this virus eradication (Wang and Charles, 1991). Various explanations have been given: absence of plasmodesm in the meristematic domes, competition between synthesis of nucleoproteins for cellular division and viral replication, inhibitor substances, absence of enzymes preiral replication, inhibitor substances, absence of enzymes present only in the cells of the meristematic zones, and suppression by excision of small meristematic domes. This last proposal could explain why some potato plant showing virus particles in the meristematic domes, could regenerate a virus free plant (Mellor and Stace Smith, 1977).

*(b) Shoot tip micrografting*

When meristem tip culture fails, it is possible to graft small meristematic domes on young seedlings growing *in vitro*. In this way, Navarro and Juarez (1977) eradicated all the virus diseases from Spanish citrus orchards.

This technique was also very successful in eliminating virus diseases from peach trees (Mosella *et al.*, 1980).

***(ii) Micropropagation techniques***

During the micropropagation process, the genetic stability of the new shoots is dependent upon their origin. Axillary shoots issue from pre-existing buds and are normally true to type; indeed the meristematic cells are genetically very stable.

Adventitious shoots, such as somatic embryos, are neoformed buds developed directly on some organs, or indirectly through a callus phase formed on this organ. So, if the mother plant presents a cell mosaic or chimaeric tissues, risks of genetic variation exists. It is similar in the case of an indirect regeneration, when the callus phase is too long.

*(a) Propagation by axillary shooting*

This technique has proved to be the most applicable and reliable method of *in vitro* propagation. Axillary shoot growth is stimulated by overcoming apical meristem dominance.

Commercial tissue culture laboratories are now able to propagate a large number of herbaceous ornamental species and several woody plants in this way.

*(b) Propagation by direct or indirect organogenesis*

Adventitious shoots could arise directly from the tissue of explants without callus formation. Several plants of the family Gesneriaceae (*Saintpaulia, Streptocarpus*) regenerate directly buds on leaf explants, likewise Lilium regenerates on scales.

However more often, like for *Ficus lyrata*, adventitious buds appear on callus. While coffee, cocoa trees, and many conifers are produced by somatic embryogenesis developed on callus or cell suspensions.

*(c) Improvement of axillary branching*

The cost of micropropagated plantlets is also an important limitation of the techniques. In New Zealand, where they produce 2-3 million micropropagated pine per annum, the relative cost of micropropagated planting stock had dropped from 13.8 times the cost of seedlings in 1988 to 6.9 by 1993 (Smith, 1997).

To reduce manpower costs, several improvements have been proposed. The more simple method was *in vitro* layering developed by Wang (1977) to clone PVX-free potato plants. The first plantlets placed on the medium in a horizontal position developed axillary shoots. They are harvested by cutting one cm above the medium surface, at 3 weeks intervals.

Ziv (1990) proposed for corm plants, Gladiolus and Nerine, a very rapid propagation system. She reduces the internodes and leaves by introduction of an anti-gibberellin agent in the medium. Finally, only aggregates of buds are formed, then they are divided and introduced in bioreactors for mass production. Similar systems were developed in Gembloux, to propagate asparagus.

*(d) Somatic embryogenesis propagation*

For genetically stable species, somatic embryogenesis offers a very fast scaling-up system, especially when it is possible to produce embryos in bioreactors. Unfortunately this production via fermenters was not as simple as first envisaged. And today, only a few model plants are successfully produced by such technology: carrot, celery. Other applications remain at the experimental stage: coffee, oil, palms, conifers, *Euphorbia pulcherrima* and several other horticultural species.

*(e) Artificial seeds*

Another very interesting possibility of the somatic embryogenesis technology has been developed during the past fifteen years by Redenbaugh (1993) and Redenbaugh *et al.* (1991). They were able to encapsulate somatic embryos by hydrogel coatings (sodium alginate), producing single embryo artificial seeds.

To date, some improvements offer the possibility to directly plant the artificial seeds in the greenhouse on special substrates (vermiculite, sans). This methodology will provide in the future a good technique to reduce the cost of transplants.

*(f) Gene bank*

Tissue culture methods offer the opportunity for *in vitro* collecting, rapid multiplication and distribution of important, elite, or rare plants that are threatened with extinction. The two major *in vitro* storage strategies are slow growth and cryopreservation.

Since the first results of Seibert (1976), who was able to initiate shoots from carnation shoot apices frozen to -196°C, this technique is now successful for many of horticultural species. Dereuddre *et al.* (1991) have provided a very simple technology to freeze encapsulated meristems in dried alginate beads. It works for pear, strawberry, eucalyptus and potato.

Some international institutions have very large collections of old and current varieties, available for exchange and introduction in crop improvement programmes. The International Potato Center (CIP) in Lima, Peru, has a large world potato collection. Germplasm of sweet potato and cassava is at the International Institute of Tropical Agriculture (IITA), Ibadan, Nigeria.

The movement of germplasm involves the risks of accidentally introducing plant quarantine pests along with the host plant material. To limit these risks, the plant material should be transferred from one country to another as *in vitro* cultures through a transit Centre, where it should be indexed. For bananas, in the framework of INIBAP, the transit Centre is the Catholic University of Leuven in Belgium where a very large *in vitro* germplasm exists.

### 8.1.2. Tissue Culture Techniques to increase genetic variability

#### *(i) Somaclonal variations*

The production of plantlets by callus regeneration, cell suspensions, protoplast cultures, could present some deviations with regard to the mother plant. This is a way to increase the genetic variability. Associated with a selective pressure (stress to toxins, pH, salinity, cold) it is used to obtain resistant lines. Indeed, after regeneration, plants can express new potentialities rarely obtained by another way. Stable and profitable variants are selected and introduced in breeding programmes. In 1976, a Pelargonium cv Velvet Rose was created by this technique (Reisch, 1983). Later, other applications were described, as in tomato and red pepper. Following Sibi (1990), extrachromosomic systems must be involved to explain the genetic behaviour of some new expressed tissue culture characteristics. She uses the term "epigenic" to evoke a whole of hereditary elements which belongs to the cytoplasmic or nuclear compartments, but are not transmitted by Mendelian rules.

#### *(ii) Haplodiploïdisation*

Many species are able to produce haploids through different *in vitro* techniques. The oldest was anther culture of androgenesis. To date, the results vary considerably from one species to another. Solanaceae (red pepper, eggplant, petunia) or crucifers (soybean, cabbage) are species easy to regenerate by anther culture. To the contrary, tomatoes, leguminous, or Compositae are recalcitrant.

Ovule culture, or gynogenesis, was a successful technique for gerbera, beet and courgette. However, the most common technique is pollination with irradiated pollen, in order to induce *in vivo* parthenogenesis. The oospheres developed in embryos without fertilization are saved by embryo rescue.

The haploid plantlets (n chromosomes) are treated with colchicine to produce fertile homozygous lines, called "double haploid lines". This haplodiploïdisation technique could give immediately new elite genotypes, hybrid parents after *in vitro* propagation, as asparagus super males (MM), or useful genetic material to establish gene mapping.

### *(iii) Protoplast culture*

Protoplasts are the smallest units able to regenerate a whole plant. Therefore protoplasts cultures can serve to enlarge genetic variability by introducing somaclonal variation. However, the main interest of protoplasts, is their capacity to fuse and to produce hybrids or cybrids, new organisms most often unknown in nature. Indeed, protoplasts allow to transgress the barrier of the botanical species or genus. Naked protoplasts can accept without rejection external elements: nuclei, cytoplasmic organelles, liposomes containing genetic information. The first protoplast fusion application was a cytoplasmic transfer from one genotype to another to induce male sterility (CMS) from mitochondrial origin. These male sterile hybrids are interesting to produce F1 hybrids (*Brassica, Cichorium*).

Another cytoplasmic transfer concerns the introduction of chloroplastic DNA to induce herbicide (atrazine) resistance, from *Solanum nigrun* to tomato for instance (Jain *et al.* 1988).

Many intergeneric and interspecific hybridizations are possible but not always profitable because these hybrids are too often infertile. This problem does not exist if we can vegetatively propagate the hybrids, as in the case of citrus rootstocks. In Florida, Grosser (1990) was able to analyze several thousand citrus hybrids.

## 8.2. Molecular biology

More and more pathologists are using molecular probes for early detection of diseases. However, the main contributions of modern biotechnology remain probably in the hands of the plant breeders. Indeed, molecular markers are of great use in detecting desirable new genes, or to identify important quantitative trait loci (QTL). This marker-assisted selection is very useful for many vegetable plants. Other applications are DNA – fingerprinting and genetic engineering. This recent technology aims to insert and to express new specific genes into a selected plants.

### 8.2.1. Marker-assisted selection

Molecular markers and genome mapping will reach a large extension in the next breeding programmes of tomatoes, peas, cabbages, melons and potatoes.

It will be possible to improve the speed and the incorporation efficiency of desirable new genes by utilizing closely linked selectable molecular markers. The number of back crosses will be reduced. Tanksley *et al.* (1981) showed that only with 12 markers, one on each tomato chromosome, the composition of the recurrent parent genome is similar to that observed in the absence of selection after the third back cross two years later. Therefore, it is easy to understand the high improvement that we can get with the 700 tomato markers already known in 1990 for this crop (De Verna and Paterson, 1990).

The utilization of molecular markers could also identify important quantitative trait loci (QTL). Recently 13 RAPD markers at eight genomic regions that were associated with QTLs affecting salt tolerance during germination in tomato have been identified.

### 8.3. DNA finger printing

The DNA polymorphism observed by RFLP, AFLP, or RAPD, allows cultivar identification of fruit trees, apple, citrus or other vegetatively propagated plants. Although some mutants, as coloured fruits, escape these identification techniques, some people would evaluate the genetic uniformity of regenerated plants in this way.

### 8.4. Genetic engineering

Genetic engineering consists of introducing a foreign gene into a plant genome to create a new function. Or at the contrary, it could be to reduce or suppress an existing gene function, as in the antisense strategy.

#### 8.4.1. Transgenic plants with new agronomic traits

Two different techniques are commonly used to introduce genetic information (DNA) inside cells, protected by pectocellulosic walls. A particle bombardment process, called ballistic, is used with success for monocots gymnosperm, squash and peas. Today, this technique allows the bombardment of meristems and avoids the difficulties of regeneration. But the mediation of *Agrobacterium* remains till now the most routine system to transform dicotyledonous plants.

The main agronomic traits introduced in horticultural plants and already commercialized are *Bt* toxin and herbicide resistance. Other studies concern virus resistance and male sterility.

Transgenic vegetable varieties suitable for processing have been developed. Bruise-resistant, delayed ripening with higher sucrose and reduced starch content tomatoes are now available. High starch content in potato, dwarf lettuce for individual serving size and broccoli with slow ripening and longer staying (green) capacity, transgenics using genes for pest and disease resistance, nutritional quality of international standard, and for development of slow ripening cvs. for PHM in tomato, brinjal, cabbage and cauliflower using anti-sense RNA have been evolved.

### 8.4.2. Antisense strategy

Calgene created in 1994 the first commercial transgenic plant, a long shelf life tomato, by the suppression of polygalacturonase activity due to an antisense gene (Smith, 1988). However, this Flavr Savr tomato variety was removed from the trade 3 years later, because of its disease susceptibility and its lack of productivity.

Later, other tomato varieties with long storage qualities were obtained by the utilization of an antisense RNA inhibition of ACC synthase or ACC oxydase, two ethylene precursors.

The antisense technique was also used to reduce the lignification of woody plants, by blocking the enzymes involved in the precursor of lignin biosynthesis. Another interesting application was the induction of white flowers in petunia and in different other ornamental plants by the suppression of chalcone synthase activity.

### 8.4.3. Transmission of the new traits

These traits induced by the transferred DNA are transmitted to the progeny as a dominant Mendelian character. Nevertheless in some cases, transgenic plants with a strong gene expression can generate progenies with only a faint no gene expression. This problem is not clearly elucidated.

### 8.4.4. Marker, reporter, promotor and expression genes

To select the transformed cells, some marker genes, very often resistant to antibiotics or to herbicides, are attached to the coding sequence, as a promoter or an expression gene. These allow the new gene to express in the whole plant or in a specific plant tissue.

Some reporter genes are used to follow the evolution of the transformed cells. The most frequent reporter genes are the gene "gus" (glucoronidase) or the gene "lux" (luciferase).

### 8.4.5. Horticultural transgenic crops already released in USA

The Flavr Savr tomato was the first genetically engineered whole food approved for commercial sale in 1994. Four other transgenic tomatoes with a delayed ripening were approved later (1995 and 1996).

In 1995, potatoes with *Bt* genes and a squash cultivar resistant to two viruses were released. Two years later, another squash cultivar resistant to 3 viruses and a papaya line also resistant to viruses were approved.

Two other horticultural crops are waiting the authority's approval: red hearted chicory (Radicchio) with male sterility and resistance to herbicide, and a tomato with Bt toxin.

If there are so few commercial transgenic plants in horticulture, it is probably due to the relatively low acreage of these horticultural crops in comparison with other agricultural crops. In 1997, in the world 5 million ha were planted with glyphosate resistant soybeans released by Monsanto alone.

Nowadays, it is true also that many transgenic plants exist in research laboratories, awaiting authorization for field testing, as they are very interesting model plants for learning physiology: lettuce with less nitrate by increasing nitrate reductase gene activity, a yeast ribonuclease gene to reduce viroid infection on potato, modification of lignin synthesis in plant to produce trees adapted to paper industry, or timber, or biomass production, regulation genes to control tree architecture.

Meanwhile, genetic engineering techniques are being applied mainly to crops which are important for the industrialized world, not crops on which the world's hungry depend. Therefore, it is unrealistic when Monsanto writes "the experts said that biotechnological innovation will increase the crop productivity without occupation of new lands, saving tropical forest of best quality and animal habitat". However, the World Bank was promising in a 1997 publication "Bioengineering of Crops" that "transgenic crops could improve food yield by up to 25 per cent in the developing countries and could help to feed an estimated additional three billion people over the next 30 years".

The objectives of Agricultural Biotechnology for Sustainable Productivity Project (ABSP project) which is managed by Michigan University are the reduction of losses due to pathogens and pests by using transgenic plants (sweet potatoes, potatoes, cucurbits, tomatoes) and by cloning commercial value-plants (bananas, pineapple, coffee).

## 8.5. Perspectives, limitations and environmental risks

Ecological impact related with the introduction of GMOs is always an open debate.

The application to agriculture of these new technologies certainly opens interesting perspectives, but also raises potential problems. The risk of crop transgene spreading has been demonstrated. A researcher of Clemson University in South Carolina reported "that in a population of wild strawberries growing within 50 meters of a strawberry field, more than 50% of the wild plants contained marker genes from the cultivated strawberries" (Kling, 1996).

**Table 8.1.** Estimated global economic, environmental, and human health benefits of *Bt* transgenic plants (Hokkanen, 1998)

| Global Market Penetration of Transgenic Plants (% use) | Reduced Use of Conventional Insecticides (US$ millions) | Cost of Transgenic Plant (US$ millions) | Economic Gain (US$ millions) | Estimated Environmental Gain (US$ millions) | Estimated Human Health Gain (US$ millions) |
|---|---|---|---|---|---|
| 1 | 90 | 45 | 45 | 91 | 16 |
| 10 | 900 | 47 | 853 | 917 | 157 |
| 25 | 2250 | 49 | 2201 | 2293 | 394 |
| 50 | 4500 | 51 | 4449 | 4585 | 787 |

A Danish team has shown a possible rapid spread of genes from oilseed rape to the weedy relative Brassica campestris.

The introduction of Bt gene allows a drastic reduction in the use of toxic chemicals for crop protection. But a poorly controlled use of *Bt*-technology can destroy more effectively the predators than the pests. Or when" many crop plants are transformed with similar effective traits, in such situation, many polyphagous pest species, which by nature are more flexible evolutionarily than those that have a narrower diet, are likely to overcome Bt resistance very quickly" (Hokkanen, 1998).

Therefore, before releasing a transgenic plant, any risk has to be weighed against the benefit of the transgenic crops. We must not forget that annually in the world 500,000 acute pesticide poisonings, with 5000 deaths, are observed. Some of these estimated environmental and human health benefits potentially provided by *Bt* crops are presented in Table 8.1.

Chapter 9

# PROTECTED CULTIVATION

Protected cultivation technology, using different kinds of structures, offers several advantages to produce vegetables and flowers of high quality with high yields and minimum risks due to uncertainty of weather. It also ensures chances for efficient use of land and other natural resources. The main purpose of protected cultivation is to create a favourable environment for the sustained growth of plant so as to realize its maximum potential even under adverse climatic conditions. There is good opportunity for greenhouse cultivation of vegetables, selected fruits and flowers by economic methods of water and fertilizer use such as fertigation. Low cost greenhouses along with micro-irrigation and fertigation techniques may be popularized in areas where evaporation exceeds precipitation during several months in a year. This becomes relevant to growers in various parts of the country who have small holdings, say less than 2 ha. They would be interested in a technology, which helps them to produce more crops each year from their land, particularly during off-season when prices are higher in nearby markets. Vegetable and flower production under location-specific and cost-effective structures may give substantial benefits even to small growers.

The protected cultivation technology is still in its preliminary stage in the country and concerted efforts are required from all concerned agencies to bring it at par with the global standards. Globalization coupled with economic liberalization will help in achieving the desired results. Work should be channelized in finding suitable and locally available construction material for low and medium cost greenhouses. Economically viable and technologically feasible greenhouse technology suitable for the Indian agro-climatic and geographical conditions is needed at the earliest. A bamboo frame green house cum rain shelter is used in NEH region. Low cost green houses have been found to be very successful under subtropical conditions for raising nurseries of vegetable crops, flowers and fruits. Environmental considerations in future will demand that the greenhouse technology of the 21st century is of the closed type with complete recycling of everything except the crop produce.

## 9.1. Protected cultivation of vegetable crops

Production of vegetable crops under protected environment has a great potential in augmenting production and quality of vegetables in off-season and maximizing water and nutrient use efficiency, under varied agro-climatic conditions of the country. By the year 2020 India needs to increase its productivity from 16 MT/ha to 25 MT/ha which calls for a better crop husbandry including the use of appropriate technology and alternative inputs. Protected cultivation of vegetables is one such area, which could be deployed effectively to tide over the short fall in vegetable production of India.

In cold arid regions, protected cultivation is the only alternative for cultivation of vegetables in winter. Of various greenhouse structures evaluated, the trench has been found to be the most suitable and economic structure for the region followed by Ladakhi polyhouse (Singh *et al.*, 1998). It retains the increased day temperatures even during night.

Vegetables can be grown in solar greenhouses in humid temperate regions. Performance of some vegetable crops like tomato, cucumber, bean, squash, bitter gourd, capsicum, brinjal, spinach, lai (vegetable mustard) and lettuce grown at Auli (9000 feet from MSL) and Harsil (10000 feet from MSL) in Uttarakhand is presented in Table 9.1 (Singh and Dhaulakhandi, 2002).

**Table 9.1.** Performance of vegetables in solar greenhouse at Auli and Harsil

| Vegetable crop | Variety/hybrid | Yield (kg/m$^2$) | |
|---|---|---|---|
| | | Auli (9000 feet) | Harsil (10000 feet) |
| Tomato | Surya Kiran | 5.45 | --- |
| | HT-8 | 6.16 | 5.90 |
| | HS-102 | 3.86 | --- |
| | Pusa Ruby | 3.86 | 6.00 |
| | Marglobe | 3.93 | --- |
| | HT-6 | 4.79 | --- |
| | Rajnee | 4.63 | --- |
| | Pusa Early Dwarf | 3.97 | --- |
| | Rupali | 4.36 | --- |
| Bean | VL-1 | 1.00 | --- |
| | VL-5 | --- | 3.00 |
| Lai (vegetable mustard) | ARU Black | 2.44 | 3.50 |
| Bell pepper | Caifornia Wonder | 3.15 | --- |
| Brinjal | ARU-2C | 4.29 | --- |
| Cucumber | Pusa Sanyog | --- | 11.00 |
| Squash | Australian Green | --- | 8.00 |
| Pea | Arkel | --- | 1.70 |
| Lettuce | Great Lake | --- | 2.10 |

Leafy vegetables can be grown in Leh (dry temperate/cold desert region). The yield of different vegetable varieties varies under different types of greenhouses is presented in Table 9.2. Trials on growing different vegetables revealed that beet leaf (palak) was a good crop for cultivation in winter. It had given yield ranging from 2.2 kg/m$^2$ to 6.3 kg/m$^2$. The maximum yield was observed in trench (6.3 kg/m$^2$) followed by local polyhouse (5.4 kg/m$^2$) (Singh and Dhaulakhandi, 2002).

**Table 9.2.** Average yield of vegetables in greenhouses at Leh during winter (October-April)

| Vegetable crop | Variety | Yield (kg/m$^2$) | | | |
|---|---|---|---|---|---|
| | | Glasshoue | Polyhouse | Local polyhouse | Trench |
| Palak | Mongol | 2.3 | 2.3 | 3.1 | 3.9 |
| Beet leaf | K. Raya | 3.7 | 2.9 | 4.8 | 5.2 |
| Fenugreek | Kasuri | 1.3 | 0.8 | 2.5 | 2.8 |
| Veg. Mustard | ARU-Black | 1.8 | 1.2 | 3.0 | 3.0 |
| Coriander | PD-1 | 1.0 | 0.7 | 1.6 | 1.9 |
| Mint | FRL-1 | --- | 1.2 | 2.3 | 2.7 |
| Orich (Dum) | Local | --- | 1.1 | 2.0 | 2.7 |
| Lettuce | Great Lake | 2.0 | 1.7 | 2.8 | 3.0 |
| Celery | Murtse-1 | 2.1 | 1.8 | 3.0 | 3.3 |
| Parsley | Leh-2 | 1.4 | 1.5 | 3.0 | 3.7 |
| Radish | Pusa Himani | 1.0 | --- | 2.8 | 2.7 |
| Turnip | Purple Top | --- | --- | 1.6 | 2.1 |
| Cabbage | Golden Acre | 2.0 | 2.2 | 2.8 | --- |
| Knol-khol | White Viena | 1.2 | 1.0 | 1.5 | 2.5 |
| Karam Sag | Kashmiri | --- | --- | 1.9 | 1.3 |

Cultivation of tomato, chilli and cucumber in greenhouse was demonstrated to the Nagaland farmers (Keitzar and Tiameren, 2003). Three crops namely; cucumber, tomato and chilli were cultivated inside the greenhouse and round–the–year–cultivation approach was adopted. Yields as affected by different planting time are given below (Table 9.3).

**Table 9.3.** Yield of crops grown in low cost greenhouse–cum–rain shelter

| Crops | Yield in kg/100 m$^2$ (average of three years) | | |
|---|---|---|---|
| | Planting time | | |
| | Feb – Mar | June – July | Oct – Nov |
| Cucumber | 227 | 222 | 212 |
| Tomato | 228 | 277 | 285 |
| Chilli | 96 | 98 | 103 |

The yield of vegetables grown under the low cost polyhouse is very encouraging (Agarwal *et. al*, 2003) (Table 9.4). Farmers of Meghalaya are using polyhouse for raising the seedlings of vegetable crops, cultivation of tomato and chilli. But other high-valued vegetables can also be grown successfully inside the polyhouse.

**Table 9.4.** Crops grown in low cost green house and their benefits

| Sl.No. | Crop | Month of planting | Area under crop ($m^2$) | Yield (kg) | Productivity (tonne $ha^{-1}$) |
|---|---|---|---|---|---|
| Crops in terrace | | | | | |
| 1. | Capsicum | May | 32.31 | 26 | 8.05 |
| 2. | Tomato | April | 32.31 | 24 | 7.43 |
| 3. | Brinjal | May | 35.0 | 38 | 10.86 |
| 4. | French Bean | June(1st week) | 19.38 | 12 | 6.19 |
| 5. | Cabbage | July | 62.70 | 63 | 10.05 |
| 6. | Cauliflower | August | 32.31 | 22 | 6.81 |
| 7. | Pea | December | 32.31 | 23 | 7.12 |
| 8. | French Bean | January | 35.0 | 25 | 7.14 |
| 9. | Tomato | March | 62.70 | 35 | 5.58 |
| Crops on riser | | | | | |
| 10. | Bitter Gourd | June | 10.16 | 9.2 | 9.06 |
| 11. | Bottle Gourd | June | 20.23 | 35 | 17.30 |
| 12. | Cucumber | July | 10.16 | 12 | 11.81 |
| 13. | Bitter Gourd | November | 10.16 | 4.6 | 4.53 |
| 14 | Cucumber | February | 30.31 | 26 | 8.58 |

There is tremendous increase in productivity of vegetables under protected cultivation in United Arab Emirates (Table 9.5).

**Table 9.5.** Productivity of different vegetable crops under greenhouse in UAE

| Vegetable crop | Yield (MT/ha) |
|---|---|
| Cucumber (long) | 100-120 |
| Cucumber (short) | 120-150 |
| Tomato | 180-200 |
| Bell pepper | 120-150 |
| Musk melon | 5 |
| Snap beans | 5 |

## 9.2. Protected cultivation of flower crops

In recent years, hi-tech commercial floriculture has become a significant economic activity. The cut flower exports indicates a tremendous increase, which was due to establishment of a large number of export oriented cut flower units around Bangalore, Pune, Delhi and Hyderabad during the last two decades. The export earnings from floricultural products though small have shown increasing trend over the years from less than Rs.10 million during 1989-90 to over Rs.1230 million by 2001-02. This shows the potential and the vast opportunity that exists for export of cut flowers. The major destinations for export of cut flowers from India are Europe (45%), Japan (30%), Australia (10%) and other countries (15%). In spite of its late entry, Indian industry made rapid strides and nearly 250 ha of land is brought under protected cultivation of floricultural crops. Nearly 90% of the area is under roses and other crops like chrysanthemum, carnation, gerbera, orchids, anthurium and lilium share the rest.

Protected cultivation has been found beneficial to produce hybrid seedlings of flower crops, export oriented production of superior quality flowers and planting material of ornamentals. There is tremendous increase in productivity of cut flowers under protected cultivation (Table 9.6).

**Table 9.6.** Yield of flower crops under protected cultivation

| Flower crop | Yield |
|---|---|
| Rose | 150-300 stems/m$^2$ |
| Chrysanthemum | 60-100 flowers/m$^2$ |
| Carnation | 200-250 flowers/m$^2$ |
| Gerbera | 250-300 flowers/m$^2$ |
| Lilium | 50-60 bulbs/m$^2$ |

## 9.3. Photoselective shade-netting

Photoselective shade-netting is an emerging approach in protected cultivation. The photoselective net products are based on the introduction of various chromatic additives, as well as light dispersive and reflective elements into the netting materials during manufacturing. They are designed to selectively screen various spectral components of solar radiation (UV, visible, and beyond), and/or transform direct light into scattered light. The spectral manipulation is aimed to specifically promote desired physiological responses, while the scattering improves the penetration of the modified light into the inner plant canopy. The photoselective netting concept was studied in bell pepper (*Capsicum annuum*) cultivation in the Negev semi-arid area in Israel during the last four years. All nets were of a similar shading capacity (30-35%) in PAR. The Red, Pearl and Yellow shade nets (ChomatiNets®) were all found to increase pepper productivity, expressed by the number of fruits produced per plant as well as MT/ha, by 15-40%, compared with the traditional black shade net.

The reduction in viral infection under the Pearl and Yellow nets resulted in higher percentage of export-quality fruit produced under these nets, relative to either the Red or Black nets. The photoselective reduction of pest infestation and viral diseases may lead to reduced pesticide application by growers.

Fruit harvested from plants grown underneath the Pearl or Yellow nets, and then stored at 7°C for 2 weeks plus 3 days at 20°C (simulating sea transport and marketing to Europe), developed significantly less decay, compared with the fruit grown under the black shade net. The results might reflect a reduced occurrence of fungal spores under the Pearl and Yellow nets, and/or an enhanced natural resistance of the plants grown under these nets to pathogens. The photoselective, light-dispersive shade nets provide a new, multi-benefit tool for crop protection.

Chapter 10

# POST HARVEST MANAGEMENT

## 10.1. Post-harvest losses

There is a considerable gap between the production and net availability of the horticultural produce due to post-harvest losses due to lack of elaborate harvesting equipment, collection centers in major producing areas, suitable containers, commercial storage plants and lack of cold chain in entire post-harvest operation.

The total post-harvest losses in fruits are estimated to be 17 to 35% amounting to nearly Rs. 2000 crores annually. In a systematic study on post harvest losses in fruits have been estimated between 17.1 to 36.7% in mango, 12 to 19% in banana, 8.3 to 30.7% in orange, 10 to 25% in apple and 23 to 90% in grapes. Considerable losses occur during transport. Minimizing these losses has, therefore, been recognized as an effective means of increasing fruit and vegetable availability without bringing additional areas under these crops. In fact, the reduction in post-harvest losses is a complementary means of increasing the productivity.

Losses in fruits can be reduced by selection of fruit cvs. having longer shelf life, adopting proper cultural practices in the orchard, use of proper harvesting, handling, packaging, storage and transport methods, use of chemicals for pre- and post-harvest treatments and proper marketing procedure to help reduce losses.

## 10.2. Pre-harvest treatments

### 10.2.1. Fruit crops

- Pre-harvest application of GA delays ripening at market stage and improves storage life in mango and guava and colour development in citrus.
- Pre-harvest sprays of Topsin-M (0.05%) or Bavistin (0.1%) at 15 days before harvest can control anthracnose and stem-end rot of mango.

- Application of thiophanate methyl (0.05%) effectively controls post-harvest losses in mango by controlling maturity and delaying ripening.
- Post-harvest decay of Nagpur mandarin can be controlled by 3 pre-harvest sprays of Benlate or Topsin-M (0.1%) at 15 days interval before harvest.
- Pre-harvest sprays of 0.6% calcium chloride 10 to 12 days prior to harvest improved shelf-life and reduces physiological loss in weight in Thompson seedless grapes.
- Application of 2.5% calcium chloride at 50 ml/sucker in the core of pineapple plant about 300 days after planting showed better ripening and improved overall quality.

### 10.2.2. Vegetable crops

- Pre-harvest application of maleic hydrazide (MH) reduces sprouting of onion and potato during storage. In onions, application of 1500 to 2000 ppm MH at 75 to 90 days after transplanting reduced sprouting during 4 to 5 months storage in ventilated structures.
- Post-harvest diseases of tomato and onion can be controlled by 3 pre-harvest sprays of 0.2% Difolatan at an interval of 10 days before harvest.
- Pre-harvest application of N-benzyladenine (growth promoter) at 10 to 20 ppm prolongs shelf-life of leafy vegetables.
- Stoppage of irrigation 10 to 15 days before harvesting onion results in enhanced storage life of bulbs.

## 10.3. Post-harvest treatments

### 10.3.1. Use of chemicals

Post-harvest losses can be reduced by use of chemicals, fungicides, waxing etc. Post-harvest application of 0.1% Bavistin or 0.1% Topsin controls storage diseases in mango. Hot water treatment with 0.1% Bavistin or 0.1% Imazil reduced storage losses of Nagpur mandarins. Spraying with aqueous emulsion of CIPC at 50 and 100 mg/kg of tubers resulted in complete inhibition of sprouting of potatoes in evaporative cool chamber for 4 and 5 months, respectively.

Protective skin coating with wax increases storage life of fruits and vegetables at ambient temperatures. This treatment protects fruits and vegetables against excessive moisture loss and higher rate of respiration. To protect against microbial spoilage, fungicides can be added to the wax emulsion.

### 10.3.2. Disinfection

Vapour heat treatment of tropical fruits at 48°C for 1 hr eliminates all stages of fruit flies.

#### 10.3.3. Irradiation

Sprouting of onion can be checked by gamma irradiation at a dose of 0.06 to 0.1 KGY. In potato, gamma irradiation at 0.1 KGY can inhibit sprouting completely and also inhibits the light induced synthesis of chlorophyll and the toxic alkaloid solanin. The irradiated potatoes could be stored successfully for 6 months at 15°C with 10% weight loss.

Fruits of climacteric class such as banana, guava, mango and papaya where irradiation in the mature but unripe pre-climacteric stage is given at dosage of 0.25 to 0.75 KGY showed improved shelf-life due to delay in the rate of ripening and senescence.

### 10.4. Grading, packaging and transport

#### 10.4.1. Grading

Systematic grading is a pre-requisite for efficient marketing system, results in quality consumption and enables the producer a competitive price. In India, fruits are graded on compulsory basis for export and voluntary basis for internal marketing.

#### 10.4.2. Packaging

Packaging is a basic tool in post-harvest management of highly perishable commodities like fruits and vegetables. Considering the long-term needs of ecosystems and to achieve an overall economy, packaging materials like corrugated fibre board boxes, corrugated polypropylene board boxes, plastic trays/crates/woven sacks, molded pulp trays/thermoformed plastic trays, stretch films and shrink wrapping have to be used.

#### 10.4.3. Transport

For successful marketing, fast and economical transport which cause minimal damage is important. For short distance transport from field to packaging houses or local market, shallow plastic crates are quite satisfactory to prevent bruising losses. For longer distances, road transport by refrigerated trucks is faster and preferred.

### 10.5. Storage

#### 10.5.1. Pre-cooling

Pre-cooling is a means of removing the field heat. Highly perishable commodities must be pre-cooled as soon as possible after harvest. It slows down the respiration of the produce, minimizes susceptibility to attack of micro-organisms, reduces water loss and eases the load on cooling system during storage and transport. Fruits like mango, grapes and vegetables like peas and okra which deteriorate fast, need pre-cooling treatment. Pre-cooling of mango also reduces incidence of spongy tissue. The methods adopted for pre-cooling before storage or loading for shipment include room cooling, forced air cooling, vacuum cooling, hydro-cooling, package icing and top icing.

### 10.5.2. Cold storage

Storage at low temperatures immediately after harvest reduces the rate of respiration resulting in reduction in build up of heat, thermal decomposition, microbial spoilage and also helps in retention of colour and quality and freshness for a longer period. The optimum refrigerated storage temperature for different fruits and vegetables is presented in Table 10.1.

**Table 10.1.** The optimum refrigerated storage temperature for different fruits and vegetables

| Fruit/vegetable Crop | Optimum Storage Temperature |
|---|---|
| Banana | 12.8°C |
| Grapes | 0 - 1.7°C |
| Guava, mango, pineapple | 8.3 - 10°C |
| Orange | 5.5 - 7.2°C |
| Cauliflower | 2 - 4°C |

At present, there is a cold storage capacity of 30 million cubic meters in India of which 90% capacity is utilized for storage of potatoes. Fruits and vegetables together utilize only about 8.1% of total cold storage capacity.

### 10.5.3. Controlled atmospheric (CA) storage

CA storage involves modification of atmosphere i.e. reduction in $O_2$ and/or elevation of $CO_2$ concentration combined with low temperature. It markedly retards respiration and delays softening, senescence and changes in colour and quality of the stored fruits. Most fruits tolerate up to 5% $CO_2$ and minimum of 2% $O_2$. This technique can be used for long term storage and for long distance transport.

CA storage can be commercially exploited for exporting mangoes by sea transport to Europe. The storage life of mango which is only 6-7 weeks at 7-8°C is increased to 10-16 weeks with CA storage at 7.5% $CO_2$ level. Mango cv. Alphanso could be kept for 35 days at 8.3-10°C, while Raspuri for 40 days at 5.5-7.2°C. For prolonged storage of oranges and banana, fruits can be packed in polythene bags to create a modified atmosphere with more $CO_2$ and less $O_2$ than air.

### 10.5.4. Low cost storage

#### Seed potato stores

CPRI, Shimla has developed a brick masonry structures of 9.1 x 4.6 x 3.7 m for storage of seed potatoes. The north and south sides are double walled with 11 cm gap filled with rice husk to provide insulation and ventilation.

### Onion stores

NHRDF, Nashik has designed structures of 4.8 x 1.5 x 1.2 m (40 quintals capacity) and 15.8 x 4.8 x 2.4 m (450 quintals capacity) with side walls of bamboo splits raised from ground level by about 20 cm for proper ventilation from the bottom. The second base is 75 cm high from the bottom base. Total loss in these structures are about 36.6% over a period of 4 months.

### 10.5.5. Zero energy cool chamber

Indian Agricultural Research Institute, New Delhi, has developed a zero energy cool chamber for short-term storage of fruits and vegetables based on evaporative cooling system. It is a double walled brick structure with sand in between which is watered at regular intervals. During summer when the outside temperature goes beyond 44°C, the temperature inside the chamber never goes beyond 28°C. The RH is maintained at 90% in the structure. A commercial size structure of 12 x 12 x 12 m has also been successfully tried.

## 10.6. Value addition

Processing of fruits and vegetables for value addition is an indispensable part of horticulture industry. In countries like USA and Brazil, nearly 60 to 70% of fruits and vegetables are used for processing as against 1 to 2% in our country. Steps should be initiated for achieving at least 10% level of processing by 2010 which would generate an investment of Rs. 1 lakh crores and create almost half a million direct employment and another 2 million indirect employment. Hence, attempts to promote processing industries in this country is very essential. In mango, banana and onion, our share is nearly 50, 15 and 8% of the total world production and hence a great scope exists for development of newer products from these commodities. Dehydrated onion and tomato products have great potential as export commodities.

Work on processing of mango all over the country and grapes in Maharashtra and Andhra Pradesh is commendable.

### 10.6.1. Grapes

Raisin production has increased from 2000 tonnes to 12000 tonnes within a decade. Processing raisins yields an additional income of 30%.

Another outstanding development in utilization of grapes is the production of champagne (sparkling wines). Two factories at Narayan Gaon (Pune) and Pimple Gaon (Nashik) together produce 15 lakh bottles of wine. However, use of grapes for juice making is yet to be exploited commercially.

### 10.6.2. Minor fruits

The processed products which have been successfully tried are as follows:

Ber: Juice, powder, candy, dehydrated ber.

Pomegranate: Juice, wine, syrup, anardhana (dried arils of sour pomegranate).

Tamarind: Pulp, juice, concentrate.

Bael: Squash.

Phalsa: Juice.

Jamun: Jam, jelly, wine, vinegar, pickles, juice.

### 10.6.3. Other fruits and vegetables

The value added products developed from other fruits and vegetables are as follows:

Mango: Jelly grade pectin (6.1%), edible fibre (5.5%), dehydrated product.

Papaya: Sweet chutney from shreds which can be stored up to 9 months without any spoilage, papain from raw papaya.

Banana: Beverages.

Watermelon: Juice.

## 10.7 Waste utilization

Large quantities of factory wastes, both solid and liquids from fruit and vegetable processing would be available for profitable exploitation into value-added byproducts.

- 0.3 million tonnes of dry mango kernels (9-13% fat content) are available annually from which 30,000 tonnes of fat valued at Rs. 200 million could be obtained.
- Citrus waste is a rich source of essential oils (mandarin and lime), pectin (lime peel), D-limone (15,000 tonnes potential availability) and other byproducts.
- Large quantities of starch can be extracted from mango seed kernels and banana pseudo stems (5% starch).
- Natural colours can be extracted from Bangalore Blue grape skin, kokum (*Garcinia indica*), phalsa and jamun.
- Animal feed can be prepared from orange waste (peels, rags, seeds), ripe banana peel, pineapple bran and mango seed kernels.
- Fruit and vegetable waste can be diverted for biogas generation.
- Central Food Technological Research Institute, Mysore has developed a machine to convert dried banana leaves into disposable cups and trays. Pineapple leaves are used to extract high quality fibre for the textile industry.

- Indian Institute of Horticultural Research, Bangalore has found that coconut waste is a suitable substrate for the cultivation of oyster mushroom.

## 10.8. Marketing

Organized marketing should be developed to reduce post-harvest losses and to improve returns to the farmers. Horticultural Produce Co-operative Marketing Society (HOPCOMS) in Bangalore and National Dairy Development Board (NDDB) in Delhi have established a large number of retail outlets for sale of fresh fruits and vegetables. Efforts should be made to develop direct link between producers and consumers by starting 'Rytu Bazars' as in the case of Andhra Pradesh and 'Apna Mandis' in case of Punjab.

Marketing of fruits and vegetables based on grading encourages the growers to produce horticultural produce of better quality. With the advancement of Information Technology, now it is possible to have marketing through internet (E-commerce).

## Chapter 11

# INTEGRATED WEED MANAGEMENT

The diverse soil types and climatic conditions in India encourage the growth of a broad spectrum of grassy (monocot) and broad leaf (dicot) weeds. All the 18 world's worst weeds (Holm *et al.*, 1977) are found in our country in association with one or the other horticultural crop (Table 11.1).

**Table 11.1.** World's worst weeds

| Sl. No. | Name of weed | Sl. No. | Name of weed |
|---|---|---|---|
| 1 | *Cyperus rotundus* | 10 | *Chenopodium album* |
| 2 | *Cynodon dactylon* | 11 | *Digitaria sanguinalis* |
| 3 | *Echinocloa crusgalli* | 12 | *Convolvulus arvensis* |
| 4 | *E. colonum* | 13 | *Avena fatua* |
| 5 | *Eleusine indica* | 14 | *Amaranthus hybridus* |
| 6 | *Sorghum halepense* | 15 | *A. spinosus* |
| 7 | *Imperata cylindrica* | 16 | *Cyperus esculentus* |
| 8 | *Eichhornia crassipes* | 17 | *Paspalum conjugatum* |
| 9 | *Portulaca oleracea* | 18 | *Rottboelia exaltata* |

Weeds harm the crops and environment in more than one way. Generally, there are two types of weeds, one that emerge before or after the crop emerges and are less competitive and can be controlled easily and the other that emerge simultaneously with the crop and are more difficult to control.

The yield loss due to unchecked weeds varies from 34.0-71.7% in fruit crops, 10-95% in vegetable crops and 25-80% in tuber crops.

Weed control represents a major cost item in horticultural crop production and has now become a part of modern horticultural technology. For centuries, man fought weeds with his hands, hoes, animal power cultivator and mechanical power. Recently, chemical control using herbicides has assumed great significance and is being adopted all over the world effectively.

Commercial growers usually aim at two types of weed control – one, where weeds are reduced and another, where weeds are completely eradicated. Owing to its difficulty and higher cost, eradication is attempted in relatively small areas, viz., nursery areas (especially container grown plants), greenhouses, in and around growing benches and in potting soil, and in certain high value ornamental beds and vegetable crops. The greatest hindrance to total eradication is the presence in the soil of long lived weed seeds and vegetative reproductive plant parts. Different methods of weed control are as follows:

## 11.1 Preventive methods

Preventive weed control is concerned with the measures taken to prevent their introduction, establishment and/or spread of specified weed species in areas not currently infested with these plant species. In foreign countries, strict laws are enforced so that noxious weeds are not carried from one region to another. But in India, such laws do not exist. As a result, *Parthenium hysterophorus*, a noxious weed, has been carried far and wide throughout the country by various means posing a national problem.

Preventive methods of weed control include weed-free crop seed, weed-free manure, clean (weed-free) harvesting equipment and ploughing implements, and elimination of weed infestations in and around irrigation channels and cultivated fields.

### 11.1.1. Weed-free crop seed

In vegetable seed production, weed seeds get mixed with crop seeds and get dispersed far and wide. Care is needed to separate weed seeds from crop seeds for better management of weeds. *Mirabilis jalapa*, a common weed of waste lands, has seeds similar to papaya and often gets mixed unintentionally resulting in poor quality papaya seeds.

### 11.1.2. Weed-free manure

To get weed-free manure is a difficult task and much of the weed problems arise due to this factor. *Parthenium hysterophorus* grows very well in and around manure pits where organic matter is in abundance. It produces innumerable number of seeds which get mixed with manure usually made up of cow dung, horse dung and poultry waste. When manure is brought from such places, the seeds are easily scattered and establish themselves in the new site along with the crop plants.

### 11.1.3. Clean (weed-free) harvesting and ploughing equipment

Harvesting and ploughing equipment are major means of dispersing underground nuts, stolons and bulbs of perennial weeds far and wide to other areas. Thus, these equipment need to be cleaned at regular intervals.

### 11.1.4. Weed infestations in and around irrigation channels

Weed seeds are also carried far and wide through irrigation channels infested heavily with weeds. Seeds of *Portulaca oleracea* are carried through water channels from long distances into vegetable fields of okra, tomato, brinjal, onion, garlic, cluster beans, knol kohl and radish in such large numbers that it becomes very difficult to control them. Same is true with seeds of *Mollugo pentaphylla* and *Urena lobata*.

To bring about preventive measures of weed control, it is necessary to enact seed laws in India.

## 11.2. Cultural methods

The cultural methods of weed control include the use of smother crops and crop rotation/intercropping/trap cropping.

### 11.2.1. Smother crops

Smother crops are highly competitive with the weed species infesting an area for light, nutrients and moisture. Examples of smother crops are barley, millet, rye, sorghum, alfalfa, clover, cowpea, sweet clover, sesbania and sunflower. Care must be taken to see that the smother crops do not become a problem by themselves.

### 11.2.2. Cover crops

Velvet bean puts up vigorous growth, accumulating greater biomass and covers the ground fairly well in a short period, smothering weeds and effectively conserving soil moisture. *Calapogonium mucunoides* is a leguminous cover crop most effective to check soil erosion and the growth of obnoxious weeds in plantations of pepper, orange and coconut. *Centrosema pubescens, Macroptilium atropurpureum* (Siratoo), *Stylosanthus hamata, Pueraria phaseoloides* (Kudzu) and *Dolichos lab lab* var. *lignosus* are the other cover crops which smother the weed growth.

### 11.2.3. Crop rotation/intercropping/trap cropping

Smother crops may be grown in rotation with less competitive crops. Intercropping is intended to protect the inter space from losses through weeds. Crop rotation is particularly effective against parasitic weeds and other crop associated weeds. The trap crops which induce germination of parasitic weeds without being parasitized also reduce weed intensity.

Certain weeds are closely associated with particular crops or cropping sequence. Important parasitic weeds are broom rape (*Orobanche* spp. parasitizes mainly solanaceous crops) and dodder (*Cuscuta* spp. parasitizes mainly leguminous crops). Parasitic weeds are host specific. Stimulus is required for triggering germination of weeds. Incidence of *Orobanche* spp. on solanaceous crops can be minimized by rotating with sorghum. Incidence of

*Cuscuta* spp. on leguminous crops can be minimized by rotating with cereal crops.

### 11.2.4. Optimum plant density and line sowing

Maintenance of plant population in any crop and its early establishment helps crops to compete better with weeds due to smothering effect. The required plant density per unit area has been worked out in most of the horticultural crops and that has to be adopted by growers for effective management of weeds. Line sowing facilitates intercultivation between plant rows.

### 11.2.5. Drip irrigation

Applying water only to rooting zone as in case of drip irrigation would curtail the growth of weeds in inter row spaces.

## 11.3. Mechanical methods

This method includes such practices as hand weeding, hoeing, mowing, flooding, smothering with non-living material, burning and machine tillage.

### 11.3.1. Tillage

Tillage operation is a common and important method of weed control in all parts of the world. The primary objective of tillage is to control weeds. Summer tillage has been age old practice to open the soil to desiccating temperatures to manage perennial weeds, soil-borne pests and diseases.

(i) **Conservation tillage:** The presence of crop residues suppressed weeds, increased organic C and total soil N in the top 5-15 cm of soil.

(ii) **Zero tillage:** Zero tillage sowing reduces the incidence of weeds such as *Phalaris minor, Chenopodium album* and *Medicago hispida* to the tune of about 40 – 50% because of less soil disturbance. The organic matter content of zero tillage soils may stay higher because of slower decomposition. In zero tillage, anchored crop residues are left standing, which promote the population of beneficial insects (natural enemies) since it provides a good habitat for their survival and thereby less insect problems.

(iii) **Rotary tillage:** Rotary tillage by rotavator is capable of cutting and incorporating the weeds as well as residues left by the previous crop. The incorporation of weeds and crop residues will increase the organic matter status of the soil and accompanying changes in the physical and chemical health of the soil resulting in the sustainability of the system.

(iv) **Ridge tillage:** It is a form of conservation tillage that appears to overcome weed control problems.

**(v) Night tillage:** Germination of weed species whose seed requires light should be impeded if a no tillage system were implemented for crop production. Alternatively, it is possible that night cultivation also could result in significant reduction in weed density. The exclusion of light during seed bed preparation has been shown to reduce weed emergence. Covering an implement to prevent light reaching the soil at the point of cultivation may be sufficient to reduce weed emergence by up to 70%.

**(vi) Deep tillage:** Deep tillage helps in controlling weeds by exposing underground parts of perennial weeds leading to desiccation and destruction during hot summer, burying weed seeds and emerged weed seedlings deep into the soil, leaving the land rough to hinder weed seed germination. Deep ploughing has to be carried out before the onset of summer when the upper soil surface is dry and lower soil layer is moist so that big clods are raised. During hot summer months exposed stolons, rhizomes of perennial weeds (*Cynodon dactylon*) die due to desiccation and lose their viability. Frequent tillage operations provide opportunity for weed seeds to germinate to be destroyed by cultivation.

**(vii) Mechanical tillage:** Machine tillage is similar to hoeing. The advantage of machine tillage is that various types of tools can be attached to machines and large areas can be weeded rapidly and economically. Cross cultivation can be done to remove much of the weeds. Machine tillage also helps in loosening of the soil which creates a suitable atmosphere for bacterial activity resulting in rapid decomposition of organic matter. Weed control by machine tillage is achieved by burial of small annual weeds with soil thrown over them, destroys close relationship between weed plant and the soil resulting in disruption of absorption of water by weed roots and death by desiccation (Fig. 11.1).

**Fig. 11.1.** Mechanical tillage for weed control.

### 11.3.2. Stale seed bed

Frequent pre-monsoon showers facilitate germination of weeds to be destroyed by cultivation. This is made use in stale seed bed where crop sowing is deliberately delayed under intense weed situation to reduce the impact of weeds.

### 11.3.3. Hand weeding

This practice is very effective for the control of young weed seedlings and established annual and biennial weed species in small garden area (nurseries and ornamentals) and in between or near crop plants grown in narrow rows where it is difficult to reach them with a hoe or cultivator (vegetable crops).

### 11.3.4. Hoeing

Hoe still remains one of the principal tools for weed control. Hoeing is effective against annual and biennial weed seedlings in crops like vegetables, ornamentals and small fruits. Weed intensity is low in line sown and transplanted crop than in broadcast sown crop. The main objectives of intercultivation are weed control, aeration, soil moisture conservation and earthing up. Wider rows permit effective intercultivation for long period with reduced hand weeding cost.

### 11.3.5. Mowing

It is primarily used to reduce weed seed production and to restrict unsightly weed growth especially in wide spaced fruit and plantation crops.

### 11.3.6. Flooding

Flooding is done to control perennial weeds. Flooding deprives the plants of air resulting in their death due to suffocation. Thus, root absorption of oxygen to the plant is prevented which is so vital for the respiration process in roots. Similarly, when shoots are immersed, it prevents absorption of carbon dioxide by leaves from atmosphere which is essential for photosynthesis.

### 11.3.7. Mulching

The principle behind this is to exclude light completely to the growing weed plants to prevent photosynthesis and further growth. Mulching also helps in moisture conservation and organic matter. Hay, manure, grass clippings, straw, saw dust, wood clips, rice hulls, paper, plastic film and even spent herbage from aromatic industry are generally used for this purpose. Black paper and black plastic films are used successfully in various horticultural crops. In Hawaiin Islands, use of black plastic film has been found to be economical and effective for unirrigated pineapple crop.

### 11.3.8. Burning

Selective burning or flaming in crop plants is done under special situations where crops attain certain heights where flames can not strike leaves or other tender parts, when their stems are woody and resistant to intense heat of flame and the weeds are small (5 cm), succulent and susceptible to the heat. The theory behind burning is to cause cell sap to expand resulting in rupture of cell walls to cause coagulation of cell protoplasm and inactivation of enzymes.

## 11.4. Soil solarization

Harvesting of solar energy through soil solarization will be the key proposition for controlling soil-borne pests including weeds. This is an eco-friendly technology used to kill weed seeds in soil. It involves mulching of the soil with clear plastic films so as to trap the solar heat in the surface soil. Sheets should be spread in such a way that there is no gap between sheet and soil surface. The resultant temperature increase would be lethal to soil pathogens, nematodes and weeds. The term 'soil solarization' denotes solar heating of the moistened soil by mulching (tarping or covering) with appropriate mulch (transparent polythene). This approach of killing weed seeds and propagules seems to have greater potential in tropical and sub-tropical regions where air temperatures goes up to 45°C during summer months. It is efficient in areas where bright sun light is available for about 4-6 weeks.

The transparent polyethylene is permeable to short wave solar radiation received from the sun which results in trapping of the heat in the soil. Polyethylene also reduces heat convection and evaporation of water from the soil. As a result of the formation of water droplets on the inner side of the mulch will trap long-wave radiation and thus eliminate radiating cooling of the soil surface.

Solarization heats the soil through repeated daily cycles. At lower soil depths, the temperature is reached to maximum later in the day and maintained for longer period of time.

A novel method of weed control has been developed using solar radiation. This technique uses intense radiation of sun to heat the soil to raise the temperature that is lethal to weed seeds. This method has future in tropical and sub-tropical countries.

Black plastic mulch has been found very effective in reducing the weed population. As sunlight does not pass through the black film, the weed growth is completely eliminated. Some of the results obtained on weed control under plastic mulch in horticultural crops are presented in Table 11.2 (Singh and Samuel, 2000).

**Table 11.2.** Effect of plastic mulch on weed management and yield of horticultural crops

| Crop | Salient findings | Name of Institution |
|---|---|---|
| Strawberry | 55% increase in yield as compared to unmulched, 85-90% weed control | Y.S. Parmar Univ. of Hort. & Forestry, Nauni, Solan (Himachal Pradesh). |
| Potato | 49% increase in yield as compared to control, 32% weed control | Orissa Univ. Agri. & Technol., Bhubaneshwar (Orissa). |
| Brinjal | 10% increase in yield with 25 $\mu$ mulch, 90% weed control | N.G. Ranga Agri. Univ., Hyderabad (Andhra Pradesh). |
| Okra | 48% increase in yield as compared to control | Kerala Agri. Univ., Tavanour, (Kerala). |

## 11.5. Biological methods

This method involves utilization of natural enemies for the control of certain weeds. The aim is to reduce weed population below the level of economic injury. This can be achieved by direct or indirect action of biological control agents. In direct action, firstly the biocontrol agent bores into plant, weakens its structure leading to its collapse and secondly, consumes and destroys the vital plant parts. In indirect action, the biocontrol agent reserves the competitive ability of weed over other plants and enhances the condition favourable for plant pathogens.

### 11.5.1. Insects

For biocontrol of weeds, the insect selected should specifically attack the target weed without harming the other plants, especially useful to man. Control of obnoxious weeds like prickly pear cactus (*Opuntia inermis*) and spiny prickly pear (*Opuntia stricta*) by *Cactoblastis cactorum*; *Hypericum perforatum* by leaf feeding beetle, *Chrysolina quadrigemina*; *Parthenium hysterophorus* by *Zygogramma bicolorata*; orobanche by flies; and cuscuta by midges (Table 11.3) are examples of biological weed control.

**Table 11.3.** Use of insects for control of parasitic weeds.

| Parasitic weed | Insects |
|---|---|
| *Orobanche* sp. | Fly-*Sipha maidis, Scotia segetum, Phytomyza orobenchia, Aphis tabae, Agriote* sp., *Lygus pratensis, Agrojrotalmia pulchella* |
| *O. crenata* | *Tropinate squarlido, Smicronix cyaneus* |
| Cuscuta | Midge – *Melar agromyza* |

### 11.5.2. Plant pathogens

Use of plant pathogens, especially fungi, to control weeds (mycoherbicides) is gaining importance in recent years (Tables 11.4 to 11.6). The potential for weed control using foliar pathogens by pre-emergent application of granular formulation of *Alternaria macrospora* spores so that biocontrol agents are present when seedlings of *Anoda cristata* emerge. This technique holds promise because reduced quantities of spores are needed when seedlings are small and emerging. Also, granular formulations can be applied on a band over the crop drill which further reduces the quantity of spores needed per unit land area.

**Table 11.4.** Mycoherbicides registered world-wide to control various notorious weeds.

| Mycoherbicide | Formulation of fungus | Weed controlled |
|---|---|---|
| De Vine | *Phytophthora palmivora* | Milk weed vine (*Morrenia odorata*) |
| Collego | *Colletotrichum gloeosporioides* f. sp. *aeschynonene* | Northern joint vetch (*Aeschynomene verginica*) |
| Caset | *Alternaria cassiae* | Sickle pod (*Cassia obtusifolia*) |
| Biomal | *C. gloeosporioides* f. sp. *malvae* | Round leaf mallow (*Malva* f. sp. *malvae pusilla*) |
| Biochon | *Chondrostereum purpureum* | Black cherry |
| Velgo | *Colletotrichum coccodes* | Velvet leaf |
| Luboa 2 | *C. gloeosporioides* f. sp. *cuscutae* | Dodder (*Cuscuta* sp.) |

Devine is a liquid formulation consisting of chlamydospores of a pathotype *Phytophthora palmivora* and is used as a post-emergent mycoherbicide against Milk weed vine (*Morrenia odorata*). Nearly 100% control of the weed is usually obtained and control lasts for 2 years. Collego is a wettable powder formulation of *Colletotrichum gloeosporioides* f. sp. *aeschynonene* and is used for controlling Northern joint vetch (*Aeschynomene verginica*). It is applied aerially or with land based equipments when the weeds have just emerged. Control reported was as high as 90%.

**Table 11.5.** Fungi possessing mycoherbicidal properties.

| Pathogen | Weed controlled |
|---|---|
| *Colletotrichum furariodes* | *Asclepias sericea* (Common milk weed) |
| *C. oriculare* | *Xanthium spinosum* (Spiny cocklebur) |
| *Alternaria* sp. | *Crisum avenae* (Canana thistle) |
| *A. crassa* | *Datura stramonium* |
| *A. helianthi* | *Xanthium strumarium* |
| *Phomopsis convolvulus* | *Convolvulus arvensis* (Field bind weed) |
| *Bipolaris halapense* | *Sorghum halepense* (Johnson grass) |
| *Puccinia abrupta* var. *parthenicola* | *Parthenium hysterophorus* |
| *P. chondrillus* | *Condrilla juncea* (Skeleton weed) |
| *P. spegazzinii* | *Mikania micrantha* |

**Table 11.6.** Use of fungi for control of parasitic weeds.

| Parasitic weed | Fungi |
|---|---|
| *Orobanche* sp. | *Fusarium oxysporum* f. sp. *orthoceras* |
| Cuscuta | *Colletotrichum gleosporoides* (Lubao II – Mycoherbicide) |

*Trichoderma* sp. has been reported to suppress *Phalaris minor*, an important weed in farmer's fields, if applied through compost.

## 11.6. Allelopathy

Inter-weed competition determined by allelopathy can be manipulated to our advantage in the natural control of weeds. Natural compounds released by some plants inhibit or prevent the growth of nearby plants. Known as allelopathic substances, they may be released from roots, leaves or other plant parts. Black walnut trees inhibit the growth of certain plants near them and thistles exude a substance that interferes with the growth of oats. Most of the inhibitors belong to phenolic group of compounds except *Oxalis corniculata* which belong to organic acids. *Parthenium hysterophorus* contains parthenin along with phenolic compounds.

Marigold flowering plant is found to suppress the growth of *Parthenium hysterophorus* when planted in the ratio of 1: 2 and 1 : 4 (*Parthenium*: Maigold). In these plots, second generation of *Parthenium* could not get re-established and very few plants left could not develop further. The recycling of marigold plants inhibits the establishment of *Parthenium*.

A few crops have been found to inhibit weeds in different cropping systems and the major sources of allelochemicals are root exudates and leachates of growing crops. Wheat, oats, peas and buckwheat suppress the growth, accumulation of above ground biomass and leaf surface of *Chenopodium album*. Oats suppressed the growth of *Erysium cheiranthoides*. When rye was used as a cover crop in no tillage cropping system, it suppressed certain weeds like *Chenopodium album* and *Amaranthus retroflexus*. Rye also inhibited other weeds like *Digitaria sanguinalis, Arena fatua, Echinochloa crusgalli* and *Ambrosia artimisifolia*. Crop residues from alfalfa and sunflower are toxic to weeds. The growing plants such as wheat, corn and soybean and their crop residues are also toxic to weeds.

## 11.7. Chemical methods

Chemical weed control was born in 1896 when Bonnet, a French grape grower, observed that the Bordeaux mixture which he used against downy mildew of grape turned the leaves of *Sinapsis arvensis* (a weed) black (Brain, 1976). Later, the weed killing properties of sulphates of ammonia, zinc, iron and other metals were soon observed. Further milestones in weed control were introduction of organic chemical, 2-methyl-4, 6-dinitrophenol (DMOC) in 1932, and discovery of hormone-type selective weed killers. Selective residual chemicals such as substituted phenyl ureas and triazines, and non-residual chemicals such as diquat and paraquat are more recent discoveries.

Increased use of herbicides have, however, resulted in multiple problems. The environment safety has been doubted with increased pollution hazards. That apart, weed species are developing resistance to chemical toxicants. A black grass strain has been identified to be resistant to chlorotoluran and diclofop methyl. Similarly, groundsel has shown resistance to simazine. In India, continuous use of isoproturan in wheat has resulted in the development of resistance in *Phalaris minor*. Added to this there is also problem of resurgence of weeds.

## 11.8. Integrated methods

Integrated weed management system (IWMS) involves co-ordinated efforts towards development of a long lasting and environment friendly practice of weed management. IWMS includes all preventive, crop husbandry and direct control measures aimed at reducing and maintaining weed population at sub-economic level. In this, all weed control methods like preventive, cultural, mechanical and biological are included in different permutation and combinations are tried to achieve maximum weed control which is not possible in any one single method. IWMS is well suited for adoption in different cropping systems of fruit crops.

In raspberry, reduced cultivation system like rotary cultivation + hoeing out- yielded non-cultivation system which depended on herbicide alone.

Chapter 12

# INTEGRATED PEST MANAGEMENT

Productivity of horticultural crops in India is relatively low compared to other countries. Of the several factors responsible for lower productivity of horticultural crops, pests (insect and mite pests) are considered as important limiting factors. Some pest problems (coconut mite, spiraling whitefly on fruit crops, coffee berry borer, diamond back moth on crucifers, serpentine leaf miners on several crops, fruit flies in fruits and cucurbits) are of national importance which causes significant losses to horticultural crops. The annual losses due to pests to all the crops in India was estimated at Rs. 60,000 million in 1983, which at today's prices could exceed Rs. 200,000 million.

## 12.1. Integrated pest management

Integrated Pest Management is an important principle on which sustainable crop protection can be based. IPM allows farmers to manage pests in a cost effective, environmentally sound and socially acceptable way. According to FAO, IPM is defined as "A pest management system that in the context of the associated environment and the population dynamics of the pest species, utilizes all suitable techniques and methods, in a compatible manner as possible and maintains the pest populations at levels below those causing economic injury".

IPM strategies are based on 3 main components/tools:

- Prevention
- Monitoring/Observation/Regulation
- Intervention

### 12.1.1. Prevention

**Aim:** To reduce initial severity of pest infestation.

Prevention of pests through indirect measures is an important part of IPM. Most of the IPM activities emphasize heavily on preventive measures especially before the crop is sown/planted *i.e.* at the field preparation or seed treatment stages since the initial pest population is largely responsible for the subsequent high pest build up. Agronomic practices such as deep summer ploughing, soil solarization, cleansing of crop refuse, elimination of weed hosts, crop rotation, water management, intercropping/mixed cropping, trap cropping, nutrition management and genetic sources of resistance along with habitat management are the major tools of prevention.

**(i) Location:** Growing crops and varieties in appropriate locations where they fit to climate, soil and topography, provides the crop with optimal growing conditions from the start.

**(ii) Crop rotation:** Growing different crops in a rotation helps to reduce the build up of certain pests, especially those in the soil, such as root feeders.

**(iii) Cropping pattern:** Planting of alternative host crops alongside each other can substantially increase pest pressure, and should be avoided if possible.

**(iv) Plant breeding:** Choice of crop variety has always been a cornerstone of crop protection, especially the growing of pest resistant varieties. These may reduce the need for frequent treatments with crop protection products, and may minimize the use-rate required of individual products. Use of resistant varieties also encourages the survival of beneficial species.

Perhaps the greatest single technological achievement is the advance in breeding crops for resistance to pests. Cultivation of resistant varieties is the cheapest and best method of controlling pests. One of the important components of IPM is the use of resistant cultivars to key pests. Under All India Co-ordinated Research Projects of Indian Council of Agricultural Research, large number of highly/moderately resistant varieties are released to the farmers (Table 12.1).

**Table 12.1.** Horticultural crop varieties resistant to pests.

| Horticultural crop | Pest | Resistant varieties |
|---|---|---|
| Brinjal | Shoot and fruit borer | Pusa Purple Long |
| Chilli | Thrips | NP 46A (T) |
| Onion | Thrips | Arka Kirthiman, Arka Pitamber, Arka Lalima |
| Field bean | Jassid, aphid & pod borer | Pusa Sem-2, Pusa Sem-3. |
| Okra | Fruit borer | Pusa A-4. |
| Pumpkin | Fruit fly | Arka Suryamukhi. |

T – Tolerant

Developments in genetic engineering will increase the potential for developing novel opportunities and choices for pest resistant plants. Transgenic plants offer a novel means of delivering biological agents, where insecticidal proteins are introduced genetically into the plant. The insect pest is killed only when it feeds on the plant and non-target organisms are unaffected.

Transgenic crop varieties in horticultural crops (tomato, potato, brinjal, beans, cabbage, cauliflower, musk melon, banana, coffee) have been developed by cloning *Bt* endotoxin genes which are cultivated in large areas. In 2006, India is the 5th largest GM crops growing country (3.8 million ha) in the world only next to USA (54.6 million ha), Argentina (18 million ha), Brazil and Canada.

**(v) Crop husbandry and hygiene:** Mechanical and physical crop protection methods can be important in promoting good crop development, and prevent or minimize pest infestation. Traditional mould board ploughing inverts the soil and buries crop residue before the preparation of a seed bed for the succeeding crop. But it can lead to increased erosion, so it should be used only when necessary. Also, it is desirable to combine it with other conservation techniques such as the use of contour ploughing and ridging. In many countries, there has been a trend towards reduced cultivation. This utilizes herbicide technology and has culminated in the practice of *direct drilling* in Europe and *no till* in the USA. As a result, soil erosion problems have been largely eliminated. Changes in soil disturbance, location of plant residue and weed ecology all influence the incidence of pests and should be taken into account in designing IPM programmes.

Good crop hygiene is also important in reducing the build up and carry-over of pest population from one crop to another. Methods are often labour intensive (e.g. removal and destruction of crop residue), but may provide essential way of reducing the survival of some pests to the next season.

**(vi) Fertilization:** Good fertilizer practice should be adopted to promote crop growth, but excessive use of nitrogen must be avoided – too much lush, vegetative growth encourages many insect pests.

**(vii) Irrigation:** Supply of water to the crop can enhance or reduce pest incidence. Flooding of some soils is important in checking soil pests. However, flood irrigation can also adversely affect the survival of some soil-inhabiting natural enemies. In vegetables this can be reduced by growing the crop on ridges or raised beds, and should be taken into account in designing IPM systems.

**(viii) Habitat management:** Protection of natural habitats within the farm environment is recognized as a means of conserving many of the natural enemies of pests. Careful management of the margins of farm land, as well as growing tree crops or hedges, are particularly important because they provide habitat, cover and refuse for beneficial insects and other animals (e.g. field bunds provide important refuge for predatory spiders which help to control several important pests; and snakes which help control rats).

**(ix) Trap crops:** Sometimes a pest can be attracted away from a valuable sensitive crop by another crop (e.g. one row of mustard sown after every 25 rows of cabbage to attract DBM, aphids, leaf webber which are controlled by spraying 4% NSKE; one row of African marigold after every 16 rows of tomato to attract fruit borer which can be controlled by spraying *Ha*NPV).

**(x) Intercropping:** Traditionally, some farmers may sow different crops in alternate rows, or under sow a crop (e.g. maize) with a legume (e.g. beans) to help improve soil fertility and help reduce pest pressures in individual crops.

Agricultural production is more sustainable in areas where proper crop-mixes and variety-mixes are adopted in a given crop season. Monocultures and overlapping crop seasons are more prone to severe outbreaks of pests and diseases. Diversity of natural enemy complexes attacking various stages of pests prevalent in poly-crop and inter-cropping systems also tend to prevent severe pest outbreaks. A poly-crop ecosystem of pulses, oil seeds, vegetables and cotton kept the populations of the polyphagous pest, the gram pod borer, *Helicoverpa armigera* and the leaf hopper, *Amrasca devastans* well below ETL for considerable time during the crop season compared to the pure crop ecosystem.

**(xi) Botanicals:** Neem products such as cake, oil, neem seed kernel extract (NSKE), neem seed powder extract (NSPE), pulverized NSPE and soaps are being used extensively to manage horticultural crop pests (bean fly, *Ophiomyia phaseoli*; serpentine leaf miner, *Liriomyza trifolii* on several crops; cucurbit fruit fly, *Bactrocera cucurbitae*; tomato fruit borer, *Helicoverpa armigera*; brinjal fruit and shoot borer, *Leucinodes orbonalis*; water melon and chilli thrips, *Thrips* spp.; chilli yellow mite, *Polyphagotarsonemus latus* and okra leaf hopper, *Amrasca biguttulla biguttulla*) (Krishna Moorthy and Krishna Kumar, 2002).

The soap sprays were highly effective on leaf hoppers, aphids, red spider mites and white flies in many vegetables, but moderately effective on thrips in water melon and chillies (Table 12.2).

**Table 12.2.** Insect pests on which soaps were found effective

| Horticultural crops | Insect pests on which soaps were found effective |
|---|---|
| Cabbage and cauliflower | DBM, leaf webber, aphids, young *Spodoptera* larva |
| Tomato | White fly, red spider mites, fruit borer (egg laying stage), leaf miner |
| Okra | Leaf hopper, white fly, aphids |
| Cucurbits | Fruit fly, leaf miner |
| Mango | Leaf hopper |
| Ornamental crops | Mites, white fly |

### 12.1.2. Monitoring/observation/regulation

**Aim:** To determine WHEN and WHAT action to take.

**(i) Crop monitoring:** Regular monitoring is very crucial to pest management as it is one of the most important decision making tools. An efficient monitoring programme can pay big dividends in lowering pest management costs. An Agro-Eco-System Analysis (AESA) based on crop health at different stages of growth, population dynamics of pests and natural enemies, soil condition and farmer's past experiences are considered for decision making. Monitoring for pests is an important part of the need to "walk" through a crop. Various tools, such as pheromone traps, diagnostics and forecasting systems are available to assist pest monitoring, minimize the time required, and achieve acceptable accuracy in determining changes in pest infestation. Economic threshold levels (ETL) are now available to assist in proper timing of IPM interventions.

**(ii) Decision support systems:** Farmers need assistance in interpreting pest monitoring data. Simple "expert systems" can be designed and made available to farmers in a range of ways, including simple peg boards or charts, special booklets, radio and television programmes or more advanced aids such as prediction models and computer-based systems. Development and provision of up-to-date information is a key factor enabling farmers to implement IPM programmes.

**(iii) Area-wide management:** IPM often requires collaborative decisions within a national or more localized area to provide effective control of pests. Some of these decisions need to be taken centrally by governments in relation to:

- Quarantine regulations and legislation.
- Provision and training of advisory services.
- Establishment of resistance management strategies against highly mobile pests on major crop systems.

Other decisions can be made by farmer associations or groups to meet localized needs. Geographical Information Systems (GIS) and remote sensing techniques can also assist in implementation of area-wide management.

### 12.1.3. Intervention – direct measures

**Aim:** To reduce the effects of economically damaging pest populations to acceptable levels.

Once it is determined that intervention must be made, decisions must be made on the type of intervention which will reduce the effects of economically damaging nematode populations to acceptable levels. Mechanical, biological, cultural and chemical control measures are applied individually or in combination, taking into consideration costs, benefits, timing, available labour force, as well as ecological and environmental effects. Some of the principal IPM interventions include cultural and physical measures, monitoring of pests, biocontrol agents (predators, parasitoids and pathogens), host plant resistance, botanicals and target specific, less hazardous chemical pesticides.

**(i) Cultural and physical control:** A number of cultural and physical methods such as hand picking of egg masses and larvae continue to be used. Recommendations should be practicable. The possibility of integrating appropriate cultural methods with judicious use of crop protection products should be explored.

**(ii) Pheromones:** Semiochemicals are more selective and biologically active ingredients which are generally target specific and used at very low application rates. Pheromones are a class of semiochemicals, which are used commercially for indirect (monitoring, quarantine and time of pest control) and direct (mass trapping) pest control. Development and availability of pheromones offers interesting new possibilities for the farmer:

- Besides selective trapping techniques to monitor the movement of pests or changes in populations during season, pheromones are also used in "lure and kill" strategies to attract the pest to localized insecticide deposits and reduce the need for overall crop spraying.
- Mating disruption to delay or reduce the need for control treatments.

**(iii) Biological control:** Research on nature's own methods of pest control is yielding new useful products and methods which can be integrated into IPM programmes. However, most biological control techniques (e.g. using beneficial insects, mites and nematodes) work best when crops are grown in controlled environments (e.g. glasshouses and plastic tunnels) to ensure consistent results. There are a few cases where control techniques with living organisms are successful in open field conditions (e.g. predatory mites against spider mites). However, biological products are generally too often unreliable or not efficient enough in the field to be commercially used on their own. Nevertheless, there is increasing interest in using other biological products (e.g. viruses, fungi and bacteria). These require similar technical expertise as chemical agents in relation to formulation, field application and resistance management.

Classical biological control (i.e. the introduction of a predator or parasitoid for the control of a particular pest species) and inundative control (i.e. repeated mass release of a control agent) programmes impose high requirements on the selectivity of chemical control agents to be combined with them. This narrows down the number of candidate chemical control agents for IPM based on classical biological and inundative control strategies.

Of late, biological suppression of pests has become an intensive area of research because of environmental concerns. About 60% of the natural control of insect pests is by the natural enemies of pests such as parasitoids, predators and pathogens (Tables 12.3 to 12.5). The Australian lady bird beetle, *Cryptolaemus motrouzieri* has been found very effective against mealy bugs infesting grapes, guava, citrus, mango, pomegranate, ber and custard apple. The encyrtid parasite, *Leptomastix dactylopii* is effective against mealy bug, *Planococcus citri* on guava, citrus, pomegranate, ber and custard apple (Mani, 2001).

*Bacillus thuringiensis* (*Bt*), a bacterium which occurs naturally, has been mass cultured and used to control pests (e.g. caterpillar pests in vegetables, vineyards and orchards) has been one of the most successful biological control measures. The insect spectrum of activity is narrow compared with synthetic chemicals, and this is useful in IPM strategies. *Bt* is effective against tomato fruit borer, okra fruit borer and diamondback moth on cabbage and cauliflower.

Several methods of enrichment and conservation of natural enemies include providing nesting boxes for wasps and predatory birds; retaining pollen and nectar bearing flowering plants like Euphorbia, wild Clover on bunds to provide supplementary food for natural enemies; placing bundles of paddy straw in fields for attracting predatory spiders. In addition, erecting perching sites, water pans, retaining bushes (Acalypha, Hibiscus, Crotons) helps in retention of predatory birds.

**Table 12.3.** Biological control of fruit crop pests.

| Fruit crop | Pest | Biocontrol agent/Dosage |
|---|---|---|
| Apple | Woolly aphid, *Eriosoma lanigerum* | *Aphelinus mali*-1000 adults or mummies/ infested tree. |
| | San Jose Scale, *Quadraspidiotus perniciosus* | *Encarsia periniciosi*-2000 adults/ infested tree. |
| | Codling moth, *Cydia pomonella* | *Chilocorus infernalis*-20 adults or 50 grubs/tree; *Trichogramma embryophagum*-2000 adults/tree; *Steinernema carpocapse* |
| Citrus | Cottony cushion scale, *Icerya purchasi* | *Rodolia cardinalis* -10 beetles/ infested plant. |
| | Mealy bug, *Planococcus citri* | *Cryptolaemus montrouzieri*-10 beetles / infested plant; *Leptomastix dactylopii*- 3000 adults/ha. |
| | Scales, *Aonidiella aurantii* | *Chilocorus nigrita*-15 adults/infested tree |
| | Scales, *Coccus viridis* | *Verticillium lecanii*-16 x $10^4$ spores/ ml + 0.05% Teepol. |
| | Leaf miner, *Phyllocnistis citrella* | *S. carpocapse* |
| Grapevine | Mealy bug, *Maconellicoccus hirsutus* | *C. montrouzieri*-2500 to 3000 beetles/ha or 10 beetles/vine. |
| Guava | Green shield scale, *Chloropulvinaria psidii* | *C. montrouzieri*-10 to 20 beetles/ infested plant. |
| | Aphid, *Aphis gossypii* | *V. lecanii*-$10^9$ spores/ml + 0.1% Teepol. |

**Table 12.4.** Biological control of vegetable crop pests.

| Vegetable crop | Pest | Biocontrol agent/Dosage |
|---|---|---|
| Beans | *Tetranychus* spp. | *Phytoseiulus persimilis*-10 adults/plant or release 1-6 leaves with predatory mites. |
| Pigeon pea | *Helicoverpa armigera* | *Ha* NPV-250 LE/ha |
| Potato | *Agrotis ipsilon, A. segetum* | *Steinernema carpocapse, S. bicornutum, Heterorhabditis indica* |

| | | |
|---|---|---|
| Tomato | *Helicoverpa armigera* | *Trichogramma brasiliensis* / *T. chilonis* / *T. pretiosum*-50,000/ha; *Ha* NPV-250 LE/ha |
| Brinjal | *Leucinodes orbonalis* | *S. carpocapse, H. indica* |
| Chilli | *H. armigera* | *Ha* NPV-250 LE/ha |
| Cabbage | *Plutella xylostella* | *S. carpocapse, S. glaseri, S. feltiae, S. bicornutum, H. bacteriophora* |
| Mushroom | *Lycoriella auripila, L. mali, L. solani, Megaselia halterata* | *S. feltiae* |

**Table 12.5.** Biological control of plantation crop pests.

| Plantation crop | Pest | Biocontrol agent/Dosage |
|---|---|---|
| Coconut | Black headed caterpillar, *Opisinia arenosella* | *Goniozus nephantidis*-3000 adults/ha. |
| | Rhinoceros beetle, *Oryctes rhinoceros* | Baculovirus-10 infected beetles/tree. |
| Areca nut | *Ischnaspis longirostris* | *Chilocorus nigrita*-20 to 50 adults/plant. |
| Coffee | Mealy bugs, *Planococcus* & *Pseudococcus* spp. | *Cryptolaemus montrouzieri*-2 to10 beetles/ infested plant. |

**(iv) Chemical control:** A very wide range of chemical products is now available. The use of chemical control agents represents in many situations the most important and widespread means of achieving effective and reliable reduction of pest infestations. As such, it is essential to reduce unnecessary user exposure to crop protection products, improve standards of work practice and hygiene, limit residues in the environment and harvested crops, and avoid potential problems of pest resurgence and pesticide resistance. Judicious application of chemical products to minimize the risk of adverse effects is integral with the principles of IPM.

To make full use of naturally occurring beneficials, great care is needed to maintain untreated refuges (e.g. hedge rows, field margins, or 'island" habitats) where natural enemies live. Improved application techniques allowing lower use-rates of chemical controls offer some way of reducing exposure of beneficials. Avoiding spray drift into refuges is essential to reduce product contamination. The main requirement is to ensure that the timing of treatment (during day and season) as well as specificity of product used, reduce adverse effects on beneficials as for as possible.

## 12.2. Biointensive integrated pest management (BIPM)

Biointensive integrated pest management is a systems approach to pest management based on an understanding of pest ecology. It begins with steps to accurately diagnose the nature and source of pest problems, and then relies on a range of preventive tactics and biological controls to keep pest populations within acceptable limits. Reduced-risk pesticides are used if other tactics have not been adequately effective, as a last resort, and with care to minimize risks.

An important difference between conventional and biointensive IPM is that the emphasis of the latter is on proactive measures to redesign the agricultural ecosystem to the disadvantage of a pest and to the advantage of its parasite and predator complex.

BIPM options may be considered as proactive or reactive.

### 12.2.1. Proactive options

Proactive options, such as crop rotations and creation of habitat for beneficial organisms, permanently lower the carrying capacity of the farm for the pest. The carrying capacity is determined by the factors like food, shelter, natural enemy complex and weather, which affect the reproduction and survival of a pest species. Cultural control practices are generally considered to be proactive strategies. Proactive practices include crop rotation, resistant crop cultivars including transgenic plants, disease-free seed and plants, crop sanitation, spacing of plants, altering planting dates, mulches, etc.

The proactive strategies (cultural controls) include:

- Healthy, biologically active soils (increasing below-ground diversity).
- Habitat for beneficial organisms (increasing above-ground diversity).
- Appropriate plant cultivars.

### 12.2.2. Reactive options

The reactive options mean that the grower responds to a situation, such as an economically damaging population of pests, with some type of short-term suppressive action. Reactive methods generally include inundative releases of biological control agents, mechanical and physical controls, botanical pesticides and chemical controls.

While developing the BIPM technologies, the following aspects have to be kept in mind:

- Economic viability. About 70% of our people depend on agriculture for their livelihood and more than 80% of them are small and marginal farmers. Hence, there is a need to develop low input crop protection technologies. The input use efficiency should be given utmost importance.
- Bringing down the pest population below the economic injury level.
- Environmental safety as enough damage has already been done to the environment by indiscriminate use of chemical pesticides. It is heartening to note that the pesticide consumption in India has come down from 75,033 MT in 1990-91 to 47,020 MT during 2001-02.
- Conservation of natural enemies that are already present in the crop ecosystem. Only safer and eco-friendly pesticides should be

recommended. In this context, avermectins, strobilurin fungicides and pheromones are safe to natural enemies.

- Conservation of biodiversity. Crop-mixes and variety-mixes should be adopted in the cropping system. In this context, use of trap crops like Indian mustard and African marigold in developing IPM technologies in cabbage and tomato, respectively are relevant.
- Not posing risk to human beings, animals, birds, honey bees, fishes and beneficial soil organisms including earthworms.
- Not coming in the way of export, since there is a growing demand for organic food in western countries. There should not be any pesticide residues in the farm produce.

Chapter 13

# INTEGRATED DISEASE MANAGEMENT

Productivity of horticultural crops in India is relatively low compared to other countries. Of the several factors responsible for lower productivity of horticultural crops, fungal, bacterial and viral/phytoplasma diseases are considered as important limiting factors. Some disease problems (bunchy top and Panama disease on banana, guava wilt, bacterial blight of pomegranate, bacterial wilt in Solanaceous vegetables, viral diseases of vegetable and fruit crops) are of national importance which causes significant losses to horticultural crops. Diseases of horticultural crops continue to cause losses of about 10% of the crop yields worth more than Rs. 15,000 crores annually. More than 9,600 MT of technical grade fungicides are used annually to manage the diseases in India.

The severity and incidence of major fungal diseases of various vegetable crops are presented in Table 13.1.

**Table13.1.** Prevalence of fungal diseases of vegetable crops.

| Crop | Disease | Pathogen | Severity (%) | Incidence (%) |
|---|---|---|---|---|
| Brinjal | Phomopsis blight | *Phomopsis vexans* | 10-70 | 32.5 |
| | Collar rot | *Sclerotium rolfsii* | 10-30 | 20.0 |
| | Leaf blight | *Alternaria alternata* | 5-35 | 25.2 |
| Tomato | Alternaria blight | *Alternaria solani* | 15-90 | 40.0 |
| | Septoria leaf spot | *Septoria lycopersici* | 5-20 | 10.0 |
| | Collar rot | *Sclerotium rolfsii* | 10-40 | 20.0 |
| | Fruit rot | *Rhizoctonia, Cladosporium, Pythium* | 5-20 | 15.0 |
| Chilli/ Capsicum | Die-back | *Colletotrichum capsici* | 15-60 | 35.0 |
| | Anthracnose | *Colletotrichum capsici* | 10-30 | 15.0 |
| | Leaf blight & | *Alternaria cineria* | 5-20 | 13.7 |

| | | | | |
|---|---|---|---|---|
| | Fruit rot | *Alternaria solani* | | |
| | Leaf spot | *Cercospora capsici* | 10-35 | 20.0 |
| Cucurbits | Downy mildew | *Pseudoperonospora cubensis* | 10-80 | 30.0 |
| | Anthracnose | *Colletotrichum lagenarium* | 30-90 | 55.0 |
| | Cercospora leaf spot | *Cercospora citrulina* | 20-50 | 30.0 |
| | Vine & Fruit rot | *Sclerotium, Rhizoctonia, Pythium, Sclerotinia* | 10-20 | 15.0 |
| | Phytophthora blight | *Phytophthora cinnamoni* | 25-60 | 35.0 |
| Cowpea/ Pea/ Beans | Leaf blight | *Cercospora cruenta, Pseudocercospora cruenta* | 20-90 | 60.0 |
| | Stem rot | *Macrophomina bataticola* | 5-20 | 10.0 |
| | Phoma leaf blight | *Phoma phaseolina* | 3-15 | 12.0 |
| | Rust | *Uromyces phaseoli* | 5-30 | 10.0 |
| | Powdery mildew | *Erysiphe pisi* | 10-90 | 50.0 |
| | Downy mildew | *Peronospora pisi* | 5-60 | 15.0 |
| | White rot | *Sclerotinia sclerotiorum* | 2-35 | 25.0 |
| Cole crops | Leaf spot | *Alternaria brassicae, A. brassicola* | 10-40 | 26.0 |
| | Downy mildew | *Peronospora parasitica* | 5-25 | 10.0 |
| Okra | Leaf blight | *Cercospora abelmoschi, C. hibisina* | 25-95 | 70.0 |
| Onion/Garlic | Purple blotch | *Alternaria porri* | 15-80 | 40.0 |
| | Leaf blight | *Stemphylium vesicarium* | 20-40 | 25.0 |
| | Downy mildew | *Peronospora destructor* | 10-25 | 10.0 |
| Vegetable nursery | Damping off | *Rhizoctonia, Alternaria, Phytophthora, Fusarium, Colletotrichum, Pythium, Sclerotium, Phomopsis* | 30-85 | 50.0 |

### 13.1. Resistant cultivars

Perhaps the greatest single technological achievement is the advance in breeding crops for resistance to diseases. Cultivation of resistant varieties is the cheapest and best method of controlling diseases. One of the important components of IDM is the use of resistant cultivars to key diseases. Under All India Co-ordinated Research Projects of Indian Council of Agricultural Research, large number of highly/moderately resistant varieties are released to the farmers (Table 13.2).

**Table 13.2.** Horticultural crop varieties resistant to diseases.

| Horticultural crop | Disease | Resistant varieties |
|---|---|---|
| Citrus | Gummosis, leaf fall, fruit rot (*Phytophthora* spp.) | Cleopatra mandarin, Rangpur lime, Trifoliate orange rootstocks. |
| Papaya | Ring spot virus | Rainbow, Sun Up. |
| Potato | Late blight | Kufri Sutlej, Kufri Badshah, Kufri Jawahar (in plains), Kufri Jyothi, Kufri Giriraj, Kufri Kanchan, Kufri Meghad (in hills). |
| Tomato | Bacterial wilt | Arka Abha, Arka Alok, Arka Shreshta, Arka Abhijit, Megha, Shakthi, Sun 7610, Sun 7611. |
| | Powdery mildew (PM) | Arka Asish. |
| | Leaf curl virus | Avinash-2, Hisar Anmol. |
| Brinjal | Bacterial wilt | Arka Nidhi, Arka Keshav, Arka Neelkant, Arka Anand, Swarna Shree, Swarna Shyamali, Surya. |
| | *Phomopsis* blight | Pusa Bhairav. |
| | Little leaf | Pusa Purple Long, Pusa Purple Cluster (Field resistant). |
| Chilli | TMV, CMV, Leaf curl | Pusa Sada Bahar, Punjab Lal, Pusa Jwala. |
| | Powdery mildew | Arka Suphala (T). |
| | Mosaic (T), leaf curl (T) | Pant C-1. |
| | Leaf curl (T) | Jawahar 218. |
| | Viruses | Arka Harita, Arka Meghana. |
| | Bacterial blight | Ujwala. |
| | Dieback (T), powdery mildew (T) | Musalwadi. |
| French bean | Angular leaf spot, Mosaic | Pant Anupama. |
| | Rust, Bacterial blight | Arka Anoop. |
| | Rust | Arka Bold, Swarna Priya, Swarna Latha, Arka Anoop. |
| | Rust, *Alternaria* leaf spot | Arka Bold. |
| Pea | Powdery mildew | Pusa Pragati, Jawahar Matar 5, Jawahar Peas 83. |
| | PM, Rust | Arka Ajit, Arka Karthik, Arka Sampoorna. |
| | *Fusarium* Wilt | JP Batri Brown 3, JP Batri Brown 4. |
| Cowpea | Bacterial blight | Pusa Komal. |
| Pigeon pea | *Fusarium* wilt | Maruti. |
| Field bean | Viral diseases | Pusa Sem-2, Pusa Sem-3. |

| | | |
|---|---|---|
| Cluster bean | PM, *Alternaria* leaf spot | Gomah Manjari. |
| Okra | YVMV | Pusa Sawani, Arka Abhay, Arka Anamika, Hisar Unnat, DVR-1, DVR-2, IIVR-10, Varsha Upkar, P-7, Pusa A-4, Parbhani Kranti (T), Punjab Kesari, Punjab Padmini. |
| Cucumber | PM | Swarna Poorna. |
| | PM, Downy mildew (DM), anthracnose, angular leaf spot | Poinsette. |
| Cabbage | Black rot | Pusa Mukta, K-1. |
| | Black leg | Pusa Drum Head. |
| Cauliflower | Black rot | Pusa Snowball K-1. |
| | Black rot & Curd blight | Pusa Shubhra. |
| | Curd blight | Pusa Synthetic. |
| | DM | Pusa Hybrid-2. |
| Onion | Purple blotch, Basal rot | Arka Pitamber, Arka Kirtiman, Arka Lalima. |
| | Purple blotch, *Alternaria porri* | Arka Kalyan. |
| Garlic | Purple blotch, *Stemphylium* sp. | Agri-found White. |
| Muskmelon | PM | Arka Rajhans, Pusa Madhuras (MR). |
| | PM, DM | Punjab Rasila. |
| | *Fusarium* wilt | Pusa Madhuras, Durgapura Madhu, Arka Jeet, Punjab Sunehari (MR), Harela. |
| Watermelon | PM, DM, Anthracnose | Arka Manik. |
| | Fusarium wilt, Anthracnose | Crimson Sweet, Petite Sweet, Super Sweet, All Sweet |
| Ridge gourd | PM, DM | Swarna Uphaar. |
| Bottle gourd | Blossom end rot | Arka Bahar (T). |
| | CMV | Punjab Komal. |
| Carrot | PM | Arka Suraj. |
| Amaranthus | White rust | Arka Arunima, Arka Suguna (MR). |
| Palak | *Cercospora* leaf spot | Arka Anupama. |
| Black pepper | *Phytophthora capsici* | IISR Shakthi. |
| Cardamom | Rhizome rot | IISR Avinash. |
| | Cardamom mosaic virus (Katte) | IISR Vijetha. |
| Ginger | Soft rot | Mahima Maran. |
| Cumin | *Fusarium* wilt | Gujarat Cumin 4. |

MR – Moderately Resistant; T – Tolerant

## 13.2. Biological control

The last decade has witnessed a tremendous break through in biological control of diseases and nematodes like *Rhizoctonia, Pythium, Fusarium, Macrophomina, Ralstonia* and *Meloidogyne* in banana, tomato, egg plant, pea, grapes, cucumber, black pepper, cardamom, ginger and turmeric, especially by using species of *Trichoderma, Pochonia, Pseudomonas* and *Bacillus* (Tables 13.3 to 13.9).

**Table 13.3.** Biological control of fruit crop diseases.

| Fruit Crop | Disease/Pathogen/s | Potential biocontrol agent/s |
|---|---|---|
| Banana | Panama wilt, *Fusarium oxysporum* f. sp. *cubense* | *T. viride, Aspergillus niger, Pseudomonas fluorescens, T. viride + P. fluorescens* – sucker treat. |
| Citrus | Root rot, *Phytophthora* spp. | *Trichoderma viride / T. harzianum* at 100 kg/ha; *Pencillium funiculosum, Pythium nunn*-soil treat. |
| | Canker, *Xanthomonas campestris* pv. *citri* | *Aspergillus niger* AN 27 |
| Strawberry | Grey mold, *Botrytis cineria* | *T. harzianum* |
| Mulberry | Leaf spot, *Cercospora moricola* | *T. viride, T. harzianum, Pseudomonas fluorescens* |
| | Cutting rot, *F. solani* | *T. virens, T. harzianum, T. pseudokoningii* |
| Grapevine | Powdery mildew, *Uncinula necator* | *Ampelomyces quisqualis*-dispersal from wick cultures at 15 cm of shoot growth & bloom |
| | Downy mildew, *Plasmopara viticola* | *Fusarium proliferatum* weekly spray starting from 15 cm of shoot growth – $10^6$ spores/ml |
| Guava | Anthracnose, *Pestalotia psidii, Colletotrichum gloeosporoides* | *T. harzianum* |
| | Wilt, *Gliocladium roseum* & *F. solani* | *Pencillium citrinum, Aspergillus niger* (AN 27), *T. harzianum* |
| Mango | Anthracnose, *Colletotrichum gloeosporoides* | *T. harzianum, Streptosporangium pseudovulgare* |
| | Powdery mildew, *Oidium mangiferae* | *S. pseudovulgare* |
| | Bacterial canker, *Xanthomonas campestris* pv. *mangiferaeindicae* | *Bacillus coagulans* |
| Apple | Scab, *Venturia inaequalis* | *Chaetomium globosum, A. pullulans, Microsphaeropsis* sp., *Cladosporium* spp., *Trichothecium roseum*-foliar spray. |

| | | |
|---|---|---|
| | *Phytophthora cactorum* | *Enterobacter aerogenes, Bacillus subtilis*-soil treatment; *T. virens*-soil treat. |
| | White root rot, *Dematophora necatrix* | *T. viride, T. harzianum, T. virens*-soil treat. |
| Pear | Blue mold, *Pencillium expansum*; Grey mold, *Botrytis cineria* | *C. infirmo-miniatus* YY6, *C. laurentii* RR87-108, *R. glutinis* HRB6-fruit spray 3 week or 1 day prior to harvest – $10^8$ cfu/ml; *Pantoea agglomerans* CPA-2-post-harvest fruit dipping in 8 x$10^8$ cfu/ml |
| | Fire blight, *Erwinia amylovora* | *Pseudomonas fluorescens*-foliar spray |
| Peach | Brown rot, *Monilia fructicola* | *Bacillus subtilis* (B-3)-post-harvest fruit line spray at 5 x $10^8$ cfu/g; *Pseudomonas syringae*-post-harvest fruit dipping in $10^7$ cfu/ml. |
| | Twig blight, *Monilia laxa* | *Pencillium frequentans*-spray shoots in early growing season – $10^{8\text{-}9}$ spores /ml |
| | Crown gall, *Agrobacterium tumefaciens* | *Agrobacterium radiobacter* K84, K1026-root dip treat. |
| Strawberry | *Botrytis cineria* | *Trichoderma* products (BINAB®TF & BINAB®T), *Bacillus pumilus, Pseudomonas fluorescens, Gliocladium roseum* –spray flowers & fruits – white flower bud to pink fruit-$10^6$ spores/ml; *G. roseum*-bee vectoring of flowers – $10^9$ cfu/g of powder. |
| Passion fruit | Collar rot, *R. solani* | *T. harzianum, Trichoderma* sp. |
| Amla | Bark splitting, *Rhizoctonia solani* | *Aspergillus niger* AN 27 |

**Table 13.4.** Biological control of vegetable crop diseases.

| Crop | Disease/Pathogen/s | Biocontrol agent/ Mode of application |
|---|---|---|
| French bean | Dry root rot, *Macrophomina phaseolina* | *Pseudomonas cepacia* UPR5C-seed treat. |
| | Wilt, *Fusarium oxysporum* f. sp. *phaseoli* | *Streptomyces* spp.- seed treat. |
| Pea | Root rot, *Aphanmomyces euteiches* | *P. fluorescens* PRA25-seed treat., *Pseudomonas cepacia* AMMD-seed treat. |
| | Damping-off, *Pythium ultimum* | *Pseudomonas cepacia* AMMD-seed treat., *P. putida* NIR-seed treat. |
| | Wilt, *F. oxysporum* f. sp. *udum* | *T. viride, T. harzianum, T. koningii* – seed treat., *Aspergillus niger* AN27 |

| | | |
|---|---|---|
| Cluster bean | Bacterial blight, *Xanthomonas axonopodis* pv. *cyamopsidis* | *Aspergillus niger* AN27 |
| Cabbage | Damping-off, *R. solani* | *T. viride, T. harzianum, T. koningii* – seed treat. |
| Cauliflower | Blight, *Alternaria brassicola* | *Streptomyces gresioviridis*-seed treat. |
| Okra | *Rhizoctonia solani* | *Bradyrhizobium japonicum, Rhizobium* spp.- seed treat. |
| Tomato | Damping-off, *Pythium aphanidermatum* | *T. viride, T. harzianum, Pseudomonas aeruginosa* 7NSK2 |
| | Wilt, *Fusarium oxysporum* f. sp. *lycopersici* | *T. viride, T. harzianum, Aspergillus niger,* non-pathogenic *F. oxysporum, F. oxysporum* f. sp. *dianthi, P. fluorescens* strains Pf1, *P. putida, Pencillium oxalicum, Pythium oligandrum, Bacillus subtilis* strains FZB-G, *Streptomyces* spp.- seed treat., seed & soil treat. |
| Potato | Black scurf, *R. solani* | *T. harzianum, T. viride-* tuber treat., *Aspergillus niger* AN27, *Verticillium biguttatum*-soil treat., *Laetisaria arvalis*-tuber treatment, Binucleate *Rhizoctonia*. |
| | Wilt, *Ralstonia solanacearum* | *Bacillus cereus, B. subtilis* |
| Bell pepper | *Phytophthora capsici* | *T. viride, T. harzianum*-fruit treat. |
| | Damping-off, *Pythium aphanidermatum* | *Streptomyces griseoviridis*-seed & soil treat. |
| Brinjal | Damping-off, Wilt, *Phytophthora* sp., *Pythium aphanidermatum, F. solani* | *T. viride, T. harzianum, T. koningii*-seed & soil treat. |
| | Collar rot, *Sclerotinia sclerotiorum* | *T. viride, T. virens, Bacillus subtilis*-soil treat. |
| Carrot | *Sclerotinia sclerotiorum* | *Coniothyrium minitans*-soil treat. |
| | *R. solani* | *T. virens* GL-21 |
| Radish | Wilt, *F. oxysporum* f. sp. *raphani* | *P. fluorescens* strains WCS374, WCS417r – soil treat. |
| | *R. solani* | *Laetisaria rosiepellis, Pythium acanthicum*-soil treat. |
| Beet root | *Pythium debarryanum, P. ultimum* | *Pencillium* spp + *P. fluorescens*-seed treat., *Pythium oligandrum* |
| Cucumber | Wilt, *Fusarium oxysporum* f. sp. *cucumerinum, R. solani* | *Colletotrichum orbiculare, F. oxysporum* f. sp. *niveum, Pseudomonas putida* 89B-27, *Serratia marcescens,* Tobacco necrosis virus. |

| | | |
|---|---|---|
| | Powdery mildew | *Ampelomyces quisqualis*-foliar spray |
| | Cucumber mosaic virus | *P. fluorescens* strain 89B-27 |
| Water melon | Wilt, *Fusarium oxysporum* f. sp. *solani, F. o.* f. sp. *niveum* | *T. viride, Aspergillus niger*-seed & soil treat., *Pencillium janczewskii.* |
| Musk melon | Wilt, *F. oxysporum, F. solani, R. solani* | *T. harzianum, Aspergillus niger* – seed treat. |
| Onion | *Sclerotium cepivorum* | *Chaetomium globosum, Trichoderma* sp. C62 –soil treat. |

**Table 13.5.** Biological control of ornamental crop diseases.

| Crop | Disease / Pathogen/s | Biocontrol agent/ Mode of application |
|---|---|---|
| Rose | Grey mold, *Botrytis cineria* | *T. viride, T. harzianum*-cutting treat. |
| Gladiolus | Yellows & corm rot, *F. oxysporum* f. sp. *gladioli* | *T. virens, T. harzianum*-corm & soil treat. |
| Chrysanthemum | Wilt, *Fusarium oxysporum* | *T. harzianum*- soil appln.-160 kg/ha |
| | *Rhizoctonia solani* | *Aspergillus niger* AN27 |
| Carnation | Wilt, *F. oxysporum* f. sp. *dianthi* | *Pseudomonas fluorescens* strain WCS 417r – soil appln; *P. putida* WCS 358r-root dip treat.; *Alcaligenes* sp., *Bacillus* sp.; *Arthrobacter* sp., *Hafnia* sp.; *Serratia liquifaciens.* |
| Gerbera | *Phytophthora cryptogea* | *Trichoderma* spp.-soil treat. |
| Narcissus | *F. oxysporum* f. sp. *narcissi* | *Streptomyces griseoviridis*-bulb treat.; *Minimedusa polyspora*-bulb treat. |
| Zinnia | *Rhizoctonia solani* | *T. virens* GL-21, *T. virens* GL-20 |
| Marigold | *Pythium ultimum* | *Glomus intraradices, G. mosseae* –soil treat. |

**Table 13.6.** Biological control of medicinal and aromatic crop diseases.

| Medicinal/ aromatic crop | Disease / Pathogen/s | Biocontrol agent |
|---|---|---|
| Opium poppy | Sclerotinia rot & blight, *Sclerotinia sclerotiorum* | *T. harzianum, T. viride, T. koningii, T. virens*- soil treat. |
| | Downy mildew, *Peronospora arborescens* | *Trichoderma* spp.-seed treat. |
| Periwinkle | *Phytophthora parasitica* | *Phytophthora parasitica* var. *nicotianae*- soil treat. |
| Jasmine | Root rot, *Macrophomina phaseolina* | *T. viride, T. harzianum*- cutting treat. |
| Chinese Rose | Wilt, *Fusarium oxysporum* | *Aspergillus niger*- soil treat. |
| Menthol Mint | Stolon decay, *Sclerotinia sclerotiorum* | *T. harzianum, T. virens*- sucker treat. |
| | *Verticillium dahliae* | *Verticillium nigrescens* |

**Table 13.7.** Biological control of tuber crop diseases.

| Tuber crop | Disease / Pathogen/s | Biocontrol agent |
|---|---|---|
| Yam | *Botrytis theobromae* | *Trichoderma viride* |
| Cassava | *Phytophthora dechsleri* | *Trichoderma viride* |
| Elephant Foot Yam | *Sclerotium rolfsii* | *Trichoderma harzianum, T. pseudokoningii* |

**Table 13.8.** Biological control of plantation crop diseases

| Crop | Disease/Causal agent | Biocontrol agents reported |
|---|---|---|
| Coconut | Stem bleeding, *Thielaviopsis paradoxa (Ceratostomella paradoxa)* | *Trichoderma virens, T. harzianum,* Phosphobacteria |
| | Basal stem rot, *Ganoderma lucidum* | *G. virens, T. harzianum* |
| Areca nut | Bud rot, *Phytophthora* spp. | *Trichoderma* spp. |
| | Fruit rot, *Phytophthora arecae, Colletotrichum capsici* | *Trichoderma* spp., *P. fluorescens* |
| | Foot rot/Anabe, *Ganoderma lucidum* | *T. harzianum* |
| Tea | Red root rot, *Poria hypolateritia* | *T. harzianum* |
| | Brown root, *Fomes noxius* | *G. virens, T. harzianum* |
| | Black root rot, *Rosellina arcuata* | *G. virens, T. harzianum* |
| Coffee | Black root, *Pellicularia koleroga* | *G. virens, T. harzianum* |
| | Brown root, *Fomes noxius* | *G. virens, T. harzianum* |
| | Santhaveri wilt, *F. oxysporum* f. sp. *coffeae* | *G. virens, T. harzianum* |
| Rubber | Brown rot, *Phellinus noxius* | *T. viride, T. harzianum, T. hamatum* |
| Betel vine | Foot & root rot, *Phytophthora parasitica* pv. *piperina* | *T. viride, T. harzianum*-soil treat. |
| | Collar rot, *Sclerotium rolfsii* | *T. harzianum, T. viride, T. koningii, T. virens*-soil treat. |

**Table 13.9.** Biological control of spice crop diseases.

| Spice crop | Disease/Causal organism | Effective biocontrol agents/Mode of application |
|---|---|---|
| Black pepper | Foot rot, *Phytophthora capsici* | *Trichoderma harzianum, T. virens, Glomus fasciculatum*-soil treat.; *Pseudomonas fluorescens, Bacillus* sp.-foliar spray |

| | | |
|---|---|---|
| | Anthracnose, *Colletotrichum gloeosporoides* | *P. fluorescens*-foliar spray |
| | Slow decline, *Radopholus similis, Meloidogyne incognita, Phytophthora capsici* | *T. harzianum, T. virens, Paecilomyces lilacinus, Pochonia chlamydosporia*-soil treat. |
| Cardamom | Damping off, *Pythium vexans* | *T. harzianum, T. viride*-soil treat. In solarized nursery beds. |
| | Clump rot/Rhizome rot, *Pythium vexans, Rhizoctonia solani, Meloidogyne incognita* | *T. harzianum*-soil treat. |
| | Capsule rot, *Phytophthora meadii, P. nicotianae* var. *nicotianae* | *T. harzianum, T. viride, T. virens, T. hamatum*-soil treat. |
| Ginger | Rhizome rot, *Pythium aphanidermatum, P. myriotylum* | *T. harzianum, T. virens*-soil solarization + soil treat.; *T. viride, P. fluorescens*-seed treat.; *Aspergillus niger* AN27-soil treat. |
| | Yellows, *Fusarium oxysporum* f. sp. *zingiberi, M. incognita* | *T. harzianum, T. virens , T. hamatum*-soil solarization + soil treat., Rhizome treat. |
| | Bacterial wilt, *Ralstonia solanacearum* | Avirulent *Ralstonia solanacearum, P. fluorescens*, Endophytic bacteria-soil treat. |
| Turmeric | Rhizome rot, *Fusarium* sp., *Pythium graminicolum, P. aphanidermatum, R. similis, M. incognita* | *T. harzianum, T. viride, T. virens* –soil treat. |
| Fenugreek | Root rot, *R. solani* | *T. viride, P. fluorescens*-seed treat. |
| Coriander | Root rot/wilt, *Fusarium oxysporum* f. sp. *corianderii* | *T. viride, T. harzianum, Streptomyces* sp.-seed treat. |
| Cumin | Wilt, *F. oxysporum* f.sp. *cumini* | *Trichoderma* spp., *T. virens*-soil treat. |
| Vanilla | Root rot, *Phytophthora meadii, F. oxysporum* f. sp. *vanillae* | *T. harzianum, P. fluorescens*-soil treat. |
| Mustard | Damping-off, *P. aphanidermatum* | *T. viride, T. harzianum*-seed & soil treat. |

## 13.3. Plant growth promoting rhizobacteria (PGPR)

PGPR are root colonizing bacteria with beneficial effects including plant growth promotion and induction of systemic resistance in plants against wilt and damping-off diseases. *Pseudomonas fluorescens, P. putida, Bacillus subtilis* and *B. pumilis* as seed treatment and soil application have been found effective against anthracnose, *Fusarium* and bacterial wilt, CMV and angular leaf spot on cucumber; halo blight of beans; damping off, CMV and root-knot nematode on tomato and *Fusarium* wilt, *Alternaria* and *Pseudomonas* leaf spots on radish (Table 13.10).

**Table 13.10.** Biocontrol of horticultural crop diseases using plant growth promoting rhizobacteria and their commercial formulations

| Rhizobacteria | Target pathogen/host | Commercial product |
|---|---|---|
| *Agrobacterium radiobacter* | Crown gall of crucifers, roses & fruit trees, *Agrobacterium tumefacians* | Galltrol-A, K-84, Norbac 84C, Nogall, Diegall. |
| *Bacillus subtilis* | Damping-off & root rots of leguminous vegetables, *Pythium ultimum, Rhizoctonia solani, Fusarium* spp. | Epic, Kodiak, Kodiak HB, Kodiak AT, Quantum, Rhizo-Plus, Rhizo-Plus Konz, Serenade, System 3, Biotok. |
| *Pseudomonas cepacia* | Damping-off of vegetables, *Pythium* spp., *Rhizoctonia, Fusarium* | Deny, Intercept. |
| *P. fluorescens* | Damping-off & wilt of vegetables, *Pythium aphanidermatum, Rhizoctonia solani, Fusarium oxysporum* | Dagger G, BlightBan A506, Conquer, Victus, Biocoat, Pant Biocontrol Agent-2, Bioshield. |
| *P. syringae* | *Botrytis cinerea, Pencillium* spp., *Mucor pyroformis, Geotrichum candidum* | Bio-save 100, Bio-save 1000, Bio-save 110 |
| *P. chlororaphis* | Leaf stripe, net blotch, spot blotch, leaf spot | Cedomon, AtEze. |
| *P. aureofaciens* Tx-1 | *Pythium aphanidermatum,* Dollar spot, anthracnose, pink snow mold | Spot-Less |
| *Ralstonia solanacearum* Nonpathogenic | Wilt of vegetables, *Ralstonia solanacearum* | PSSOL |
| *Streptomyces griseoviridis* K61 | Ornamentals & vegetables *Pythium, Fusarium* | Mycostop |

## 13.4. Foliar spray therapy

Spray application of $K_2HPO_4$ or $KH_2PO_4$ at 3.5 g/litre of water has been reported to control powdery mildew in rose and carnation. Similarly, the above treatment was also found effective for the management of powdery mildew on mango, grapes and cucurbits.

## 13.5. Non pathogenic strains (NPS)

A novel strategy of using non-pathogenic strains of fungi is gaining importance in the management of plant diseases. NPS of *Fusarium* (Fusaclean Fo 47) is being used widely to control *Fusarium* wilt in tomato, melons and carnation over thousands of hectares in Europe. In Japan, NPS of *Fusarium* is successfully used for the management of *Fusarium* wilt in sweet potato. In Italy, NPS of *Fusarium* (251/2) is being successfully used for the management of *Fusarium* wilt in carnation and other horticultural crops (Table 13.11).

**Table 13.11.** Management of wilt diseases using non-pathogenic strains

| Horticultural crop | Disease | Non-pathogenic strain |
|---|---|---|
| Cucumber | Wilt, *Fusarium oxysporum* f. sp. *cucumerinum* | Non-pathogenic *Fusarium oxysporum*-soil treat. |
| Water melon | Wilt, *Fusarium oxysporum* f. sp. *solani, F. o.* f. sp. *niveum* | Avirulent *F. o.* f. sp. *niveum* |
| Carnation | Wilt, *F. oxysporum* f. sp. *dianthi* | Non-pathogenic *F. oxysporum* |
| Sweet potato | Wilt, *F. oxysporum* f. sp. *batatas* | Non-pathogenic *Fusarium oxysporum* |

NPS are known to compete with pathogens for nutrients in the rhizosphere, compete for infection sites on the rhizoplane and induce systemically acquired resistance in the host.

## 13.6. Strobilurin fungicides

Strobilurin fungicides are also called Qo inhibitors as they act on cytochrome Qo of the fungi. The Basidiomycetous fungus, *Strobilurus tenacellus* produces antibiotics to ward off competition from other fungi. Based on this principle, several fungicides have been developed namely Azoxystrobin, Kresoxy methyl, Metominostrobin, Trifloxystrobin, Picoxystrobin, Pyraclostrobin, Famoxadone and Fenomidone during 1996-2001. Within 4 years, sale of these fungicides totaled $620 million, accounting for 10% of total fungicide market in the world. This success is unparalleled in the history of fungicides.

Strobilurin fungicides are naturally occurring compounds and hence eco-friendly, highly systemic, have unique mode of multisite action, hence development of resistance is remote. They have broad spectrum of activity on all groups of fungi and registered in 72 countries on 84 crops representing over 400 crop/disease systems.

## 13.7. Biotechnological approaches

Biotechnological intervention in crop protection has been through tissue culture, genetic engineering, diagnostic probes and molecular markers.

One of the applications of tissue culture technique is the production of disease-free planting material. The use of apical meristem as an explant source generated virus and fungus-free planting material in banana and citrus.

With the advent of recombinant DNA technology, genetic engineering is being used to generate transgenic plants resistant to diseases. Molecular biology techniques are used to identify disease resistance genes. Transgenic virus resistance has been developed in potato, tomato, cucumber, melon and papaya. Papaya transgenic cvs namely

'Rainbow' and 'Sun Up' have been developed by Hawaiian scientists who were honoured with 'Von Humboldt Award' recently.

The defense proteins chitinase and β-1,3-glucanase have been used to develop transgenic fungal resistance. These enzymes catalyze the hydrolysis of chitins and β-1,3- glucans, which form the major components of the fungal cell wall, leading to inhibition of mycelial growth and preventing further spread of the pathogen. Transgenic fungal resistance has been developed in tobacco against *Phytophthora infestans* using glucanase and against *Rhizoctonia solani* using chitinase. Transgenic rice resistant to *R. solani* has also been developed using chitinase.

Lysozyme, cecropins and defensins genes isolated from arthropods and mammals have been used to develop transgenic bacterial resistance in potato against soft rot and tobacco against bacterial wilt.

Molecular probes helps us to rapidly screen germplasm for several diseases simultaneously with high precision. These probes can detect pathogens occurring in extremely low quantities. Molecular markers for disease resistance genes have been identified in citrus for tristeza virus, in beans for rust, in lettuce for downy mildew, in peas for *Fusarium* wilt and powdery mildew, in potato for PVY and late blight and in tomato for *Fusarium* wilt, spotted wilt virus, powdery mildew and bacterial wilt.

Chapter 14

# INTEGRATED NEMATODE MANAGEMENT

Productivity of horticultural crops in India is relatively low compared to other countries. Of the several factors responsible for lower productivity of horticultural crops, plant parasitic nematodes are considered as important limiting factors. Some nematode problems (root-knot and reniform nematodes on vegetables; burrowing nematode on banana, plantation and spice crops; and cyst nematodes on potato and cowpea) are of national importance which causes significant losses to horticultural crops.

## 14.1. Economic importance

Estimated overall average annual yield loss of the world's major horticultural crops due to damage by plant parasitic nematodes is 13.54% (Table 14.1). For the 7 horticultural crops (left-hand column) that stand between man and starvation (life sustaining horticultural crops), an estimated annual yield loss of 12.77% is reported. Monetary losses due to nematodes on 10 horticultural crops, six of which are life sustaining, were estimated at US$ 19.37 billion annually based on 1984 production figures and prices.

**Table 14.1.** Estimated annual yield losses due to damage by plant parasitic nematodes – world basis (Sasser and Freckman, 1987).

| Life sustaining horticultural crops | Loss (%) | Economically important horticultural crops | Loss (%) |
|---|---|---|---|
| Banana | 19.7 | Cocoa | 10.5 |
| Cassava | 8.4 | Citrus | 14.2 |
| Coconut | 17.1 | Coffee | 15.0 |
| Field bean | 10.9 | Cowpea | 15.1 |
| Potato | 12.2 | Eggplant | 16.9 |
| Sugar beet | 10.9 | Grapes | 12.5 |
| Sweet potato | 10.2 | Guava | 10.8 |

| | | | |
|---|---|---|---|
| | | Melons | 13.8 |
| | | Okra | 20.4 |
| | | Ornamentals | 11.1 |
| | | Papaya | 15.1 |
| | | Pepper | 12.2 |
| | | Pineapple | 14.9 |
| | | Tea | 8.2 |
| | | Tomato | 20.6 |
| | | Yam | 17.7 |
| Average | 12.77 % | Average | 14.31 % |
| Overall average – 13.54% | | | |

The 16 economically important horticultural crops (right-hand column) that represent a miscellaneous group important for food or export value were reported to have an estimated annual yield loss of 14.31%.

In India, the avoidable yield losses due to plant parasitic nematodes in horticultural crops have been presented in Table 14.2.

**Table 14.2.** Avoidable yield losses in horticultural crops due to plant parasitic nematodes.

| Crop | Nematode(s) | Yield loss (%) |
|---|---|---|
| Banana | *Radopholus similis* | 32.00-41.00 |
| | *Meloidogyne incognita* | 30.90 |
| Sweet orange | *Tylenchulus semipenetrens* | 69.00 |
| Lemon | *T. semipenetrans* | 29.00 |
| Sweet lime | *T. semipenetrans* | 19.00 |
| Grapevine | *M. incognita* | 55.00 |
| | *M. javanica* | 53.00 |
| Papaya | *Rotylenchulus reniformis* | 28.00 |
| Pomegranate | *Meloidogyne* sp. | 24.64-27.45 |
| Peach | *Macroposthonia xenoplax* | 33.00 |
| Plum | *M. xenoplax* | 10.00 |
| Potato | *incognita* | 42.50 |
| | *Globodera rostochiensis* | 99.50 |
| Tomato | *incognita* | 30.57-46.92 |
| | *M. javanica* | 77.50 |
| | *R. reniformis* | 42.25-49.02 |
| Brinjal | *M. incognita* | 27.30-48.55 |
| Chilli | *Meloidogyne* sp. | 24.54-28.00 |
| Okra | *M. incognita* | 28.08-90.90 |
| | *M. javanica* | 20.20-41.20 |

| | | |
|---|---|---|
| French bean | *M. incognita* | 19.38-43.48 |
| | *M. javanica* | 30-40 |
| Cowpea | *M. incognita* | 28.60 |
| | *R. reniformis* | 13.20-32.00 |
| Peas | *M. incognita* | 20.20-50.61 |
| | *R. reniformis* | 15.8 |
| Carrot | *M. incognita* | 56.64 |
| Bitter gourd | *M. incognita* | 36.72 |
| Pointed gourd | *M. incognita* | 30-40 |
| Water melon | *M. incognita* | 18-33 |
| Mushroom | *Ahelenchoides sacchari* | 40.6-100.0 |
| | *A. composticola* | 35-60 |
| | *A. avenae* | 25.8-53.5 |
| Crossandra | *M. incognita* | 21.64 |
| Tuberose | *M. incognita* | 13.78 |
| Patchouli | *M. incognita* | 47.00 |
| Davana | *M. incognita* | 50 |
| *Cymbopogon* | *M. incognita* | 20 |
| Carnation | *M. incognita* | 27 |
| Gerbera | *M. incognita* | 31 |
| *Coleus forskohlii* | *M. incognita* | 70.2 |
| Chethikoduveli (*Plumbago rosea*) | *M. incognita* | 29.00-43.96 |
| Kacholam (*Kaempferia galanga*) | *M. incognita* | 18-64 |
| Colocasia | *M. incognita* | 24 |
| Menthol mint | *M. incognita* | 30 |
| Cardamom | *M. incognita* | 32-47 |
| Betel vine | *M. incognita* | 21.1-38 |
| Ginger | *M. incognita* | 29.60-74.10 |
| | *R. similis* | 39-73 |
| Turmeric | *M. incognita* | 18.6-25.0 |
| | *R. similis* | 46-76 |
| Coriander | *M. incognita* | 51 |
| Cumin | *M. incognita* | 34 |
| Fennel | *M. incognita* | 39 |
| Black pepper | *R. similis & M. incognita* | 38.5-64.6 |
| | *R. similis* | 59 |
| | *M. incognita* | 46 |
| Coconut | *R. similis* | 30 |

Realizing that nematodes cannot be eradicated, and that we must live with them, the overall goal is to keep the population density as low as possible. The economic threshold is the density of nematodes where the losses incurred exceed the cost of nematode management.

Horticultural ecosystem provides a congenial ecological niche for one or the other nematodes, the population of which if unchecked, may reach alarming proportions resulting in serious losses. Nematode management is therefore important for high yields and quality that are required by the high cost of modern crop production. The direct and indirect benefits of nematode management include – increased quantity and quality of the produce, improved health of plants thereby reducing their susceptibility to plant pathogens and increasing the ability to withstand adverse growing conditions, and a better utilization of nutrients and moisture.

In the field, several nematode management methods like regulatory, physical, cultural, chemical, biological and host resistance can be effectively employed to keep nematode population to a minimum level. It is not advisable to depend on a single method to control nematodes. Efficient management requires the carefully integrated combinations of several practices.

## 14.2. Integrated nematode management

Integrated nematode management (INM) is an economically viable and socially acceptable approach to crop protection. INM is a system that, in the context of the associated environment and the population dynamics of the nematode species, utilizes all suitable techniques and methods in as compatible manner as possible and maintains the nematode populations below those causing economically unacceptable damage or loss. The criteria for judging the effectiveness of INM are productivity, stability, sustainability and equity. Sound knowledge of nematodes, cropping systems and the environment are important for the implementation and success of INM.

INM is an important principle on which sustainable crop protection can be based. The goal of INM is to achieve long-term sustainable systems of crop protection and production. The ideal objective of INM should be to maximize profit, while minimizing human health and environmental risks.

INM is the best combination of cultural, biological, chemical and regulatory measures that provides the most effective, environmentally sound and socially acceptable method of managing nematodes under the circumstances in which they work. Conserving the natural enemies of nematodes is recognized as an important part of INM programmes.

There is an urgent need to shift from the 'chemical intensive' to 'bio-intensive' cropping system with a view to restore the viability, sustainability and health of our agro-ecosystems. There is a need for development and adoption of INM in agriculture because of the economic and environmental benefits it brings to us. Major emphasis should be given to non-chemical methods of control, such as the biological, cultural, physical and regulatory.

Nematode-resistant varieties of important crops must be evolved as these would be most useful tools of INM for resource poor farmers.

To alleviate nematode problems, INM procedures are implemented based on the principles of prevention, population reduction and tolerance. In practice, integration of different methods of nematode control is the rule than the exception. In most cases, practical control of a nematode disease involves integration of several diverse control measures. In this context, biological control agents form ideal components in the INM along with botanicals and to limited extent chemicals.

Integrated nematode management (INM) includes research, development, technology transfer and implementation needed to integrate two or more control measures to manage one or more nematode species. The aim of INM is to reduce and stabilize the nematode pest population below damaging levels, resulting in favourable long term socio-economic and environmental consequences. The components of INM systems include (a) biological monitoring; (b) environmental monitoring; (c) agricultural production systems; (d) nematode crop ecosystem models; (e) system design; and (f) implementation. The proceedings available for INM system design and implementation are based on the principles of exclusion, population reductions and tolerance.

INM strategies are based on 3 main components/tools:

- Prevention
- Monitoring/Observation/Regulation
- Intervention

### 14.2.1. Prevention

Prevention of nematodes through indirect measures is an important part of INM. The aim of prevention is to reduce the initial severity of nematode infestation. Most of the INM activities emphasize heavily on preventive measures especially before the crop is sown/planted *i.e.* at the field preparation or seed treatment stages since the initial nematode inoculum is largely responsible for the subsequent high nematode build up. Agronomic practices such as deep summer ploughing, soil solarization, cleansing of crop refuse, elimination of weed hosts, crop rotation, water management, inter-cropping/mixed cropping, trap cropping, nutrition management and genetic sources of resistance along with habitat management are the major tools of prevention.

(i) **Site selection:** Avoid growing a crop in nematode-infested area/sick plot.

(ii) **Crop sanitation:** Reduces the nematode load.

(iii) **Removal of alternate hosts:** Breaks the nematode life cycle.

(iv) **Land preparation:** Deep summer ploughing or soil solarization kills vulnerable stages of nematodes due to high temperatures.

(v) **Crop rotation:** Growing of non-host crops reduces the nematode build up. Basically, crop rotation involves depriving the nematode of a suitable host and thereby reducing the nematode population by starvation. Probably the most common cultural practices used in nematode management is crop rotation.

(vi) **Use of organic amendments:** Use decomposed FYM/ vermicompost which enriches natural population of bioagents.

(vii) **Timely sowing:** Avoidance of peak activity of the nematode e.g. planting during winter helps to avoid nematode attack.

(viii) **Need based irrigation:** Avoid excessive irrigation as it favours nematodes.

(ix) **Biodiversity or habitat management:** Crop mixes and variety mixes should be adopted in the cropping system. Follow intercropping/mixed cropping/trap cropping.

(x) **Host resistance:** The use of resistant and tolerant cultivars is economically attractive management tactics because most infested fields have a dominant or *target* nematode. Furthermore, nematodes spread slowly, persist in the soils for long periods of time, and management by physical and chemical means is relatively expensive. Available information indicates that plant resistance to nematodes is genetically controlled by one gene in some cases, and by more than one gene in other plant-nematode combinations. Practical results have been obtained in breeding for resistance to a number of major nematodes (Table 14.3).

**Table 14.3.** Nematode-resistant cultivars of horticultural crops.

| Crop | Nematode | Resistant cultivars |
|---|---|---|
| Banana | *Radopholus similis* | Kadali, Pedalimoongil, Peykunnan, Ayiramkapoovan, Kunnan, Pisang Seribu, Tongat, Vennettu Kunnan, Anaikomban, Gold Finger |
| Citrus | *Tylenchulus semipenetrans* | Trifoliate orange, Swingle Citrumello |
| | *Pratylenchus coffeae* | *Poncirus trifoliata*<br>*Microcitrus* selections |
| Grapevine | *Meloidogyne incognita* | Black Champa, Dogridge, 1613, Salt Creek, Cardinal, Banquabad, Zinfundel, Kandharia, Kishmish Chorni, Redvine, Convert Large Black, Luglienga |
| | *Rotylenchulus reniformis* | Dakshi, Joazbeli, Mukchilani |
| Papaya | *R. reniformis* | Solo, Washington |
| Passion Fruit | *M. incognita* | Yellow, Kaveri |
| Tomato | *Meloidogyne* spp. | Nematox, SL-120, NTR-1, SL-12, Patriot, VFN-8, VFN Bush, Piersol, Radiant, Nemared, Ronita, |

| | | |
|---|---|---|
| | | Anahu, Bresch, Helani, Campbell-25, Punuui, Arka Vardan, Pelican, Hawaii-7746, Hawaii-7747, Hisar Lalit, Punjab NR-7 |
| | *R. reniformis* | Healani, VFN Bush, Radiant, SL-120, Patriot, Kalyanpur Selection I, Kalyanpur Selection III, Petropride, Peto 95, Peto 108, Nema 1400, ACE 55 |
| Brinjal | *M. incognita* | Giant of Banaras, Black Beauty, Gola, SM – 67, Gachha Baigan, Syamala (MR) |
| Chilli | *M. javanica* | 579, CAP-63, Pusa Jwala Wonder Hot, Teja, LCA 206 |
| Capsicum | *R. reniformis* | Sweet Bell, California Wonder |
| Potato | *incognita* | Kufri Dewa |
| | *Globodera rostochiensis* | Kufri Swarna |
| | *Pratylenchus penetrans* | Katahdin, Kenebec, Superior |
| | *R. reniformis* | La Rouge, Red Aboda |
| Onion | *R. reniformis* | Evergreen |
| Sweet potato | *M. incognita* | Sree Bhadra, Sree Vardhini, Sree Nandini, Kanjangad |
| | *R. reniformis* | Gold Rush |
| Cowpea | *M. incognita* | Barsati Mutant, Iron, Clay, Chinese Red, Groit, New Era, Missisipi Purple, Missisipi Silver, New Selection, Viktor K 798, Broweye 7, Brabham, Moneetta, Catjang, Columbia, Gujarat Cowpea 1 (GAU-1), C-152, 92-1-B, IC 9642-B, TVU 2430-P |
| | *Heterodera cajani* | CO 6 |
| French bean | *M. incognita* | Banat, Blue Lake Stringless, Bountiful Flat, Brown Beauty, Cambridge Countess, Gallaroy, Kenya-3, Pinto W5-114, Seafarer, Suttan's Masterpiece |
| Okra | *M. incognita* | Parbhani Kranthi (MR) |
| Muskmelon | *incognita* | Scarsol |
| | *M. javanica* | S-445 |
| Watermelon | *M. incognita* | Shehjanpuri |
| Ridge gourd | *M. incognita* | Panipati, Meerut Special |
| Ash gourd | *M. incognita* | Jaipuri, Agra |
| Pumpkin | *M. incognita* | Jaipuri, Dasna |
| China aster | *M. incognita* | Shashank-Resistant<br>Poornima-Mod. Resistant |
| Tuberose | *M. incognita* | Shringar-Resistant<br>Suvasini-Tolerant |
| Mentha | *M. incognita* | Kukrail, Arka, Neera, SS-1-4, SS-2-7, SS-18 |

| | | |
|---|---|---|
| Black pepper | *M. incognita* | Pournami (Tolerant) |
| Ginger | *Meloidogyne* spp. | Mahima Maran |
| Yam | *M. incognita* | Sree Latha, Sree Keerthi |
| Cassava | *M. incognita* | Sree Sahya, H-97, Narayaniyakappa |
| Coffee | *R. reniformis* | *Coffeae robusta, C. canefora* |

### 14.2.2. Monitoring/observation/regulation

The aim of monitoring/observation/regulation is to determine when and what action to take. Regular monitoring is very crucial to nematode management as it is one of the most important decision making tools. An efficient monitoring programme can pay big dividends in lowering nematode management costs. An Agro-Eco-System Analysis (AESA) based on crop health at different stages of growth, population dynamics of nematodes and natural enemies, soil condition and farmer's past experiences are considered for decision making. Soil sample analysis for nematodes is usually employed as monitoring tool. Economic threshold levels (ETL) are now available to assist in proper timing of INM interventions.

- Quarantine against introduction of exotic nematodes.
- Domestic quarantine against further distribution e.g. potato cyst nematode in Nilgiri Hills of Tamil Nadu.
- Regulation on supply of quality bioagents.

If a harmful species of nematode is not present in an area, legal quarantine is the most effective method for prevention. Numerous attempts have been made to prevent the introduction of nematodes into countries or provinces by means of plant quarantine. Quarantines are established by legislative or other government actions and usually give quarantine authorities power to develop and enforce regulations to accomplish the purpose. Such regulations prohibit bringing infected plants into protected areas where similar crops are vulnerable to nematode infection.

Nematode surveillance and decision making ought to be done either visually or through soil sampling. Similar monitoring of natural enemies is also desirable.

### 14.2.3. Intervention

Once it is determined that intervention must be made, decisions must be made on the type of intervention which will reduce the effects of economically damaging nematode populations to acceptable levels. Mechanical, biological, cultural and chemical control measures are applied individually or in combination, taking into consideration costs, benefits, timing, available labour force, as well as ecological and environmental effects. Some of the principal INM interventions include cultural and physical measures, monitoring of nematodes, biocontrol agents (predators, parasitoids and pathogens), host plant resistance, botanicals and target specific, less hazardous chemical pesticides.

**(i) Mechanical and physical methods:** a number of cultural and physical methods such as collection and destruction of nematode infected plants continue to be used. Recommendations should be practicable. The possibility of integrating appropriate cultural methods with judicious use of crop protection products should be explored (instead of broadcast application of nematicide, use bare-root dip of seedlings before transplantation, spot/row application).

Heat treatment is one of the oldest methods of nematode control. Plant parasitic nematodes can be killed by heat, desiccation, irradiation, high osmotic pressure, etc., but it is more difficult to employ physical methods in killing nematodes in soil and planting materials. It is virtually impossible to kill nematodes when they are in growing plants. Field scale treatment of soil is also difficult. Nematodes can be killed at 44 to 48°C. The nematode enzymes are inactivated at short time exposure to about 50°C *in vitro*. Bare root dip in hot water has to be specifically determined for different nematode species. Solarization is a modern technique to be developed for each situation.

**(ii) Biological methods:** Biological suppression of plant parasitic nematodes with antagonistic fungi, mycorrhizae, bacteria and Actinomycetes is gaining increasing importance as a result of realization that many environmental and health hazards are linked with the use of chemicals. Standardization of methods for effective utilization of biocontrol agents is very important for evolving ecologically sound integrated nematode management strategies.

Bioagents can be used for seed treatment or soil application with *Trichoderma viride, T. harzianum, Paecilomyces lilacinus, Pochonia chlamydosporia, Pseudomonas fluorescens* for the management of nematodes.

Avermectins are the secondary metabolites of an Actinomycete, *Streptomyces avermitilis,* which are highly effective at a very low concentration against plant parasitic nematodes and mite pests. Scientists at the Indian Institute of Horticultural Research, Bangalore, for the first time in India, have isolated 6 strains of *S. avermitilis* and showed their effectiveness for the management of root-knot nematodes infecting tomato, egg plant, chilli, carnation and gerbera (Parvatha Reddy and Nagesh, 2002) (Table 14.4).

**Table 14.4.** Management of horticultural crop nematodes using avermectins

| Horticultural crop | Nematode | Avermectin/dose |
|---|---|---|
| Citrus | *T. semipenetrans* | A monthly rate of 1.1 kg a.i. per ha |
| Tomato | *M. incognita* | Aqueous solution of avermectins (250 ml of 0.001%/m$^2$ nursery bed); charcoal formulations of *S. avermitilis* at 100 g/m$^2$ nursery bed |
| Carnation and gerbera (commercial polyhouses) | *M. incognita* | Post-plant treatment at 250 ml/m$^2$ at two intervals (6 and 12 months after planting) |

**(iii) Botanicals:** Botanicals and their products play an important role in the reduction of nematode population in soil. A large number of reports have been published showing that growing of antagonistic plants (marigold, mustard, sesame, asparagus, etc.) or addition of organic materials (green manures, oil cakes, crop residues, cellulosic soil amendments, etc.) to the nematode infested soils results in definite reduction of several plant parasitic nematodes.

**(iv) Chemical methods:** Nematode monitoring should be done for scheduling the chemical nematicide application. Nematicides should be used in right dosages, time and through appropriate equipment.

The use of chemicals on a field scale for control of plant parasitic nematodes was not possible until early 1940's when effective and economical soil fumigants like D-D and EDB were discovered which made it possible to provide growers with spectacular differences in growth and yield through the effective control of nematodes and other soil pests. Several effective nematicides belonging to organophosphate and carbamate groups were developed and improvements in methods of application provided more economic control. In many parts of the world today, soil treatment for nematode control is an established farm practice.

The primary advantage of chemical control over other methods is the quick reduction of nematode population within days after the chemical is applied. The grower is able to plant a crop soon after treatment or, in some cases, at the time of treatment. Most crops are especially vulnerable to nematode attack during the seedling stage when the young root system is becoming established. Crops planted in treated soil develop extensive root systems, and the crop, especially annuals, mature before residual population of nematodes can increase to a damaging level.

## 14.3. Biointensive integrated nematode management (BINM)

Biointensive integrated nematode management is a systems approach to pest management based on an understanding of nematode ecology. It begins with steps to accurately diagnose the nature and source of nematode problems, and then relies on a range of preventive tactics and biological controls to keep nematode populations within acceptable limits. Reduced-risk nematicides are used if other tactics have not been adequately effective, as a last resort, and with care to minimize risks.

An important difference between conventional and biointensive INM is that the emphasis of the latter is on proactive measures to redesign the agricultural ecosystem to the disadvantage of a nematode and to the advantage of its parasite and predator complex.

BINM options may be considered as proactive or reactive.

### 14.3.1. Proactive options

Proactive options, such as crop rotations and creation of habitat for beneficial organisms, permanently lower the carrying capacity of the farm for the nematode. The carrying capacity is determined by the factors like food, shelter, natural enemies complex and weather, which affect the reproduction and survival of a nematode species. Cultural control practices are generally considered to be proactive strategies. Proactive practices include crop rotation, resistant crop cultivars including transgenic plants, disease-free seed and plants, crop sanitation, spacing of plants, altering planting dates, mulches, etc.

The proactive strategies (cultural controls) include:

- Healthy, biologically active soils (increasing below-ground diversity).
- Habitat for beneficial organisms (increasing above-ground diversity).
- Appropriate plant cultivars.

### 14.3.2. Reactive options

The reactive options mean that the grower responds to a situation, such as an economically damaging population of nematodes, with some type of short-term suppressive action. Reactive methods generally include inundative releases of biological control agents, mechanical and physical controls, botanical pesticides and chemical controls.

## 14.4. Low input sustainable INM strategy

### 14.4.1. *Summer*

- Two or three deep summer ploughings with a soil turning harrow at fortnightly intervals, preferably with a light irrigation between two ploughings.
- Soil solarization of nursery beds or pit soil using clear thin polyethylene mulch for 3-6 weeks before sowing.
- Rabbing of nursery beds (burning of crop residues over the soil surface), preferably after turning of soil, before sowing.
- Application of FYM/compost enriched with *Trichoderma harzianum* / *Paecilomyces lilacinus* / *Pochonia chlamydosporia* to nursery beds.
- Growing of non-host green manure/antagonistic crops/trap crops and incorporating in the soil at the time of flowering.
- Ploughing back non-commercial crop residues into the soil before monsoon.

### 14.4.2. *Monsoon*

- Growing of non-host green manure/antagonistic crops/trap crops and incorporating in the soil at the time of flowering.
- Application of organic amendments (FYM/compost/oil cakes/crop residues) to the soil while preparing the land.

### 14.4.3. *Kharif*

- Growing of non-host or nematode-resistant crop cultivars.
- Application of FYM/compost enriched with *Trichoderma harzianum* / *Paecilomyces lilacinus* / *Pochonia chlamydosporia* to nursery beds.
- Uprooting and burning of roots of host crops and weeds after harvest.

### 14.4.4. *Rabi*

- Delaying sowing/planting to mid-November (when soil temperature falls to 15-18°C).
- Restricting of growing susceptible or tolerant crops to *rabi* season with a suitable period of rotation.
- Removal and burning of non-commercial but disease-free crop residues.

*Section II*

# CROP-WISE PRODUCTIVITY ENHANCING TECHNOLOGIES

Chapter 15

# HORTICULTURAL CROPS

## 15.1. Importance of horticultural crops

Agriculture is the mainstay of the Indian economy, and horticulture is a crucial component thereof. Horticulture development has been accorded high priority in the XI[th] Five Year Plan. The impact of enhanced investment in horticulture has been highly encouraging in terms of vastly improved quality production and export potential-an increase from 96.1 million MT in 1990-91 to 275.4 million MT during 2007-08.

**Table 15.1.** Area, production and productivity of horticultural crops in India (2007-08)

| Horticultural crops | Area ('000 ha) | Production ('000 MT) | Productivity (MT/ha) | % Share of hortil. crops |
|---|---|---|---|---|
| Vegetables | 7803 | 125887 | 16.1 | 60.84 |
| Fruits | 5775 | 63503 | 11.0 | 30.69 |
| Plantation crops | 3226 | 12045 | 3.7 | 5.82 |
| Spices | 2603 | 4103 | 1.6 | 1.98 |
| Flowers | 161 | 870 | 5.4 | 0.42 |
| Aromatic & Medicinal plants | 386 | 325 | 0.8 | 0.16 |
| Nuts | 132 | 177 | 1.3 | 0.09 |
| Mushrooms | --- | 37 | --- | --- |
| Honey | --- | 65 | --- | --- |
| Total | 20087 | 207012 | 10.3 | |

India is the second largest producer of fruits next only to China contributing 12.0% of the total world production. Total production of fruits during 2007-08 has been estimated at 63.503 million MT from 5.775 million ha (Bijay Kumar, 2009) (Table 15.1). India is the largest producer of mango, banana, and papaya. India produces 41.0% of the world mangoes, 29.0% of the bananas and 30.0% of papayas. In grapes, India has recorded the highest productivity per unit area in the world (26 MT/ha). The per capita consumption of fruits has increased from 40 to 85 g/day within the last one decade.

India is the second largest producer of vegetables next only to China contributing 13.0% of the total world production. Vegetables occupy an area of 7.803 million ha with a production of 125.887 million MT (Bijay Kumar, 2009). India produces 18.0% of the world onions. In vegetables, India occupies the first position in the production of peas and okra, second in onion, brinjal, cabbage, cauliflower and third in potato and tomato. The per capita consumption of vegetables has increased from 96 to 175 g/day within the last one decade.

Floriculture is estimated to cover an area of 161,000 ha with production of 870,000 MT of loose flowers. The area under cut flowers has increased significantly in recent years and the production has reached 4341 million numbers.

India has emerged as the largest producer of spices, tea, cashew and areca nut. Coconuts sustain an area of 1.940 million ha with a production of 108.94 million MT. Areca nut is cultivated in 0.386 million ha with a production of 0.476 million MT. Cashew nuts are grown in an area of 0.868 million ha with a production of 0.665 million MT. The annual production of spices is 4.103 million MT from 2.603 million ha.

Horticulture has emerged as an important sector for diversification of agriculture. Presently, horticulture has established its credibility in improving income through increased productivity, generating employment and in enhancing exports besides providing protective food, household nutritional security, precious raw material for various processing and ancillary industries and finally in enhancing the aesthetic aspects of life. The focused attention on investment in horticulture during the last decade has been rewarding in terms of increased production and productivity of horticultural crops with manifold export potential. Horticulture has played a significant role in women empowerment, providing employment opportunities through mushroom cultivation, floriculture, processing, nursery raising, vegetable seed production etc.

## 15.2. Export of horticultural produce

Horticulture in India is fast emerging as a major commercial venture, because of higher remuneration per unit area and the realization that consumption of fruits and vegetables is essential for health and nutrition. In the last one decade, export potential of horticultural crops has significantly increased attracting even multinationals into floriculture, processing and value added products.

The present annual growth rate of horticulture is more than 6.5% which contributes 24.5% from the mere 8% of area to the GDP and 54.55% to the export earnings in the agriculture sector (Tables 15.2 and 15.3).

**Table 15.2.** Export of horticultural produce from India (2007-08) (Bijay Kumar, 2009)

| Horticultural product | Export quantity (MT) | Value (Rs. Crores) | % Share of value |
|---|---|---|---|
| **Floriculture & seeds** | | | |
| Floriculture | 42016.60 | 64983.50 | 13 |
| Fruit & vegetable seeds | 8031.09 | 12146.68 | 2 |
| Total | 1654282.66 | 159332.63 | |
| **Fruits and vegetables** | | | |
| Fresh onions | 1378373.17 | 116330.57 | 25 |
| Other fresh vegetables | 275909.49 | 43002.06 | 9 |
| Walnuts | 5062.06 | 11803.06 | 2 |
| Fresh mangoes | 79060.88 | 14193.95 | 3 |
| Fresh grapes | 85897.79 | 30192.45 | 6 |
| Other fresh fruits | 159569.71 | 25643.64 | 5 |
| Total | 549949.22 | 173163.18 | |
| **Processed fruits and vegetables** | | | |
| Dried and preserved vegetables | 118787.46 | 42406.20 | 9 |
| Mango pulp | 156835.52 | 50582.79 | 10 |
| Pickle and chutneys | 145216.00 | 29359.48 | 6 |
| Other processed fruits and vegetables | 129110.24 | 50814.71 | 10 |
| Grand Total | 2754181.00 | 505659.00 | |

**Table 15.3.** Export of Horticultural produce from India (2007-08) (Bijay Kumar, 2009)

| Horticultural produce | Quantity in MT | Value in lakh Rs. |
|---|---|---|
| Apple | 32665 | 3330 |
| Banana | 16662 | 2607 |
| Oranges | 29261 | 2726 |
| Grapes | 96723 | 31706 |
| Gauva | 2495 | 417 |
| Litchi | 161 | 63 |
| Mango | 54350 | 12741 |
| Papaya | 10879 | 1123 |
| Pineapple | 4194 | 339 |
| Sapota | 2150 | 503 |
| Brinjal | 338 | 191 |
| Cauliflower | 70 | 2 |
| Onion | 100860648 | 103577 |

| | | |
|---|---|---|
| Peas | 814 | 270 |
| Tomato | 134845 | 15290 |
| Potato | 78450 | 4142 |
| Sweet potato | 346 | 35 |

The major destinations for export of different horticultural products and potential international markets for fresh and processed products are presented in Table 15.4.

**Table 15.4.** Major destinations for export of different horticultural products (Bijay Kumar, 2009)

| Horticultural crops | Major destinations for export |
|---|---|
| Fruits | |
| Fresh grapes | UK, UAE, Netherlands, Bangla Desh, Germany, Sri Lanka, Norway, Belgium, Nepal, Oman, Russian Republic |
| Pomegranate | UAE, Saudi Arabia, Sri Lanka, Kuwait |
| Fresh guava | Saudi Arabia, Netherlands, UAE, Kuwait, Yaman Arab Republic, Netherlands, Nepal, Sudan, Canada, Oman, Japan |
| Fresh Gooseberries | UAE, Qatar |
| Fresh mango | UAE, UK, Bangladesh, Nepal, Saudi Arabia, Kuwait, USA, Bahrain, Singapore, Germany |
| Fresh apple | Bangladesh, Nepal, China, USA, Saudi Arabia, Germany, UAE |
| Fresh Banana | UAE, Saudi Arabia, Bahrain, Kuwait, Qatar, Nepal, Oman, Maldives, UK, Bangladesh |
| Fresh orange | Bangla Desh, Nepal, USA, Canada, Netherlands, Oman, Spain, China, UAE, Malaysia |
| Fresh litchi | Nepal, UAE, UK, Bahrain, Bahamas, Canada, Oman |
| Fresh papaya | UAE, Saudi Arabia, Qatar, Bahrain, France, Nepal, Kuwait, Bahamas, Germany, Netherlands |
| Fresh pineapple | Nepal, UAE, Maldives, Saudi Arabia, Spain, Pakistan, Kuwait, USA, Bangladesh, Armenia |
| Fresh sapota | UAE, UK, Bahrain, Saudi Arabia, Qatar, Canada, Bahamas, USA, Oman, Kuwait |
| Mango pulp | UK, Canada, Bangladesh, Bahrain, Netherlands, Japan, Saudi Arabia |
| Mango slices | UK, Saudi Arabia, Japan, USA |
| Dried grapes | UK, Malaysia, Netherlands, UAE |
| Mango pickles | Saudi Arabia, UAE, UK, Canada |
| Mango chutney | Oman, USA |
| Guava jam & jelly | UK, Germany, Hong Kong, USA, Netherlands, Saudi Arabia |
| Dried tamarind | Egypt, UAE, Pakistan, Saudi Arabia, UK |
| Dried figs | Nepal, Bangladesh |

| | |
|---|---|
| Vegetables | |
| Fresh brinjal | UK, France Saudi Arabia, Germany, Hong Kong, Portugal, Sweden, Canada, Belgium, UAE |
| Fresh cauliflower | Nepal, UAE, Maldives, Canada |
| Fresh onion | UAE, Singapore, Sri Lanka, Bangladesh, Malaysia, Nepal, Bahrain, Saudi Arabia, Mauritius, Oman |
| Fresh peas | Saudi Arabia, UAE, Kuwait, USA, Qatar, Nepal, Comoros, Oman, Singapore, Liberia |
| Fresh tomato | Pakistan, UAE, Bangladesh, Nepal, Maldives, Taiwan, China, Oman, Kuwait, Tanzania Republic |
| Fresh potato | Nepal, Mauritius, Sri Lanka, Maldives, Seychelles, UAE, Iran, Bangladesh, Pakistan, Malaysia |
| Fresh sweet potato | UAE, Maldives, Israel, France, Kuwait, Oman, Bahrain, Canada, Qatar |
| Gherkin | France, USA, Spain, Belgium |
| Dried vegetables | UAE, Sri Lanka, USA, UK, Singapore |
| Dehydrated onion | Germany, Romania, Sri Lanka, USA, Malaysia, Netherlands, UK |
| Vegetable seeds | USA, Netherlands, France, Italy, Japan, Pakistan |
| Flowers | |
| Flowers | USA, Japan, Netherlands, UK, Italy |

Horticulture today is not merely a means of diversification but forms an integral part of food and nutritional security, as also an essential ingredient of economic security. Adoption of horticulture both by small and marginal farmers brought prosperity in many regions of the country of which Maharashtra, Karnataka and Andhra Pradesh are the prime examples. A Golden Revolution through significant production increases in horticultural crops is already on the horizon.

Since India is now a signatory member of WTO, it will require its produce and products to be competitive both in the domestic and export markets. This would call for the use of improved horticultural practices such as use of HYV/F1 hybrids, recommended agro-techniques and plant protection measures, hi-tech intervention and scientific post-harvest management.

Chapter 16

# FRUIT CROPS

Fruits are amongst the most cherished bounties of nature. They are the major sources of Vitamins A and C and minerals like calcium and iron. Fresh fruits have low calorific values and are low in fats. A large variety of fruits are grown in India. Of these, banana, citrus, pineapple, papaya, sapota, jack fruit and avocado (tropical fruits); mango, grape, guava, litchi and loquat (subtropical fruits); apple, pear, peach, plum, apricot, almond and walnut (temperate fruits) and amla, ber, pomegranate, annona, fig and phalsa (arid zone fruits) are important. The leading fruit growing states are Maharashtra, Andhra Pradesh, Tamil Nadu, Gujarat, Karnataka, Uttar Pradesh, Bihar, West Bengal, Kerala, Jammu and Kashmir, Assam, Orissa, Madhya Pradesh and Punjab.

**Table 16.1.** Major fruit producing countries of the world (2007-08) (Bijay Kumar, 2009)

| Country | Area (in '000 ha) | Production (in '000 MT) | Productivity (in MT/ha) | % share of world production |
|---|---|---|---|---|
| China | 9587 | 94418 | 10 | 19 |
| India | 5775 | 63503 | 11 | 12 |
| Brazil | 1777 | 36818 | 21 | 7 |
| USA | 1168 | 24962 | 21 | 5 |
| Italy | 1246 | 17891 | 14 | 3 |
| Spain | 1835 | 15293 | 8 | 3 |
| Indonesia | 846 | 11615 | 14 | 2 |
| Mexico | 1100 | 15041 | 14 | 3 |
| Iran | 1256 | 12102 | 10 | 2 |
| Turkey | 1049 | 12390 | 12 | 2 |
| Others | 22841 | 208036 | 9 | 42 |
| Total (World) | 48481 | 512070 | | |

India is the second largest producer of fruits next only to China contributing 12.0% of the total world production. The present area under fruit crops is estimated to be about 5.775 million hectares with an estimated production of 63.503 million MT (Tables 16.1 to 16.3). The nut crops are grown in 0.1209 million hectares with an annual production of 0.1335 million MT (Tables 16.4 and 16.5). Fruit crops and nuts contribute 31.81% share to the total production of horticultural crops. India is the largest producer of mango, banana and papaya. India produces 41.0% of the world mangoes, 29.0% of the bananas and 30.0% of world papaya. In grapes, India has recorded the highest productivity per unit area (26.0 MT/ha) in the world. The per capita consumption of fruits has increased from 40 to 85 g/day within the last one decade.

**Table 16.2.** Crop-wise area and production of fruit crops in India (2007-08)

| Fruit Crops | Area ('000 ha) | Production ('000 MT) | Productivity (MT/ha) | % share of India production |
|---|---|---|---|---|
| Apple | 264 | 2002 | 7.6 | 3.2 |
| Banana | 647 | 23205 | 35.9 | 36.5 |
| Citrus | 843 | 7574 | 9.0 | 11.9 |
| Grapes | 64 | 1677 | 26.0 | 2.6 |
| Guava | 178 | 1975 | 11.1 | 3.1 |
| Litchi | 69 | 418 | 6.0 | 0.7 |
| Mango | 2205 | 13792 | 6.3 | 21.7 |
| Papaya | 80 | 2686 | 33.4 | 4.2 |
| Pineapple | 80 | 1216 | 15.1 | 1.9 |
| Pomegranate | 122 | 858 | 7.0 | 1.4 |
| Sapota | 150 | 1238 | 8.2 | 1.9 |
| Others | 1071 | 6862 | 6.4 | 10.8 |
| Total | 5775 | 63503 | 11.0 | |

**Table 16.3.** State-wise area, production and productivity of fruit crops (2007-08) (Bijay Kumar, 2009)

| State | Area ('000 ha) | Production ('000 MT) | Productivity (MT/ha) | % share of India production |
|---|---|---|---|---|
| Maharashtra | 1432.3 | 11047.6 | 7.7 | 18 |
| Andhra Pradesh | 853.0 | 10772.3 | 12.6 | 17 |
| Tamil Nadu | 292.3 | 7530.0 | 25.8 | 12 |
| Gujarat | 306.9 | 5849.7 | 19.1 | 9 |
| Karnataka | 279.0 | 4588.0 | 16.4 | 7 |
| Uttar Pradesh | 316.0 | 3757.5 | 11.9 | 6 |
| Bihar | 286.2 | 3252.4 | 11.4 | 5 |
| West Bengal | 194.3 | 2766.7 | 14.2 | 4 |

| | | | | |
|---|---|---|---|---|
| Kerala | 323.3 | 2579.8 | 8.0 | 4 |
| Jammu & Kashmir | 194.9 | 1435.9 | 7.4 | 2 |
| Assam | 122.7 | 1410.8 | 11.5 | 2 |
| Orissa | 265.3 | 1275.1 | 4.8 | 2 |
| Madhya Pradesh | 46.7 | 1237.0 | 26.5 | 2 |
| Punjab | 61.6 | 1055.4 | 17.1 | 2 |
| Others | 800.5 | 3924.8 | 4.7 | 8 |
| Total | 5775 | 62503 | 11.0 | |

**Table 16.4.** Crop-wise area, production and productivity of nut crops in India (2007-08) (Bijay Kumar, 2009)

| Nut crop | Area ('000 ha) | Production ('000 MT) | Productivity (MT /ha) |
|---|---|---|---|
| Almond & walnut | 132 | 177 | 1.3 |

**Table 16.5.** State-wise area, production and productivity of nut crops (2007-08) (Bijay Kumar, 2009)

| State | Area ('000 ha) | Production ('000 MT) | Productivity (MT/ha) | % share of India production |
|---|---|---|---|---|
| Jammu & Kashmir | 98.9 | 156.8 | 1.585 | 88.8 |
| Uttaranchal | 18.6 | 16.3 | 0.876 | 9.2 |
| Himachal Pradesh | 11.0 | 3.3 | 0.300 | 1.9 |
| Arunachal Pradesh | 3.8 | 0.1 | 0.026 | 0.16 |
| Total | 132.3 | 176.5 | | |

India has been exporting fruits and fruit products since several decades. However, in the recent past export promotional activities received greater attention. Mango and grapes are the main items in fresh fruit export and some other fruits like pomegranate, sapota, banana, citrus, litchi, apple and strawberry also entered fresh fruit export market in limited quantities.

## 16.1. BANANA, *MUSA* SPP.

### 16.1.1. High yielding varieties

**Table 16.6.** Variety-wise average yield of banana (MT/ha.)

| High yielding varieties | Yield (MT/ha) |
|---|---|
| Basrai, Rasthali | 40-50 |
| Robusta | 60-70 |
| Shrimanti | 70 |
| Grand Naine | 65 |
| Ardhapuri, Meanyham | 55 |
| Hirsal, Safed Velchi, Red banana, Lal Velchi | 45 |

| | |
|---|---|
| Poovan | 40-50 |
| Monthan | 30-40 |
| Dwarf Cavendish, Robusta Champa, Chini Desi | 50-60 |
| Nendran | 30-35 |
| Udhayam (NRC for Banana, Trichy) | 45-50 |
| CO 1 (TNAU, Coimbatore) | 83 fruits/bunch (10.57 kg/bunch) |

### 16.1.2. High density planting

The normal spacing provided for Robusta and Dwarf Cavendish is either 2.1 m x 2.1 m (2267 plants/ha) or 1.8 m x 1.8 m (3086 plants/ha) with yield levels of 60-65 MT per ha. In recent years, the emphasis has been on increasing the productivity per unit area by following closer spacing. A closer spacing can be adopted under good management conditions using micro-irrigation and fertigation techniques. However, a major limiting factor in banana high density planting will be the sunlight which affects flowering, crop duration and maturity and also on the performance of ratoon crop. A net profit of Rs. 3,19,714/ha could be realized from the main and ratoon crops of banana.

Due to availability of dwarf clones in banana, HDP is gaining popularity. For the highest possible yields of good quality fruit, an optimum plant density (2.1m x 2.1m of 2,267 plants/ha) for poor soils and 1.5 to 2.1m for Dwarf Cavendish, 2.4m x 2.4m for Chakkarakeli and 2.4 to 3.0m for Bontha (Monthan) has been tried. Results revealed that the Dwarf Cavendish and Robusta are fit to be planted under high density planting concept at a planting distance of 1.2m X 1.2m (6,944 plants/ha) in a square system of planting under a proper nutrient management practice. There is minimum weed growth, effective utilization of the solar radiation and appropriate utilization of the inputs, which are the other major factors, attributed to the success of such planting systems. Yield realization varies from 86 MT/ha in Basrai at 1.5 x 1.5m to 174 MT/ha in Robusta planted at 1.2m x 1.2m spacing.

Robusta, Dwarf Cavendish and Lacatan are good cultivars for high density planting. Yield response of various cultivars planted in different densities resulted in yields of 86.6 to 174.39 MT/ha at 4444 to 10000 plants/ha (Table 16.7). A plant population of 4444/ha produced fruit yield of 120 MT/ha (Awasthi and Mehta, 2000).

**Table 16.7.** Yield of banana under high density orcharding

| Variety | Spacing (m) | Plants/ha | Yield (MT/ha) | Reference |
|---|---|---|---|---|
| Traditional system | 2.0 x 2.0 | 2500 | 35-36 | Chattopadhyay *et al.*, 1985 |
| Robusta | 1.2 x 1.2 | 6944 | 174.39 | Chacko & Reddy, 1981a |
| | 1.5 x 1.5 | 4444 | 145.44 | Chacko & Reddy, 1981b |

| | | | | |
|---|---|---|---|---|
| | 1.8 x 1.8 | 4630 (3 plants/pit) | 110.35* | Kumar & Nalina, 2001 |
| Basrai | 1.2 x 1.2 | 6944 | 130.13 | Chundawat *et al.*, 1982 |
| | 1.3 x 1.3 | 5917 | 105.90 | Chundawat *et al.*, 1982 |
| | 1.5 x 1.5 | 4444 | 86.60 | Chundawat *et al.*, 1982 |
| Lacatan | 1.2 x 1.2 | 1944 | 113.30 | Satyanarayana & Rao, 1985 |
| Poovan | 1.0 x 1.0 | 10000 | 99.10 | Chattopadhyay *et al.*, 1985 |
| Giant Governor | 1.0 x 1.0 | 10000 | 100.45 | Chattopadhyay *et al.*, 1985 |
| Nendran | 2.0 x 2.0 | 7500 (3 plants/pit) | 58.65** | Aphsara & Sathiamoorthy, 2003 |

* 86.41 MT/ha in conventional planting

** 26.38 MT/ha in conventional planting

Banana cv. Rasthali planted under high density planting (planting of three suckers per pit at a spacing of 2 x 3m with 5000 plants per hectare) significantly registered higher bunch yield of 38.31 tonnes per hectare than that of normal planting (planting of one sucker per pit at a spacing of 2 x 2m with a total population of 2500 plants per hectare) which recorded only 28.21 tonnes per hectare of bunch yield.

### 16.1.3. Micro irrigation

Studies conducted at the University of Agricultural Sciences, Dharwad, Karnataka on banana have shown that the yield in case of drip irrigated banana was 31.0 MT/ha as compared to 24.7 MT/ha in case of basin irrigated system (Table 16.8).

**Table 16.8.** Water saving and yield in drip irrigated banana cv. Poovan

| ET replenishment | Water saved over ring basin (%) | Yield (MT/ha) | Increase over ring basin (%) |
|---|---|---|---|
| 60% (drip) | 63.07 | 29.5 | 19 |
| 80% (drip) | 51.77 | 30.0 | 21 |
| 100% (drip) | 40.47 | 31.0 | 25 |
| Ring basin | --- | 24.7 | --- |

Irrigation at 40% depletion gave higher bunch weight than at 60 or 80% depletion (Krishnan and Shanmugavelu, 1979). Further increase in frequency (20% depletion) gave the highest bunch weight of 35 kg and yield of 89.375 MT/ha as a result of increase in total consumptive use of water and water use efficiency (Krishnan and Shanmugavelu, 1980).

Evaporation replenishment of 100% moisture gave maximum fruit yield in banana cv. Robusta (91 MT/ha). However, 80% replenishment also gave comparable yield (90 MT/ha) and thus saving 20% water (Table 16.9).

**Table 16.9.** Fruit yield and water use in banana cv. Robusta under drip irrigation

| Evaporation replenishment | Yield (MT/ha) | Water used (mm) | Water use efficiency (kg/ha/mm) |
|---|---|---|---|
| 20% | 68 | 638 | 1092 |
| 40% | 76 | 1002 | 761 |
| 60% | 87 | 1362 | 644 |
| 80% | 90 | 1728 | 511 |
| 100% | 91 | 2093 | 432 |
| 120% | 83 | 2452 | 370 |

Drip irrigation in banana at Jalagaon recorded 62% increase in yield and 40-45% reduction in water used. Drip irrigation also brought the uniformity in harvesting and reduced the days taken for the harvest by 20-25 days (Table 16.10).

**Table 16.10.** Impact of drip irrigation in banana

| Particulars | Flood/furrow irrigation | Drip system | Difference |
|---|---|---|---|
| 1. Height of plant (cm) | 153.66 | 169.94 | + 16.28 |
| 2. Girth of plant (cm) | 60.39 | 71.28 | + 11 |
| 3. Average no. of leaves (Nos./plant) | 34.11 | 35.29 | + 1.18 |
| 4. Average flowering period (Days) | 342 | 310 | - 32 |
| 5. Average harvesting period (Days) | 433 | 397 | - 36 |
| 6. Average weight (kg) | 14 | 17 | + 3 |
| 7. Plants harvested (%) | 75 | 97 | + 22 |
| 8. Yield (MT/ha) | 46.40 | 69.72 | + 23.32 |
| 9. Increase in yield (%) | - | 46 | - |
| 10. Water saved (%) | - | 58 | - |

*Source:* Extension Publication No.298, 2000, Mahatma Phule Krishi Vidyapeeth, Rahuri, Maharashtra

### 16.1.4. Fertigation

With the advent of drip irrigation system and availability of liquid fertilizers, the nutrient application technology is considered to be the most efficient. Results available in banana indicate that 20 to 60% saving of nutrients without adversely affecting the yield and leaf nutrient status of plants (Tables 16.11 to 16.13).

**Table 16.11.** Studies on fertilizer application in banana (pooled data for 3 years 1986-88)

| Irrigation method and mulch | N-level, % of recommended dose | Yield (MT/ha) | % increase over control |
|---|---|---|---|
| Drip | 100 | 61.2 | 8.7 |
| Drip | 80 | 66.1 | 17.4 |
| Drip | 60 | 60.5 | 7.5 |
| Drip | 40 | 54.9 | (-) 2.5 |
| Drip + STM | 100 | 68.9 | 22.4 |
| Drip + STM | 80 | 66.7 | 18.5 |
| Drip + STM | 60 | 65.0 | 15.5 |
| Drip + STM | 40 | 54.2 | (-) 3.7 |
| Surface (control) | 100 | 56.3 | --- |

**Table 16.12.** Response of banana cv. Elakki bale to nitrogen fertigation

| Nitrogen/dose | Fruit yield (MT/ha) |
|---|---|
| 100 g/plant through drip | 29.9 |
| 200 g/plant through drip | 30.0 |
| 300 g/plant through drip | 29.7 |
| 100 g/plant through soil application | 27.2 |
| 200 g/plant through soil application | 29.3 |
| 300 g/plant through soil application | 28.4 |

**Table 16.13.** Response of banana cv. Robusta to nitrogen and potash fertigation

| NK levels | Fruit yield (MT/ha) under different NK ratios | | |
|---|---|---|---|
| | 1: 1.0 | 1: 1.5 | 1: 2.0 |
| 50 g/plant | 51 | 57 | 61 |
| 100 g/plant | 77 | 78 | 79 |
| 150 g/plant | 84 | 86 | 87 |
| 200 g/plant | 87 | 88 | 89 |

**Table 16.14.** Fertigation studies in banana cv. Robusta

| Treatment | Yield (MT/ha) |
|---|---|
| 250% PE, RDF (600: 90: 900 g/plant), water soluble fertilizers | 132.50 |
| 250% PE, RDF (600: 90: 900 g/plant), normal fertilizers | 116.00 |
| 250% PE, RDF (450: 67.5: 675 g/plant), water soluble fertilizers | 78.65 |
| 250% PE, RDF (450: 67.5: 675 g/plant), normal fertilizers | 91.50 |
| 200% PE, RDF (600: 90: 900 g/plant), water soluble fertilizers | 122.55 |
| 200% PE, RDF (600: 90: 900 g/plant), normal fertilizers | 112.65 |
| 200% PE, RDF (450: 67.5: 675 g/plant), water soluble fertilizers | 94.65 |
| 200% PE, RDF (450: 67.5: 675 g/plant), normal fertilizers | 86.90 |

The experiment conducted to investigate the effects of fertilizer sources on the yield of banana cv. Robusta grown under fertigation revealed that the treatment [250% PE, RDF (600: 90: 900 g/plant), water soluble fertilizers] recorded the highest fruit yield of 132.50 MT/ha (Table 16.14) (Kavino *et al.*, 2002).

In an experiment on nitrogen and potassium fertigation in banana, 200 g of N and K per plant gave higher fruit yield of 88.5 MT/ha (Table 16.15). Application of 150 g of N and K per plant also gave results at par with the higher dose.

**Table 16.15.** Nitrogen and potassium fertigation in banana

| Treatment N and K (g/plant) | Yield (MT/ha) |
|---|---|
| 50 | 56.4 |
| 100 | 78.2 |
| 150 | 85.7 |
| 200 | 88.5 |
| CD (P = 0.05) | 3.48 |

Fertigation experiment conducted with cv. Robusta (AAA) revealed that fertigation treatment with 50 to 75 % of recommended NPK (200: 30: 300g NPK/plant) registered the maximum bunch weight, more number of hands and fingers both under normal and high density planting system (Table 16.16) (Mahalakshmi *et al.* 2001).

**Table 16.16.** Influence of fertigation on banana cv. Robusta (AAA)

| Water + Fertilizer | Yield (MT/ha) | % increase over conventional |
|---|---|---|
| Normal planting system (Single plant/pit) | | |
| Plant crop 25 LPD + 100: 30: 150 g NPK* | 95.00 | 61.07 |
| Ratoon crop 25 LPD + 150: 30: 225 g NPK ** | 111.05 | 61.07 |
| High density planting system (3 plants/pit) | | |
| 50 LPD + 450: 90: 675 g NPK** | 174.88 | 196.51 |
| Conventional (Single plant/pit) | | |
| 200: 30: 300 g NPK/ plant | 58.98 | --- |

* 50 percent recommended dose ** 75 percent recommended dose

The effect of fertigation on the growth, yield, quality and physiological parameters of banana cv. Red banana (AAA) revealed that planting of one plant/pit along with 100% of RDF (110: 35: 300 g NPK) through fertigation resulted in higher bunch weight (22.55 kg) (Suganthi, 2002).

The fertigation treatment (25 litres/day/plant + 200: 30: 300 g NPK/ plant) produced 88.66% increase in bunch weight (44.53 kg) and productivity (111.33 MT/ha) over conventional method (23.59 kg bunch weight and 58.98 MT/ha productivity) (Mahalakshmi *et al.*, 2006).

### 16.1.5. Integrated nutrient management

Banana generally requires high amount of mineral nutrients for proper growth and production. Studies conducted by different authors indicated that banana crop requires more of K than N and P (Table 16.17). These nutrients are also varyingly required at different growth stages, more during the fruit bud initiation and differentiation stage.

**Table 16.17.** Variations in the uptake of nutrients by banana

| Nutrient uptake (kg/ha) | | | Yield (MT/ha) | Reference |
|---|---|---|---|---|
| N | P | K | | |
| 30.00 | 50-75 | 15-20 | 175-200 | Jacob & Vonvexkull (1960) |
| 25.00 | 50 | 12.50 | 150 | Martin-Prevel (1964) |
| 25.00 | 250 | 60 | 1000 | Montagut & Martin-Prevel (1965) |
| 28.00 | 250 | 90 | 350 | Martin-Prevel *et al.* (1968) |
| 16.75 | 38 | 8 | 285 | Joseph (1971) |
| 30.00 | 300 | 80 | 800 | Shanmugam & Velayutham (1972) |
| 25.00 | 300 | 80 | 800 | Veeraraghavan (1972) |
| 57.50 | 321.84 | 72.86 | 1179.67 | Veerannah *et al.* (1976) |
| 77.00 | 180-450 | 49-309 | 964-2440 | Martin-Prevel (1992) |

Shanthi (2004) found that application of 50% K through MOP + 50% K through Cement Kiln Flue Dust (CKFD) had a positive response on bunch weight in banana cvs. Karpooravalli and Poovan (Table 16.18).

**Table 16.18.** Influence of Cement Kiln Flue Dust on yield characters of banana

| Treatment | Bunch weight (kg) | |
|---|---|---|
| | Karpooravalli | Poovan |
| 100 % K through MOP | 19.42 | 15.83 |
| 50 % K through MOP + 50 % K through CKFD | 23.46 | 19.14 |
| CD (P = 0.05) | 2.12 | 1.65 |

The efficiency of different sources of nitrogen on banana cv. Ney Poovan was assessed at Tamil Nadu Agricultural University, Coimbatore and the result showed that combined application of 25 per cent N as CAN + 25 per cent N as Urea + 50 per cent of N as Ammonium sulphate had resulted in better vegetative growth, physiological attributes, soil and leaf nutrient status culminating in an increased yield in terms of bunch weight and other economical traits (Table 16.19) (Keshavan, 2004).

**Table 16.19.** Influence of different 'N' sources on yield characters of banana cv. Ney Poovan

| Treatment | Bunch weight (kg) |
|---|---|
| Control | 8.44 |
| 25 per cent of N as CAN + 25 per cent of N as urea + 50 per cent of N as ammonium sulphate | 12.10 |
| CD (0.05) | 0.81 |

Integrated nutrient management involving supplying 50% of nutrients (0.5% urea, 0.5% SOP, 0.2% Zn $SO_4$ and 0.1% Boric Acid) by spray on leaf and bunch (four sprays on leaf from 5th month to 8th month and two sprays on bunches and leaves in 10th and 11th month) and improving root health by application of *Trichoderma* enriched FYM resulted in yield increase to 30 kg/bunch from 25kg in farmers practice in cultivar Robusta. The technology was once more refined by filling the pit at planting with tank silt clayey in texture but with more than 55% water-holding capacity. This resulted in mean bunch yield increase to 40-45 kg.

Tissue-cultured (TC) banana cultivation is expanding in India as farmers are now realizing its advantages due to its rapid growing, early and high yielding characters. As they produce significantly high yields than the conventional suckers, the requirements of macro as well as micronutrients might be more. Application of 150% of recommended doses of NPK (*i.e.* 165: 52.50: 495 g) in four splits was found essential to increase the plant growth and development, yield and quality in the plant and ratoon crops of TC banana (Table 16.20) (Nalina, 2002).

**Table 16.20.** Response of tissue culture derived banana to increased dose of nutrients

| Treatments | Bunch weight (kg) |
|---|---|
| Recommended dose (110: 35: 330 g N, $P_2O_5$, $K_2O$) in 3 split doses | 26.87 |
| 150% of recommended dose (i.e. 165: 52.5: 495 g N, $P_2O_5$, $K_2O$) in 4 split doses | 35.18 |
| CD (P = 0.05) | 1.09 |

The entire plant canopy was sprayed including the developing bunches twice with sulphate of potash (SOP), initially after the opening of the last hand (*i.e.* 7th month after planting) and 30 days later. Foliar spray concentration had a significant and positive impact on bunch weight, total number of fingers and finger weight (Ramesh Kumar and Kumar, 2007) (Table 16.21).

**Table 16.21.** Effect of foliar spraying of SOP on bunch traits of banana cv. Ney Poovan

| Treatment | Bunch weight (kg) | TSS (%) |
|---|---|---|
| Control | 10.80 | 24.40 |
| 0.5% SOP spray | 11.53 | 27.90 |
| 1.0% SOP spray | 12.63 | 27.90 |
| 1.5% SOP spray | 14.27 | 28.90 |
| CD (P = 0.05) | 1.02 | 2.06 |

Vidhya (2004) found that application of 100 per cent of the recommended dose of fertilizer (RDF) along with 50 g each of *Azospirillum*, phosphate solubilizing bacteria and arbuscular mycorrhizal fungi (AMF) in banana cv. Robusta resulted in better yield traits (Table 16.22).

**Table 16.22.** Influence of bio fertilizers on yield characters of banana

| Treatment | Bunch weight (kg) |
|---|---|
| 100 percent of recommended dose of fertilizers alone | 24.47 |
| 100 per cent of the (RDF) along with 50 g each of *Azospirillum*, phosphate solubilizing bacteria and arbuscular mycorrhizal fungi | 26.10 |
| CD (P = 0.05) | 0.75 |

Application of 25% N from FYM and 75% of N from inorganic sources recorded significantly higher fruit yield of banana (77 MT/ha) (Palwe *et al.*, 1999b) (Table 16.23).

**Table 16.23.** Effect of organic and inorganic fertilizers on productivity of banana

| Treatment | Fruit yield (MT/ha) |
|---|---|
| 200 g N inorganic | 68.90 |
| 25% N through FYM + 75% N through inorganic | 77.03 |
| 25% N through neem cake + 75% N through inorganic | 68.93 |
| 25% N through FYM + 25% N through neem cake + 50% N through inorganic | 73.93 |
| 25% N through FYM + 50% N through neem cake + 25% N through inorganic | 70.90 |
| 25% N through FYM and green manure + 75% N through inorganic | 76.70 |
| CD (P = 0.05) | 4.00 |

Application of 200 g N in four splits (30, 75, 120, 165 DAP) during vegetative phase and 50 g N in three splits (210, 255 and 300 DAP) during reproductive phase gave maximum banana fruit yield of 77.01 MT/ha (Shinde *et al.*, 1999) (Table 16.24).

**Table 16.24.** Effect of nitrogen levels and time of application on yield of banana cv. Basrai

| Nitrogen levels | | Time of application | | Fruit yield ( MT/ha) |
|---|---|---|---|---|
| Vegetative phase | Reproductive phase | Vegetative phase | Reproductive phase | |
| 200 g N | --- | 4 splits (30, 75, 120 & 165 DAP) | --- | 72.42 |
| 150 g N | 50 g N | 4 splits (30, 75, 120 & 165 DAP) | 3 splits (210, 255 & 300 DAP) | 69.89 |
| 100 g N | 100 g N | 4 splits (30, 75, 120 & 165 DAP) | 3 splits (210, 255 & 300 DAP) | 64.66 |
| 200 g N | 50 g N | 4 splits (30, 75, 120 & 165 DAP) | 3 splits (210, 255 & 300 DAP) | 77.01 |
| 200 g N | 100 g N | 4 splits (30, 75, 120 & 165 DAP) | 3 splits (210, 255 & 300 DAP) | 73.83 |
| 200 g N | 150 g N | 4 splits (30, 75, 120 & 165 DAP) | 3 splits (210, 255 & 300 DAP) | 71.29 |
| CD (P = 0.05) --- | | --- | --- | 2.53 |

Application of 200 g $K_2O$ per plant in four equal splits at planting, 90, 135 and 210 DAP with 40 g $P_2O_5$ and 200 g N per plant for Basrai variety of banana is an appropriate proposition to maximize the productivity per unit area of banana plantation (Palwe *et al.*, 1999a) (Table 16.25).

**Table 16.25.** Effect of potassium levels on yield of Basrai banana

| Level of $K_2O$ | Fruit yield (MT/ha) | % increase over control |
|---|---|---|
| 0 (Control) | 38.14 | --- |
| 100 | 60.61 | 58.91 |
| 200 | 66.39 | 74.06 |
| 300 | 65.08 | 70.63 |
| 400 | 62.76 | 64.55 |
| 500 | 61.78 | 61.98 |
| 600 | 60.76 | 58.34 |
| CD (P = 0.05) | 2.82 | --- |

Banana plants inoculated with *Glomus fasciculatum* recorded higher yield (33.88 MT/ha) as compared to uninoculated ones. Similarly, these plants registered less number of days for shooting (205.96); harvest after shooting (124.69) and total crop duration (329.97) when compared to non AMF plants (Table 16.26).

**Table 16.26.** Effect of *Glomus fasciculatum* on the yield of banana.

| Treatment | Yield ( MT/ha) | |
|---|---|---|
| | With *G. fasciculatum* | Without *G. fasciculatum* |
| 75% RDF + Microbial consortia-I* | 37.29 | 33.44 |
| 50% RDF + Microbial consortia-I | 33.18 | 25.98 |
| Microbial consortia-I | 24.95 | 21.09 |
| RDF (Recommended Dose of Fertilizer) | 36.96 | 33.69 |
| Microbial consortia-II** | 37.04 | 33.41 |
| CD (P = 0.05) | 3.772 | 3.772 |

***Microbial consortia-I**: Mixture of *Azotobacter chroococcum, Azospirillum brasiliense, Pseudomonas striata and Trichoderma harzianum.*

****Microbial consortia-II**: Mixture of 15 local isolates of microbes in cow dung powder.

Growing and incorporation of cowpea along with application of arbuscular mycorrhizal fungi (AMF) recorded the highest bunch weight (8.22 kg) which was on par with *Azospirillum* + AMF and AMF + vermicompost (Geetha, 2006).

The treatments 80% NPK (160: 24: 320) + phosphobacteria (25 g/plant) + rice husk ash (25 kg/plant) and 80% NPK (160: 24: 320) + phosphobacteria (25 g/plant) + vermicompost (2 kg/plant) recorded the highest bunch weight of banana cv. Rasthali, which was 29% more than that of 100% inorganic NPK fertilizers only (Jeyabaskaran *et al.*, 2006).

Banana plants receiving 100% RDF were superior for obtaining the highest fruit yield (21.29 MT/ha and 34.13 MT/ha in plant and ratoon crops, respectively) (Athani *et al.*, 2006b).

Supplying ¼ NPK through soil application during third month and remaining ¾ nutrient needs as 12 foliar sprays at fortnightly intervals has resulted in higher pseudo stem girth, higher leaf N and K, optimum P and 20% higher yield over the farmer's practice of supplying entire NPK (300: 100: 300g NPK/plant) needs through soil application (Edward Raja *et al.*, 1999) (Table 16.27).

**Table 16.27.** Effect of soil vs. foliar application of NPK to banana

| Treatment | Pseudo stem girth (cm) | Leaf nutrient (%) (7th month) | | | Yield (MT/ha) |
|---|---|---|---|---|---|
| | | N | P | K | |
| Soil appln. of entire 300g N, 100g $P_2O_5$ and 300g $K_2O$ (Farmer's practice) | 35 | 2.8 | 0.68 | 2.9 | 58 |
| ¼ NPK through soil + ¾ NPK as 12 foliar sprays | 42 | 3.4 | 0.32 | 3.4 | 68 |

### 16.1.6. Bioregulators

Cavendish group of bananas are known to be very vigorous and foliar spray of certain growth regulators play a significant role in improving the yield through increasing the fertilizer use efficiency. Foliar spray of chlormequat chloride at 1000 ppm during 4th and 6th month after planting in addition to the soil application of NPK at 110: 35: 330g/plant/year resulted in improved physiological efficiency, nutrient status and fruit yield including fruit quality (Jeyakumar *et al.*, 2003) (Table 16.28).

**Table 16.28.** Influence of chlormequat chloride application on banana

| Treatment | Bunch weight (kg) | TSS (° Brix) |
|---|---|---|
| Control | 24.80 | 20.00 |
| Chlormequat chloride 1000 ppm | 33.62 | 23.60 |
| CD (P = 0.05) | 1.76 | 1.40 |

Sea weed based bioregulators are widely used in the recent years to increase the nutrient use efficiency in various agricultural and horticultural crops. Foliar spray of 0.2 % plantozyme during fourth and sixth month after planting, in addition to the fertilizer dose of 200: 40: 200 g N: $P_2O_5$: $K_2O$ per plant per year revealed the influence of nutrients and amino acids in plantozyme on higher cell wall plasticity and dry matter accumulation resulting in better yield besides improving the leaf nutrients status (Jeyakumar and Kumar, 2002) (Table 16.29).

**Table 16.29.** Effect of bioregulator application on yield and quality of banana

| Treatment | Bunch weight (kg) | TSS (%) |
|---|---|---|
| Control | 18.0 | 19.7 |
| Plantozyme 0.2% | 26.3 | 25.7 |
| CD (P = 0.05) | 1.56 | 1.22 |

Spraying of 100 ppm $GA_3$ twice at 25 and 50 days after the shooting resulted in an increase of 32.5%, 23.9% and 16.98% in respect of length, girth and weight of the fingers, respectively in banana cv. Rajapuri (Rao, 1997).

### 16.1.7. Mulching

The fruit yield obtained in Robusta banana were 112.9, 95.5 and 85.6 MT/ha, respectively by using black polythene, sugarcane trash at 15 MT/ha and banana trash at 10 MT/ha covers compared to 76.8 MT/ha in control and thus resulting in the highest cost-benefit ratio by sugarcane trash mulch (Bhattacharya and Rao, 1985).

Polythene covering of banana bunch increased the bunch weight by 30.3% and reduced the period of cropping by 15 days. Covering of banana bunches with white polythene during winter was found beneficial as it helped in increasing the yield of bunches in winter by 15-20%. Black and white colour polythene bags used for sleeving did not show any difference in the response.

### 16.1.8. Tissue cultured plants

Use of four months old tissue cultured Basrai banana seedlings as planting material flowered and came to harvest early and gave maximum bunch weight of 19.74 kg. Use of under sized rhizomes (below 500 g) and two month-old tissue cultured seedlings were not found suitable as planting material (Birhade *et al.*, 1999) (Table 16.30).

**Table 16.30.** Performance of tissue cultured seedlings versus normal suckers.

| Planting material | Days to flowering | Days to harvest | Bunch weight (kg) |
|---|---|---|---|
| 4-month old issue culture seedlings | 272 | 372 | 19.74 |
| 2-month old issue culture seedlings | 294 | 396 | 16.07 |
| Big suckers (500-700 g) | 278 | 379 | 18.84 |
| Small suckers (below 500 g) | 299 | 402 | 13.54 |
| CD (P = 0.05) | 19.39 | 21.58 | 1.11 |

The tissue cultured plants of Robusta and Grand Naine produced 10-14 % higher yield than their sucker raised plants. The cycle duration of tissue cultured plants was shorter by one and half months.

### 16.1.9. Integrated weed management

Weeds in banana plantations can cause yield reduction up to 40 %. Weeding up to 9 months stage is vital for obtaining good yield from the plantations.

The significantly highest banana fruit yield was obtained in weed-free treatment (60.84 MT/ha) which was on par with weed-free till 9th month (58.24 MT/ha), traditional weeding practice (58.23 MT/ha) and no weeding till 3rd month, then weed-free (56.21 MT/ha) treatment giving 40.48, 34.47, 34.45 and 29.79% increase in yield over untreated control, respectively (Patel *et al.*, 1999) (Table 16.31).

**Table 16.31.** Effect of different weeding practices on productivity and economics of banana

| Treatment | No. of weeding operations | | | Fruit yield (MT/ha) | Benefit: cost ratio |
|---|---|---|---|---|---|
| | Hand weeding | Tillage | Weedicidal spray | | |
| No weeding | --- | --- | --- | 43.31 | --- |
| Weed free | 3 | 3 | 1 | 60.84 | 8.08 |
| Traditional weeding practice | 3 | 3 | -- | 58.23 | 7.77 |

| | | | | | |
|---|---|---|---|---|---|
| No weeding till 3rd month then weed free | 2 | 1 | 1 | 56.21 | 10.00 |
| No weeding till 6th month then weed free | 2 | 1 | 1 | 51.51 | 6.36 |
| Weeding till 3rd month, no weeding 3rd to 9th month | 2 | 1 | 1 | 50,61 | 5.66 |
| Weeding till 6th month, no weeding 6th to 9th month, then weed free | 2 | 1 | 1 | 51.66 | 6.47 |
| Weeding till 9th month, no weeding 9th to 12th month | 3 | 2 | 1 | 58.24 | 7.74 |

## 16.1.10. Integrated disease management

Banana hybrid BRS-1 developed by Banana Research Station, Kannara, Kerala, is resistant to leaf spot (*Cercospora musicola*) and wilt (*Fusarium oxysporum* f. sp. *cubense*)

## 16.1.11. Integrated nematode management

### i) Burrowing nematode, *Radopholus similis*

*R. similis* causes 'rhizome rot' or 'toppling' or 'black head' disease of banana and is becoming a serious problem. It causes retarded growth and extensive root and rhizome necrosis. The burrowing nematode is responsible for 30.76 to 41 per cent yield loss in banana.

**Table 16.32.** Effect of different isolates of *Pseudomonas fluorescens* on plant growth, yield and management of *Radopholus similis* infecting banana

| Treatment | Plant height (cm) (270 DAP) | Pseudo stem girth (cm) (270 DAP) | Bunch weight (kg) | Root lesion index |
|---|---|---|---|---|
| *P. fluorescens* – PfB 13 | 328.2 | 57.4 | 15.6 | 14 |
| *P. fluorescens* – PfB 19 | 287.3 | 46.4 | 13.4 | 32 |
| *P. fluorescens* – PfB 24 | 270.3 | 47.3 | 11.4 | 28 |
| *P. fluorescens* – PfB 32 | 266.8 | 41.2 | 11.2 | 42 |
| *P. fluorescens* – PfB 35 | 287.8 | 45.2 | 13.1 | 46 |
| *P. fluorescens* – PfB 39 | 278.6 | 46.1 | 10.1 | 36 |
| *P. fluorescens* – Pf 1 | 275.4 | 45.3 | 14.3 | 18 |
| Carbofuran | 270.9 | 43.1 | 13.2 | 19 |
| Control | 210.9 | 39.7 | 10.9 | 54 |
| CD (P = 0.05) | 26.59 | 6.99 | 0.96 | 1.37 |

Field application of *P. fluorescens* (isolate PfB 13) at 10 g/plant significantly reduced nematode population in soil and roots and root lesion index; and increased plant height, pseudo stem girth, leaf area, number of leaves and fruit yield (Table 16.32).

Integration of neem cake at 200 g/plant along with *Glomus mosseae* at 100 g/plant (containing 25-30 chlamydospores/g of inoculum) was most effective in reducing the *R. similis* population both in soil and roots, while karanj cake with *G. mosseae* gave maximum increase in fruit yield of banana. Mycorrhizal root colonization and number of chlamydospores of *G. mosseae* were maximum in neem cake amended soil (Parvatha Reddy *et al.*, 2002) (Table 16.33).

**Table 16.33.** Effect of *Glomus mosseae* and oil cakes on population of *Radopholus similis* and yield of banana.

| Treatment | Dose (g)/ plant | Popn. of *R. similis* | | Yield(kg)/ plant |
|---|---|---|---|---|
| | | Roots (10 g) | Soil (250 ml) | |
| *G. mosseae* | 200 | 112 | 122 | 8.64 |
| Castor cake | 400 | 146 | 132 | 8.18 |
| Karanj cake | 400 | 118 | 128 | 10.34 |
| Neem cake | 400 | 118 | 112 | 8.91 |
| *G. mosseae* + Castor cake | 100 + 200 | 90 | 108 | 12.68 |
| *G. mosseae* + Karanj cake | 100 + 200 | 76 | 80 | 16.61 |
| *G. mosseae* + Neem cake | 100 + 200 | 48 | 62 | 14.80 |
| Control | --- | 218 | 184 | 5.45 |
| CD (P=0.05) | | 11.97 | 8.31 | 0.84 |

**Table 16.34.** Effect of paring of suckers and dipping in insecticide solution on nematode population and yield of banana

| Treatment | Nema popn in 200 ml soil | Nema popn in 5g roots | Yield (MT/ha) | Benefit: cost ratio |
|---|---|---|---|---|
| Unpared sucker | 365 | 27 | 57 | --- |
| Paring + drying for 72 hr | 209 | 23 | 59 | 1.17 |
| Double paring + drying for 72 hr | 187 | 21 | 59 | 1.72 |
| Paring + dipping in 0.5% monocrotophos for 45 min | 109 | 18 | 61 | 2.01 |
| Paring + dipping in 0.75% monocrotophos for 45 min | 89 | 16 | 60 | 0.30 |
| Double paring + dipping in 0.5% monocrotophos for 45 min | 85 | 15 | 63 | 2.92 |
| Double paring + dipping in 0.75% monocrotophos for 45 min | 76 | 12 | 63 | 2.22 |
| Paring + pralinage with carbofuran at 40g/sucker | 76 | 12 | 63 | 0.15 |

Double paring of banana suckers along with dipping in 0.5% monocrotophos for 45 min gave maximum yield (63.283 MT/ha) and recorded higher benefit: cost ratio (2.92) (Patil *et al.*, 1999b) (Table 16.34).

Integration of paring of banana suckers, dipping in 0.5% monocrotophos solution for 45 min along with intercropping with marigold or sunnhemp gave higher fruit yield (62.838 and 61.816 MT/ha, respectively) and benefit cost ratio (1.70 and 1.28, respectively) (Patil *et al.*, 1999a) (Table 16.35).

**Table 16.35.** Effect of paring of suckers, dipping in insecticide solution and intercropping on nematode population and yield of banana

| Treatment | Nema popn in 200 ml soil | Nema popn in 5g roots | Yield (MT/ha) | Benefit: cost ratio |
|---|---|---|---|---|
| Untreated control | 454 | 29 | 58.439 | --- |
| Paring + dipping in 0.5% monocrotophos | 283 | 18 | 60.927 | 1.20 |
| Paring + dipping in 0.5% monocrotophos + incorporation of sunnhemp at 10g/m$^2$ | 252 | 14 | 61.816 | 1.28 |
| Paring + dipping in 0.5% monocrotophos + incorporation of cowpea at 10g/m$^2$ | 274 | 17 | 60.483 | 1.04 |
| Paring + dipping in 0.5% monocrotophos + incorporation of marigold at 2g/m$^2$ | 239 | 15 | 62.838 | 1.70 |
| Carbofuran at 1.25g a.i./plant | 240 | 14 | 61.860 | 1.58 |

Soil application of FYM enriched with *Paecilomyces lilacinus* and *Pseudomonas fluorescens* at 2 kg/plant at the time of planting and subsequent application for four times at an interval of six months reduced the root population of *R. similis* and *M. incognita* by 64.5 and 75%, respectively and increased the fruit yield by 21% (Rao *et al.*, 2010).

Banana hybrid BRS-1 developed by Banana Research Station, Kannara, Kerala, is resistant to the burrowing nematode.

### 16.1.12. Integrated management of disease complex

#### i) Root-knot nematode, Meloidogyne incognita and wilt, Fusarium oxysporum f. sp. cubense, Ralstonia solanacearum disease complex

Soil application of neem cake + *T. viride* + carbendazim was found effective in reducing the burrowing nematode (*R. similis*) and wilt (*F. o.* f. sp. *cubense*) disease complex and in increasing the banana fruit yield (15,147 kg/ha). This treatment also gave minimum lesion index (1.1) and root-knot index (1.0) as compared to control (4.0) with cost: benefit ratio of 1: 2.72 (Ravi *et al.*, 2001).

**ii) Burrowing, spiral and root-knot nematode complex**

*Radopholus similis, Helicotylenchus multicinctus* and *Meloidogyne incognita* cause nematode complex in banana.

Integration of paring and hot water treatment (at 55°C for 20 min) of banana suckers along with application of carbofuran 3G (at 16.6 g/plant) and neem cake (at 1 kg/plant) at the time of planting reduced the soil and root population of nematodes besides increasing growth, development and yield of banana (Table 16.36).

**Table 16.36.** Management of nematode complex in banana

| Treatment* | Fruit yield/ plant (kg) | Cost: benefit ratio | Nema popn/ 200 ml soil** | | | Nema popn/ 5g roots | | |
|---|---|---|---|---|---|---|---|---|
| | | | *Rs* | *Hm* | *Mi* | *Rs* | *Hm* | *Mi* |
| Untreated check | 4.2 | 1: 1.49 | 213 | 150 | 133 | 62 | 51 | 3 |
| Paring + hot water treat. + carbofuran | 5.8 | 1: 2.01 | 97 | 78 | 40 | 30 | 19 | 21 |
| Paring + hot water treat. + carbofuran + neem cake | 7.6 | 1: 2.12 | 74 | 53 | 48 | 20 | 11 | 10 |
| CD (P = 0.05) | 0.6 | --- | 10.6 | 10.6 | 13.9 | 3.0 | 3.5 | 3.3 |

* Hot water treatment at 55°C for 20 min, Carbofuran 3G at 16.6 g/plant, Neem cake at 1 kg/plant.

** *Rs – Radopholus similis, Hm – Helicotylenchus multicinctus, Mi – Meloidogyne incognita.*

## 16.2. CITRUS, *CITRUS* SPP.

### 16.2.1. High yielding varieties

**Table 16.37.** High yielding varieties of acid lime

| Acid lime varieties | Yield | Developed by |
|---|---|---|
| Jai Devi (PKM-1) | High yielder 934 fruits/tree/ year (36.975 kg/tree) | HC&RI, TNAU, Periyakulam |
| Vikram | High yielder | MAU, Parbhani |
| Pramalini | High yielder | MAU, Parbhani |

### 16.2.2. High density planting

High density planting has been aimed with the use of dwarfing viral agents, use of inters tocks selected from citrus related genera and the use of dwarfing Trifoliate Orange rootstock. Flying Dragon Trifoliate Orange is capable of reducing growth by 75 per cent as compared to standard rootstocks. Kinnow mandarin has been grown successfully under high density planting using Troyer Citrange, Karna Khatta and Soh Sarkar rootstocks at a planting distance of 6' x 6' (2,990 plants/ha), 8' X 8' (1,682 plants/ha) on

Karna Khatta and 10' x 10' (1,076 plants/ha) on Soh Sarkar rootstocks, respectively. Similarly, Cleopatra mandarin rootstock was ideal for Mosambi sweet orange. For Khasi mandarin, a spacing of 5m x 5m (400 plants/ha) was found ideal.

- Flying Dragon TO – only genetically dwarfing rootstock.
- Capable of reducing growth by 75%
- Spacing: 2m x 2m.
- Most productive rootstock.
- Swingle citrumello and Troyer citrange-6 x 4 ft.
- Kinnow on Troyer citrange-6 x 6 ft.
- Kinnow on Karna Khatta-8 x 8 ft.
- Budded plants of Kinnow are allowed to grow with multiple shoots above bud union. It helps in keeping the terminal growth in check, provides adequate fruiting area.

**Advantages**

- Start bearing after 3rd year.
- Give very high initial production and net returns.
- Weed problem is minimized.
- Better moisture conservation, reduce number of irrigations.
- Cost of inputs and intercultural operations is reduced.
- Bright orange-coloured juicy fruits of excellent quality are produced.
- Rootstocks useful in staggering the fruit maturity. Kinnow on Troyer citrange come to harvest by the end of November which is very early. Kinnow on Karna Khatta come to harvest up to mid January.
- Life span – 10 to12 years. Afterwards remove alternate rows.

Varying plant densities in sweet orange (1250-2020 trees/ha) yielded 31.5 to 39.5 MT/ha. Similarly yields are high in mandarin (33-72 MT/ha) with 555-5000 trees/ha. Grape fruit yielded 58 MT/ha at planting density of 2020 trees/ha (Table 16.38).

**Table 16.38.** Yield of citrus under high density orcharding

| Citrus variety | Rootstock | Spacing (m) | Density (plants/ha) | Yield (MT/ha) | Tree age (years) | Reference |
|---|---|---|---|---|---|---|
| Traditional system | --- | 6-8 x 6-8 | 400-227 | 10-15 | --- | Singh, 1995 |
| Orange cv. St. Inves Valencia | Trifoliate orange | 3.2 x 2.5 | 1250 | 135.0 | 1-7 | Hutton & Cullis, 1982 |
| Orange cv. Hamlin | Mortan | 3.3 x 1.5 | 2020 | 39.5 | 7 | Wheatan *et al.*, 1991 |

| | | | | | | |
|---|---|---|---|---|---|---|
| Orange cv. Valencia | Carrizo citrange | --- | 1318 | 19.5 | 4 | Bevington, 1994 |
| Orange cv. Navelina | Sour orange | --- | 1200 | 493.0 | 0-13 | Tribulato *et al.*, 1994 |
| Orange cv. Tarocco | Sour orange | --- | 1200 | 832.0 | 0-13 | Wheatan *et al.*, 1991 |
| Orange cv. Valencia | Sour orange | 3.3 x 1.5 | 2020 | 31.5 | 7 | Wheatan *et al.*, 1991 |
| Mandarin cv. Miyugawa wase | --- | 2.0 x 1.0 | 5000 | 70.2 | 6 | Hirano *et al.*, 1982 |
| Mandarin cv. Miyugawa wase | --- | 2.0 x 2.0 | 2500 | 72.5 | 10 | Hirano *et al.*, 1982 |
| Mandarin cv. Kinnow | --- | 6.0 x 3.0 | 555 | 33.3 | 9 | Sharma *et al.*, 1992 |
| Grape fruit cv. Red blush | Koethen x Rubidoux | 3.3 x 1.5 | 2020 | 58.2 | 7 | Wheatan *et al.*, 1991 |
| Grape fruit cv. Red blush | Mortan | 3.3 x 1.5 | 2020 | 58.0 | 7 | Wheatan *et al.*, 1991 |

### 16.2.3. Micro irrigation

Studies conducted at the University of Agricultural Sciences, Dharwad, Karnataka on acid lime have shown that the yield in case of drip irrigated acid lime was 11.15 MT/ha as compared to 7.89 MT/ha in case of basin irrigated system (Table 16.39).

**Table 16.39.** Water saving and yield in drip irrigated acid lime with recommended dose of fertilizers (500: 300: 500 g/plant)

| ET replenishment | Water saved over ring basin (%) | Yield (MT/ha) | Increase over ring basin (%) | Benefit: cost ratio |
|---|---|---|---|---|
| 60% (drip) | 64.40 | 10.76 | 36 | 2.32 |
| 80% (drip) | 53.85 | 10.85 | 37 | 2.35 |
| 100% (drip) | 43.72 | 11.15 | 41 | 2.44 |
| Ring basin | --- | 7.89 | --- | 1.89 |

**Table 16.40.** Effect of irrigation levels on fruit yield of acid lime

| Irrigation levels | Fruit yield (kg/tree) | % increase |
|---|---|---|
| - 20 kPa | 13.35 | 43.54 |
| - 40 kPa | 14.24 | 53.10 |
| - 80 kPa | 14.87 | 60.00 |
| No irrigation | 9.30 | --- |

In acid lime, irrigations at – 20, - 40 and – 80 kPa of soil water potential increased fruit yield over non-irrigation mainly because of significant increase in size and number of fruits per tree as a result of increased tree vigour (Singh, 1985) (Table 16.40).

### 16.2.4. Fertigation

Shirgure *et al.* (2001) reported higher fruit yield (26.1 MT/ha), TSS (8.12°Brix) and leaf nitrogen content (2.42%) in Nagpur mandarin, when irrigation was scheduled at 20% depletion of available water and fertigating plants at 500, 140 and 70 g NPK/plant compared to other irrigation schedules and fertigation levels (Table 16.41).

**Table 16.41.** Effect of irrigation and fertigation on yield and quality of Nagpur mandarin

| Treatment | Fruit yield (MT/ha) | TSS (°Brix) | Leaf N content |
|---|---|---|---|
| Irrigation at 10% depletion of available water (600: 200: 100 g NPK/plant) | 18.0 | 7.93 | 2.31 |
| Irrigation at 10% depletion of available water (500: 140: 70 g NPK/plant) | 17.8 | 7.25 | 2.12 |
| Irrigation at 10% depletion of available water (400: 80: 40 g NPK/plant) | 17.0 | 7.27 | 2.10 |
| Irrigation at 20% depletion of available water (600: 200: 100 g NPK/plant) | 18.2 | 7.95 | 2.28 |
| Irrigation at 20% depletion of available water (500: 140: 70 g NPK/plant) | 26.1 | 8.12 | 2.42 |
| Irrigation at 20% depletion of available water (400: 80: 40 g NPK/plant) | 16.5 | 7.97 | 2.20 |

Shirgure *et al.* (1999) reported that acid lime fruit yield per plant was higher with 80% nitrogen fertigation (49.41 kg) as compared to 100% nitrogen supplied through band placement (39.89 kg) (Table 16.42).

**Table 16.42.** Effect of N fertigation in acid lime

| Treatment | Fruit yield/plant (kg) |
|---|---|
| 60% N fertigation | 38.03 |
| 80% N fertigation | 49.41 |
| 100% N fertigation | 41.17 |
| 100% N – Band placement | 39.89 |

### 16.2.5. Integrated nutrient management

Studies conducted at Tinsukia (Assam) by Borah *et al.* (2001) revealed that maximum yield with appreciable tree vigour and fruit quality of Khasi mandarin could be obtained from balanced nutrition of the plants through combinations of organic (neem cake) and inorganic fertilizers (Table 16.43).

**Table 16.43.** Effect of organic and inorganic nutrition on yield and quality of Khasi mandarin plants

| Treatments | No of fruits /plant | Yield (kg/plant) | Juice (%) | Ascorbic acid (mg/100g) | TSS (°Brix) |
|---|---|---|---|---|---|
| 600: 300: 600g NPK /plant | 805 | 118.01 | 46.33 | 48.27 | 14.35 |
| 600: 300: 600g NPK /plant + Neem cake at 15 kg/plant | 1072 | 203.55 | 55.66 | 57.26 | 15.26 |
| CD (P = 0.05) | 19.66 | 10.45 | 1.05 | 3.50 | 0.21 |

Results of a study conducted on 20 year old seedling trees of sweet orange cv. Sathgudi revealed that the maximum yield, cost benefit ratio with better fruit quality could be obtained by the balanced nutrition of 400: 150: 300g NPK/plant/year along with oil cake (castor cake at 7.5 kg) (Seshadri and Madhavi, 2001) (Table 16.44).

**Table 16.44.** Effect of organic and inorganic fertilizers on yield of sweet orange

| Treatments | Fruit number / plant | Yield/ plant (kg) | Weight of fruit (g) | Juice (%) | TSS (°Brix) |
|---|---|---|---|---|---|
| T 1. (Inorganic fertilizers at 800: 300: 600 g NPK/ plant/year) | 1960 | 296.27 | 160.76 | 42.04 | 12.86 |
| T 2. (Castor cake at 7.5 kg/plant/year + 50 % of T 1 ) | 2539 | 399.87 | 170.71 | 41.68 | 13.28 |

Integrated nutrient management studies conducted at Andhra Pradesh revealed that 50% of the recommended dose of fertilizers besides 100g each of *Azospirillum* and Phosphobacteria + 5 kg of vermicompost resulted in higher fruit yield (15.48 MT/ha) as against 13.98 MT/ha with farmer's practice (Umajyothi *et al.*, 2004).

**Table 16.45.** Effect of bioinoculants in combination with organic manure and inorganic fertilizers on yield and fruit quality of mandarin orange

| Treatments | Fruit weight (g) | No. of fruits/tree | Yield/ tree (kg) | TSS (°Brix) | Ascorbic acid (mg/100g) |
|---|---|---|---|---|---|
| RDF (600: 200: 400 g NPK/ plant) | 91.50 | 100.97 | 9.00 | 8.89 | 21.80 |
| 100 % RDF + Bioinoculants* | 84.50 | 147.95 | 12.90 | 9.44 | 23.13 |
| 75 % RDF + Bioinoculants * | 86.00 | 140.95 | 12.00 | 9.51 | 22.47 |
| CD (P = 0.05) | 1.75 | 8.06 | 0.08 | 0.26 | 0.62 |

* Arbuscular mycorrhiza (500 g/plant) + *Azospirillum lipoferum* (100 g/plant) + Phosphate solubilizing bacteria (100 g/plant) + *Pseudomonas fluorescens* (100 g/ plant).

At Lower Pulney Hills, Saravanan (2007) investigated the effect of bioinoculants viz., *Azospirillum lipoferum,* phosphate solubilizing bacteria (*Bacillus megaterium*), Arbuscular mycorrhizae and *Pseudomonas fluorescens* in combination with farmyard manure and inorganic fertilizers on growth, yield and quality of mandarin orange during 2005-2007 and found 100 per cent RDF along with all bioinoculants registered higher fruit number, fruit yield with good quality fruits than the pure inorganically fertilized trees (Table 16.45).

With the maximum K-fertilizer rate of 420 kg $K_2O$/ha, the cumulative yield of tangerine for three years was 15% higher as compared with the control treatment in which only N- and P-fertilizers were applied. As regards the positive effect of K on fruit quality, farmers harvested up to 21% more large-sized fruits (>6 cm) due to K application. Moreover, juice content in fruits increased by 10-12% when K-fertilizer was applied.

### 16.2.6. Mulching

In Doon valley, mulching with black polythene and grass has been found to improve fruit yield in Eureka lemon (Arora and Mohan, 1985) (Table 16.46).

**Table 16.46.** Effect of mulching micro catchments in Doon valley on fruit yield in Eureka round lemon

| Mulch | Fruit yield (kg/tree) | % increase |
|---|---|---|
| Black polythene (full run-on length) | 7.735 | 1127.77 |
| Black polythene (half run-on length) | 2.100 | 333.33 |
| Grass | 3.654 | 480.00 |
| Flat plot + no mulch | 0.630 | --- |

### 16.2.7. Integrated disease management

Acid lime cv. Rasraj (Indian Institute of Horticultural Research, Bangalore) is resistant to bacterial canker, while Sai-Sharbati (Mahatma Phule Krishi Vidyapeeth, Rahuri) is tolerant to canker and tristeza virus.

### 16.2.8. Integrated nematode management

#### i) Citrus nematode, *Tylenchulus semipenetrans*

#### a) Chemical control

Mukhopadhyaya and Dalal (1971) achieved 97.6% control by ethoprophos when applied at 4 kg a.i./ha and 39.9% increase in yield of sweet lime during second year. Chhabra *et al.* (1977) found fensulfothion very effective at 3 kg a.i./ha in controlling the nematodes on pumello and increased the yield by 68-76% (Table 16.47).

**Table 16.47.** Effect of different nematicides on *Tylenchulus semipenetrans* infecting different *Citrus* spp.

| Citrus spp. | Nematicide & Dosage | Type of application | Effectiveness |
|---|---|---|---|
| Grape fruit | Aldicarb – 5.7 to 11.4 kg a.i./ha | Soil application | Reduced nema popn, improved fruit size, quality & fruit yield by 15.7-108% |
| Naval orange | Carbofuran, Ethoprophos, Phenamiphos-2.2 kg a.i./ha | Soil application | Reduced larval popn. in soil, increased fruit yield by 108-137% |
| Orange | Aldoxycarb, Carbofuran, Carbosulfan – 1.1 kg a.i./ha | Through drip irrigation | Reduced female nemas on roots and increased fruit yield |
| | Phenamiphos – 10 ml/m$^2$ | With irrigation water | Reduced larval popn., increased uptake of K, Ca and Cu, gave extra yield of 6.1-7.7 MT |
| Sweet lime | Ethoprophos – 4.0 kg a.i./ha | Soil application | Reduced nema popn. by 98%, checked popn. build-up for 16 months, increased fruit yield at 12 kg/tree |
| Sweet orange | Aldicarb – 6 kg a.i./ha | Soil application | Reduced nema popn. by 84%, increased fruit yield by 224% with benefit: cost ratio of 11 |
| | Carbofuran – 6 kg a.i./ha | Soil application | Reduced nema popn. by 76%, increased fruit yield by 152% with benefit: cost ratio of 8 |

## 16.3. PAPAYA, *CARICA PAPAYA*

### 16.3.1. High yielding varieties

**Table 16.48.** High yielding varieties of papaya

| High yielding varieties of papaya | Yield (MT/ha) | Developed by |
|---|---|---|
| Pusa Nanha | 50 | IARI, New Delhi |
| Punjab Sweet | 50 kg/plant | PAU, Ludhiana |
| CO 1 | 100 kg/plant | TNAU, Coimbatore |
| CO 2 | 120-180 kg/plant | TNAU, Coimbatore |
| CO 5 | 112-128 kg/plant | TNAU, Coimbatore |
| CO 6 | 160-200 kg/plant | TNAU, Coimbatore |
| CO 7 | 340.9 kg/plant | TNAU, Coimbatore |
| Surya | 60 kg/plant (2x 2 m) | IIHR, Bangalore |
| Sunrise Solo | 25 - 37.5 kg/plant | IIHR, Bangalore |

Other high yielders include Coorg Honey Dew (IIHR, Bangalore), Pusa Dwarf (IARI, New Delhi) and Pant-1(GBPUAT, Pantnagar).

### 16.3.2. High density planting

With the development of dwarf varieties like Pusa Dwarf, Pusa Nanha and Ranchi Dwarf, it is now possible to grow papaya under the HDP concept. Normally papaya varieties are planted at a distance of 2.5m x 3.0m or 2.5m x 2.5 m, which accommodates 1,333 to 1,600 plants/ha, while Pusa Nanha may be planted at a distance of 1.25 x 1.25m (6,400 plants/ha). Such orchards may give 3 to 4 times higher yields (60 to 65 MT/ha) in comparison to the yield of the traditional orchards (15 to 20 MT/ha). Other cultivars like CO-1, CO-2, Pusa Dwarf and Honey Dew-1 are also suitable for high density planting.

A plant population of 2500/ha is recommended for high density planting (Awasthi and Mehta, 2000). Biswas *et al.* (1989) planted Ranchi Dwarf at 2922 plants/ha and harvested 98.05 MT/ha fruits.

### 16.3.3. Micro irrigation

Daily drip irrigation with 2 emitters/plant placed midway between trunk and skirt line gave maximum fruit yield along with maximum water use efficiency (Table 16.49).

**Table 16.49.** Yield and water use efficiency in relation to irrigation frequency, number of emitters and their placement under drip irrigation

| Treatment | Fruit yield (MT/ha) | Water use efficiency (kg/ha/mm) |
|---|---|---|
| Irrigation frequency | | |
| Drip daily | 116.5 | 42.4 |
| Drip alternate day | 110.2 | 36.9 |
| Number of emitters | | |
| 1 emitter/plant | 98.2 | 32.9 |
| 2 emitters/plant | 115.6 | 38.7 |
| Emitter placement | | |
| Skirt line | 101.4 | 33.9 |
| Midway between trunk and skirt line | 120.5 | 40.4 |
| CD (P =0.05) | 13.6 | --- |

**Table 16.50.** Water saving, yield and return in drip irrigated kagzi lime

| ET replenishment | Water saved over ring basin (%) | Yield (MT/ha) | Benefit: cost ratio |
|---|---|---|---|
| 60% (drip) | 64.40 | 10.76 | 2.32 |
| 80% (drip) | 53.85 | 10.85 | 2.35 |
| 100% (drip) | 43.72 | 11.15 | 2.44 |
| Ring basin | --- | 7.89 | 1.89 |

Studies conducted at UAS, Dharwad on kagzi lime have shown that the benefit cost ratio in case of drip irrigated lime was 2.44 as compared to 1.89

in case of basin irrigated system with recommended dose of fertilizer (500: 300: 500 g NPK/plant) as shown in Table 16.50.

### 16.3.4. Fertigation

In fertigated papaya plants there was an increase in number of fruits per tree by 24%, fruit weight by 30.6%, fruit length by 12.5% and TSS by 10.7% over plants which are not fertigated (Table 16.51) (Jeyakumar *et al.*, 2001).

**Table 16.51.** Comparison of soil application of fertilizers and fertigation in papaya

| Parameter | Soil application | Fertigation | % increase |
|---|---|---|---|
| Plant height (cm) | 288.3 | 327.7 | 13.6 |
| No. of fruits/plant | 76.4 | 94.7 | 23.9 |
| Av. fruit weight (kg) | 1.86 | 2.43 | 30.6 |
| Yield/plant (kg) | 142 | 230 | 62.0 |
| Fruit length (cm) | 26.3 | 29.6 | 12.5 |
| Fruit circumference (cm) | 43.6 | 51.4 | 17.8 |
| TSS (°Brix) | 11.2 | 12.4 | 10.7 |

Chaudhari *et al.* (2001) reported higher fruit yield (64.7 MT/ha) with 50% recommended dose of liquid fertilizers accompanied by drip irrigation.

### 16.3.5. Integrated nutrient management

Papaya being a perennial crop and the nutrient is required at all stages of growth continuously, split application of fertilizer is found beneficial. Results of experiments conducted at Tamil Nadu Agricultural University, Coimbatore revealed that six split applications (Irulappan *et al.*, 1984) was found to be good while Ravichandran *et al.* (2002) recommended twelve splits instead of six as it resulted in higher yield and quality of fruits in CO2 papaya (Table 16.52).

**Table 16.52.** Effect of split application of nutrients on yield of papaya cv. CO 2

| Treatment | Yield of fruits/plant (kg) |
|---|---|
| 300 g each of NPK/plant/year in six split doses | 167.01 |
| 300 g each of NPK/plant/year in 12 split doses | 213.64 |
| CD (P = 0.05) | 16.55 |

Micronutrient deficiencies in papaya may affect the growth and development of papaya crop if soils are found deficient. Kavitha *et al.* (2000a, b) studied the influence of foliar spraying of Zinc and Boran and found significant effect on yield and quality of papaya (Table 16.53).

**Table 16.53.** Effect of certain micro nutrients spray on yield and fruit quality of papaya cv. CO 5

| Treatment | Yield of fruits/ plant (kg) | Latex yield/ fruit (g) | TSS (°Brix) |
|---|---|---|---|
| Control | 230.17 | 12.54 | 12.30 |
| Foliar spraying of Zn 0.5 % + B 0.1 % at 4, 8, 12, 16th month after planting | 330.68 | 15.23 | 14.80 |

The maximum fruit yield of papaya cv. Pusa Delicious was obtained in the treatment comprising application of 200 g N/plant (126 MT/ha) and spacing of 1.25 m x 1.25 m (105.8 MT/ha) (Shukla and Singh, 2006).

### 16.3.6. Integrated disease management

Papaya cv. Pant-1 developed by G.B. Pant University of Agriculture and Technology, Pantnagar is resistant to anthracnose disease.

### 16.3.7. Integrated nematode management

#### i) Root-knot nematode, *Meloidogyne incognita*

Soil application of *Pochonia chlamydosporia* at 5 kg/ha increased plant height and fruit yield of papaya and reduced nematode population both in soil and roots (Kumar, 2009) (Table16.54).

**Table 16.54.** Effect of talc formulation of *Pochonia chlamydosporia* for the management of *Meloidogyne incognita* infecting papaya

| Treatment | Plant height (cm) | Nematode population | | Yield (kg/plant) |
|---|---|---|---|---|
| | | Soil | Roots | |
| *P. chlamydosporia* at 1 kg/ha | 22.6 | 162 | 12.4 | 24.7 |
| *P. chlamydosporia* at 2 kg/ha | 23.9 | 155 | 11.3 | 29.6 |
| *P. chlamydosporia* at 3 kg/ha | 22.4 | 140 | 9.4 | 31.8 |
| *P. chlamydosporia* at 4 kg/ha | 26.2 | 134 | 8.6 | 34.4 |
| *P. chlamydosporia* at 5 kg/ha | 36.8 | 122 | 7.3 | 38.1 |
| Control | 20.7 | 305 | 19.8 | 20.5 |
| CD (P = 0.05) | 4.9 | 8.1 | 0.2 | 2.7 |

## 16.4. PINEAPPLE, *ANANAS COMOSUS*

### 16.4.1. High density planting

A few decades ago, the planting density followed for pineapple in India was only fifteen to twenty thousand plants per hectare, with a productivity of less than twenty tonnes fruit per hectare, resulting in high cost of production. Under such low density planting, 20 to 25% fruits become unmarketable due to sunburn in the tropical region. Therefore, extensive studies were undertaken at the Indian Institute of Horticultural Research, Bangalore, during the 1970's and 80's to develop high density planting of pineapple for increasing the productivity and fruit quality and reducing the cost of production. These studies proved the benefits of increased planting density in the range of 53,000 to 63,500 plants per hectare.

Kew, Queen and Smooth Cayenne cultivars are found highly suitable for HDP using double row method of the planting suckers (500 to 600 g) or slips (350-400gm) spaced at 25 to 30 cm in the rows at a distance of 45 to 60 cm with a spacing of 90 to 105 cm between the beds (63,000 plants/ha). There is ample scope for expanding area under HDP for the benefit of the fruit growers, so that more and more area could be brought under pineapple cultivation. Yield realization varies from 60-70 MT/ha with use of high density in Kew pineapple.

Two-row trench system of planting has been found to be the best for high density planting in the plains, whether the crop is grown with or without irrigation. The field is laid out into 22.5 to 30.0 cm deep trenches alternating with mounds. In each trench, two shallow furrows about 10 to 15 cm deep and 15 cm inside from the edge of the trench are opened and suckers or slips are planted in these furrows, so arranged that two plants will not be exactly opposite to each other (Fig. 16.1). Spacing of pineapple plants depend on the growth of the plant. Soil moisture and fertility influences plant growth and indirectly determine spacing required per plant and eventually planting density. In overall analysis, plant-to-plant spacing of 22.5 to 25.0 cm and row-to-row spacing of 45 to 60 cm are ideal. Where pineapple plants grow luxuriantly with long leaves, a wider spacing of 90 cm between the trenches is required, but in places where growth of the leaf is moderate, a trench-to-trench spacing of 75 cm is adequate. A plant density of 63,000 plants/ha (22.5 cm x 60 cm x 75 cm) has been found ideal in semiarid mild tropical areas. In hot and humid tropical areas, plant density of 53,000 plants/ha (25 cm x 60 cm x 90 cm) performs well. However, decrease in fruit weight was quite evident when plant-to-plant spacing was reduced to 20 cm and row-to-row to 40 cm irrespective of the spacing between beds.

The yields recorded under high-density planting are in the range of 85 to 105 tonnes/ha, which are 55 to 85 tonnes more than the conventional planting densities, without adversely affecting the fruit size, quality and canning recovery. It is associated with other advantages like less weed infestation, protection to fruits from sunburn and increased production of suckers and slips per unit area, and non-lodging of plants. Close planting also saves on the upright orientation of the apical leaves, and eventually results in uniformly coloured lustrous fruits, which are cylindrical with low taper ratio, giving more canning recovery. Another advantage of high-density planting is the overlapping of basal leaves forming a sort of natural covering over the soil, preventing evaporation losses and thereby resulting in moisture conservation. Under dense planting a microclimate with high humidity will be created around the plant, which is congenial for growth and fruiting. Two successive ratoon crops harvested at twelve monthly intervals amounted to

50.7 and 53.8% of the plant crop yield at high-density planting under good management. Highest cost: benefit ratio of 1: 4.2 was observed in planting densities of 53,000 and 59,200 plants/ha (Tables 16.55 and 16.56).

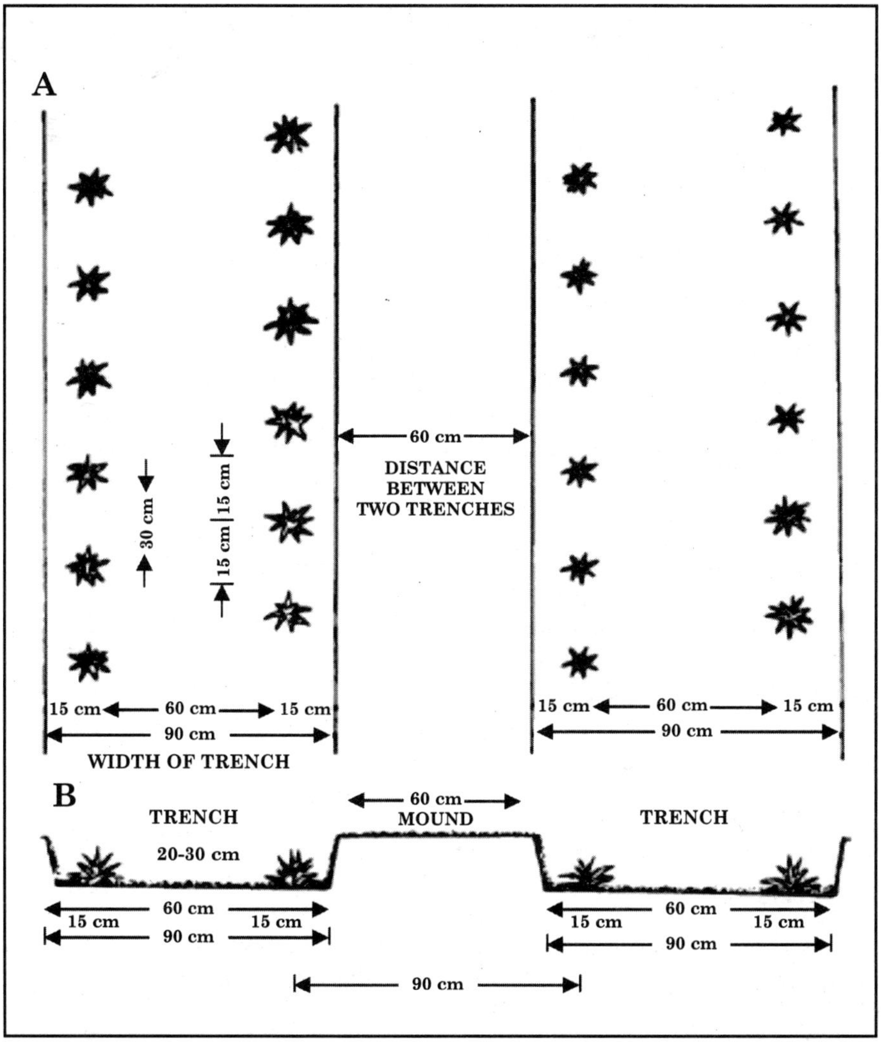


**Fig. 16.1.** Layout of pineapple plot for population density of 43,500 plants/ha

**Table 16.55.** The spacing required for different plant populations per hectare.

| Plant population/ha | Distance (cm) | | |
|---|---|---|---|
| | Plant-to-plant within a row | Row-to row | Trench-to-trench |
| 43,500 | 30.0 | 60.0 | 90.0 |
| 53,300 | 25.0 | 60.0 | 90.0 |
| 63,700 | 22.5 | 60.0 or 45.0 | 75.0 or 90.0 |

**Table 16.56.** Cost of cultivation and returns from one hectare of Kew pineapple with high density planting.

| Particulars | Single crop cycle – plant crop only | | Three crop cycle – plant crop + two ratoons | |
|---|---|---|---|---|
| | Traditional* | High density** | Traditional* | High density** |
| Fruit yield (MT ha-1) | 24 | 70 | 50 | 147 |
| Expenditure (Rs.ha-1) | 11,400 | 33,723 | 17,600 | 56,649 |
| Gross returns (Rs.ha-1) | 36,000 | 1,05,000 | 75,000 | 2,20,500 |
| Net returns (Rs.ha-1) | 24,000 | 71,277 | 57,400 | 1,63,851 |
| Benefit cost ratio | 2.16 | 2.11 | 3.26 | 2.89 |

* Traditional planting density of 19,000 plants ha$^{-1}$ (25 x 60 x 90 cm)

** High density planting of 53,000 plants ha$^{-1}$ (45 x 90 x 135 cm)

A planting density of 63750 plants/ha on proven management practices (two row system of planting, use of Kew, Queen and Smooth Cayenne cultivars, use of suckers weighing 500-600 g and slips weighing 350-400 g for planting) has increased yield from 15-20 MT/ha to 70-80 MT/ha (Awasthi and Mehta, 2000) (Table 16.57). Fruits obtained from high density planting are cylindrical and less tapering, besides low sun scald problem.

**Table 16.57.** Yield of pineapple cv. Kew under high density orcharding

| Density (plants/ha) | Yield (MT/ha) | Reference |
|---|---|---|
| Traditional system 27777 | 49.4 | Das *et al.*, 1987 |
| High density planting 55555 | 50.7 | Chadha, 1977 |
| 63788 | 79.3 | Chadha *et al.*, 1974 |
| 53300 | 139.8 | Kumar, 1977 |
| 72000 | 139.8 | Lee,1997 |
| 75923 | 102.2 | Das *et al.*, 1987 |

### 16.4.2. Mulching

Mulching with paddy husk or banana straw conserved soil moisture and improved fruit yield of pineapple even in high rainfall areas (Anon, 1985) (Table 16.58).

**Table 16.58.** Effect of mulching on fruit yield of pineapple

| Mulch | Fruit yield (MT/ha) | % increase over control |
|---|---|---|
| Paddy straw | 41.95 | 43.46 |
| Paddy husk | 46.01 | 57.35 |
| Banana leaves | 40.79 | 39.50 |
| Control | 29.24 | --- |

Singh and Samuel (2000) reported higher fruit yield of pineapple (76.73 MT/ha) under 50$\mu$ plastic mulching as compared to unmulched control (35.18 MT/ha).

## 16.5. SAPOTA, *MANILKARA ACHRAS*

### 16.5.1. High yielding varieties

**Table 16.59.** High yielding varieties of sapota

| High yielding varieties | Fruit yield (kg/tree) | Developed at |
|---|---|---|
| CO-1 | 175-200 | TNAU, Coimbatore |
| CO-2 | 175 (11.8 MT/ha) | TNAU, Coimbatore |
| CO-3 | 157 (40-50 MT/ha) | TNAU, Coimbatore |
| PKM-1 | 236 | HC&RI, TNAU, Coimbatore |
| PKM-2 | 344 | HC&RI, TNAU, Coimbatore |
| PKM-3 | 14 MT/ha | HC&RI, TNAU, Coimbatore |
| PKM-4 | 20.8 MT/ha | HC&RI, TNAU, Coimbatore |
| PKM-5 | 18.7 Mt/ha | HC&RI, TNAU, Coimbatore |

### 16.5.2. High density planting

Sapota is normally planted at a wide spacing of 8 x 8 m. Results have shown that high plant density of 312 plants per ha (8m x 4m) is possible yielding 15.35 MT/ha after 15 year of planting in PKM-1.

### 16.5.3. Fertigation

With the advent of drip irrigation system and availability of liquid fertilizers, the nutrient application technology is considered to be the most efficient. Results available in sapota indicate that 26 to 40% saving of nutrients without adversely affecting the yield and leaf nutrient status of plants (Table 16.60).

**Table 16.60.** Impact of the fertilizer application on the yield of sapota cv. Kalipatti

| Treatment | Fruit yield (kg/plant) | % increase over control |
|---|---|---|
| Normal fertilizers 100% rain fed (control) | 30.4 | --- |
| Normal fertilizers 100% + basin irrigation | 43.1 | 41.77 |
| Normal fertilizers 100% + drip irrigation | 48.3 | 58.88 |
| Water soluble fertilizers 60% + fertigation | 34.9 | 14.80 |
| Water soluble fertilizers 80% + fertigation | 58.0 | 90.78 |
| Water soluble fertilizers 100% + fertigation | 57.1 | 87.82 |

Application of 320: 130: 360 g NPK/tree/year being 80% in the form of water soluble fertilizers through fertigation at monthly intervals is recommended for bearing sapota trees (more than 10 year-old) for getting higher productivity and better quality of fruits (Khan *et al.*, 2001) (Table 16.61).

**Table 16.61.** Productivity of sapota cv. Kalipatti as influenced by sources and level of fertilizers

| Treatment | Fruit yield (MT/ha) |
|---|---|
| Normal fertilizers 100% + Rained | 28 |
| Normal fertilizers 100% + Basin irrigation | 37 |
| Normal fertilizers 100% + Drip irrigation | 40 |
| Water soluble fertilizers 60% Fertigation | 48 |
| Water soluble fertilizers 80% Fertigation | 65 |
| Water soluble fertilizers 100% Fertigation | 61 |

### 16.5.4. Integrated nutrient management

The maximum sapota (cv. Kalipatti) fruit yield (92 kg/plant) was obtained in combined application of 50% nutrients applied through FYM (50 kg/plant) + 50% nutrients through chemical fertilizers (N-500g, P-200g, K-200g/plant) (Hiwale *et al.*, 2006).

### 16.5.5. Bioregulators

In order to obtain higher sapota fruit yield from unit area during the formative years, planting of sapota at 5 x 5m spacing is beneficial up to 13 years. Further those trees were to be pruned (10 x 5 x 5m or 8 x 5 x 5m canopy) and applied with paclobutrazol at 4-8 ml/tree to sustain the productivity (Table 16.62).

**Table 16.62.** Effect of pruning and paclobutrazol on production of sapota cv. PKM- 1 (age 14 years) at Periyakulam

| Treatment | Trees/ha | Yield (MT/ha) |
|---|---|---|
| 10 x 10m spacing, tree topping at 5m height | 100 | 3.2 |
| 8 x 8m spacing, tree topping at 5m height | 156 | 6.2 |
| 10 x 5 x 5m canopy + 8 ml paclobutrazol/tree | 200 | 9.0 |
| 8 x 5 x 5m canopy + 4 ml paclobutrazol/tree | 312 | 11.4 |

## 16.6. MANGO, MANGIFERA INDICA

### 16.6.1. High yielding varieties

**Table 16.63.** High yielding varieties of mango

| High yielding varieties | Fruit yield (kg/tree) | Developed by |
|---|---|---|
| Prabhasankar | 23.5 | BAC, RAU, Sabour |
| Mahmood Bahar | 23.0 | BAC, RAU, Sabour |
| Sundar Langra | 25.0 | BAC, RAU, Sabour |
| Alfazli | 21.0 | BAC, RAU, Sabour |

| | | |
|---|---|---|
| Jawahar | 28.0 | BAC, RAU, Sabour |
| Sabri | 22.9 | BAC, RAU, Sabour |
| Subhash | 22.5 | BAC, RAU, Sabour |
| Menaka | 22.5 | BAC, RAU, Sabour |
| Paiyur 1 | 22.3 | HRS, TNAU, Paiyur |
| PKM 1 | 102.7 | HC&RI, TNAU, Coimbatore |
| Arka Puneet | 10-11 MT/ha | IIHR, Bangalore |
| Alphanso | 10.06 MT/ha | |

Other high yielders include Manjeera, AU Rumani (FRS, APAU, Sangareddy); Mallika, Amrapali, Pusa Surya (IARI, New Delhi) and Konkan Ruchi (RFRS, Vengurla).

### 16.6.2. High density planting

Mango yields worldwide are generally poor, ranging from 4 to 9 MT/ha in the major mango producing countries. This is partly attributable to wide tree spacing, which is traditionally based on expected eventual tree size. Canopies of these mango trees often take more than 10 years to fill the allocated space in the orchard row. Little consideration is generally given to canopy size maintenance once the overcrowding eventually occurs. Yields for more closely spaced trees of well-managed orchards in Florida and Puerto Rico, where annual or biennial hedging is employed to maintain canopy size, generally exceed 20 MT/ha. Most of the commercial varieties of the mango are irregular bearers and generally planted at a distance of 10 to 15 m, accommodating about 70 to 100 plants/ha, while Amrapali variety has been recommended by Indian Agricultural Research Institute, New Delhi, to be planted at 2.5 x 2.5 m (triangular method) and accommodating 1,600 plants/ha (as against 70-100 plants/ha under traditional planting) density yielded up to 22 MT/ha. The high density orchard provides several times higher yields than the traditional densities. Dashehari mango at 2.5m x 3.0m (1,333 plants/ha) can also be raised under HDP with pruning and dehorning after the harvesting after 10$^{th}$ year of planting. In Maharashtra, HDP has been achieved with application of paclobutrazol. Successful HDP (666 plants/ha) plantations of different commercial varieties namely Kesar, Alphanso, Tommy Atkins, Maya, etc. have been demonstrated by the Reliance Agro Ltd. at Jamnagar (Gujarat) under arid agro-climate. A solid plantation of over 80,000 plants has been established using different hi-tech interventions like fertigation, use of growth regulator (Paclobutrazol), uniform clonal (Polyembryonic) root-stock, modern packaging unit, etc. It is exclusively an export-oriented plantation and is likely to revolutionize mango production and productivity in India. This model is proposed to be replicated in different mango belts for hi-tech cultivation of fruit crops.

Yield commences with about 5 MT/ha in third year of planting which increases to 30-40 MT/ha after about 10 tears of planting (Table 16.64). Dashehari is prolific bearer but in alternate years and has surpassed yield of unpruned Amrapali (a regular bearer) (Ram *et al.*, 1997). In Israel, productivity of mango has been doubled adopting high density planting technology producing quality fruits which are being exported to European countries. They plant mango trees at 3 x 5 m or 3 x 6 m and use pruning machines mounted on tractor. However, Indian farmers may adopt hand pruning and closer planting (3 x 3 m) to increase the productivity of mango.

**Table 16.64.** Fruit yield of mango under traditional and high density plantings

| Orcharding system/Cultivar | Spacing (m) | Density (plants/ha) | Yield (MT/ha) | Tree age (years) | Reference |
|---|---|---|---|---|---|
| Traditional | | | | | |
| Dashehari | 12.5 x 12.5 | 64 | 4.25 | 14 | Ram *et al.*, 2001 |
| Dashehari | 10.0 x 10.0 | 100 | 6.00 | 14 | Ram *et al.*, 2001 |
| Dashehari | 7.5 x 7.5 | 178 | 9.00 | 14 | Ram *et al.*, 2001 |
| Dashehari | 5.0 x 8.0 | 400 | 18.00 | 14 | Ram *et al.*, 2001 |
| High density | | | | | |
| Tommy Atkins | 4.5 x 9.0 | 247 | 18.00 | 6 | Nel, 1993 |
| Tommy Atkins | 3.0 x 9.0 | 370 | 22.00 | 6 | Nel, 1993 |
| Tommy Atkins | 2.0 x 9.0 | 555 | 35.00 | 6 | Nel, 1993 |
| Irwin | 4.5 x 9.0 | 247 | 18.00 | 7 | Nel, 1993 |
| Irwin | 3.0 x 9.0 | 370 | 40.00 | 7 | Nel, 1993 |
| Irwin | 2.0 x 9.0 | 555 | 44.00 | 7 | Nel, 1993 |
| Kent | 6.0 x 6.5 | 256 | 25.00 | 7 | Ram, 2000 |
| Kent | 5.0 x 3.0 | 667 | 28.00 | 7 | Ram, 2000 |
| Mallika | 4.0 x 3.0 | 833 | 10.48 | 12 | Pandey & Majumdar, 1988 |
| Dashehari | 3.0 x 2.5 | 1333 | 5.00 | 5 | Ram, 1996 |
| Dashehari | 3.0 x 2.5 | 1333 | 19.30 | 10 | Ram, 1996 |
| Dashehari | 3.0 x 2.5 | 1333 | 23.16 | 15 | Ram *et al.*, 1997 |
| Dashehari | 3.0 x 2.5 | 1333 | 42.86 | 20 | Ram *et al.*, 2001 |
| Dashehari | 3.0 x 2.5 | 1333 | 34.74 | 24 | Ram, 2000 |
| Dashehari | 2.5 x 2.5 | 1600 | 25.92 | 9 | Ram *et al.*, 1997 |
| Amrapali | 2.5 x 2.5 | 1600 | 24.46 | 12 | Pandey & Majumdar, 1988 |
| Amrapali | 2.5 x 2.5 | 1600 | 20.64 | 10 | Ram *et al.*, 1997 |
| Amrapali | 2.5 x 2.5 | 1600 | 11.50 | 4 | Majumdar *et al.*, 1982 |

### Double hedge row planting

Double hedge row system (10 x 5 x 5m) can accommodate 222 plants/ha compared to 100 trees/ha in conventional system of planting (10m x 10m). The increased population is around 122% (Fig. 16.2).

```
*— 5 m— *        *——10 m——*        *— 5 m —*
5 m                                       5 m
*— 5 m — *       *——10 m—— *       *— 5 m —*
5 m                                       5 m
*— 5 m — *       *——10 m—— *       *— 5 m —*
```

Spacing: 10 x 5 x 5 m

Fig. 16.2. Double hedge row planting

**Benefits**

- More plants can be accommodated (122% increase in population over the square system of planting) per unit area.
- More yield can be obtained (9.45 MT/ha) compared to square system (5.05 MT/ha).

**Economics**

- The increased yield in double hedge row system is (88%) over conventional system.
- The yield in double hedge row system is 9.45 MT/ha with a benefit cost ratio of 5.82 compared to conventional system (5.05 MT/ha with a benefit cost ratio of 5.05).

Among the different plant densities in mango cv. Kesar, 10 x 2.5m recorded significantly higher fruit yield (5.056 MT/ha) followed by 10 x 5m (4.287 MT/ha) (Mistry and Patel, 2006).

### 16.6.3. Micro irrigation

Significantly higher fruit number (357) and yield (66.62 kg/tree) were recorded at 75% evaporation replenishment in 8 year-old mango (cv. Arka Anmol) trees (Kumar and Pandey, 2006a).

The fruit number (228) and yield (50 kg/tree) were highest in organic mulching + 50% evaporation replenishment in 6 year-old mango (cv. Lat Sundari) trees (Kumar and Pandey, 2006b).

**Fruit retention:** The maximum number of fruits per plant during 2001-02, 2002-03, 2003-04 and 2004-05 was recorded as 446, 385, 550 and 728, respectively, in case of treatment (100 percent dose of fertilizer under drip irrigation at 20 kPa soil metric tension with mulch). Against this, the lowest number of fruit set per plant was recorded as 144, 124, 250, and 545 in case of surface irrigation in combination with 50 percent level of conventional fertilizer application.

**Fruit yield:** The experimental results indicated that 8 years young mango plants under drip irrigation, irrigated at 50 kPa soil moisture tension + 100 micron black plastic mulch + fertigation at 100% of recommended dose was found best as far water saving, yield (10 MT/ha) and quality of produce are concerned. The crop under conventional system of irrigation and fertilizer application yielded 6 MT/ha.

The water saving in case of drip irrigation at 20, 30, and 60 kPa soil moisture level and 50% soil moisture regime were calculated as 11.52, 30.56 and 54.63 per cent, respectively over surface irrigated treatments in mango cv. Dashehari.

### 16.6.4. Fertigation

Application of 100% recommended dose of N and K through fertigation gave maximum fruit yield in mango cvs. Alphanso and Arka Puneet (Table 16.65).

**Table 16.65.** Response of mango to NK fertigation

| NK fertigation | Fruit yield (kg/plant) | |
|---|---|---|
| | Alphanso | Arka Puneet |
| 100% recommended dose | 20.4 | 17.0 |
| 75% recommended dose | 11.4 | 12.5 |
| 50% recommended dose | 9.1 | 9.0 |

An experiment was conducted in mango cv. Ratna planted under HDP during 2005-07 to study the influence of N and K nutrients applied through fertigation. The results showed that fruit weight, number of fruits per tree and per tree yield were improved by application of 100 per cent recommended dose of nutrients through fertigation (Table 16.66).

**Table 16.66.** Effect of fertigation on yield characters in mango cv. Ratna

| Treatments | Number of fruits tree-1 | Mean fruit weight (g) | Fruit yield (kg tree-1) | Estimated fruit yield (t ha-1) |
|---|---|---|---|---|
| 100% of RDF as soil application | 116.5 | 364.6 | 40.8 | 17.98 |
| 100% N + 100% P + 50% K of RDF through fertigation | 126.7 | 340.8 | 40.2 | 17.69 |
| 100% N + 100% P + 75% K of RDF through fertigation | 142.7 | 436.3 | 54.0 | 23.78 |
| 100% N + 100% P + 100% K of RDF through fertigation | 160.0 | 465.3 | 59.8 | 26.34 |
| CD (P = 0.05) | 6.14 | 12.50 | 1.43 | 0.63 |

RDF: (800: 400: 800 g NPK/plant/year)

The highest mango fruit yield (11.73 MT/ha) was obtained when drip irrigation was coupled with 50% of the recommended nitrogen fertigation. Under fertigation treatment, fruit yield ranged from 8.97 to 13.56 MT/ha, while without nitrogen fertigation fruit yield ranged from 8.13 to 9.50 MT/ha. Mango trees irrigated to the replenishment of 90, 70 and 60% of open pan evaporation resulted in saving of irrigation water by 15.0, 31.4 and 40.5%, respectively (Bhriguvanshi and Singh, 2006).

Soil application of fertilizer recorded significantly lower number of fruits (98) and fruit weight (14 kg/plant) as compared to fertigation with 100% of RDF (fruit number-120 and fruit weight-22 kg/plant) (Srinivas, 2006).

### 16.6.5. Integrated nutrient management

Application of recommended dose of N and P fertilizers with 1 kg $K_2O$ through potassium sulphate (SOP) + 1% potassium nitrate applied through foliar spray and paclobutrazol at 0.75 g a.i./m canopy produced maximum yield of Alphanso mango fruits (78.5 kg/tree) which was on par with treatment receiving 0.5 kg $K_2O$ through SOP; and the treatment receiving 2.0 kg $K_2O$ through SOP (Dabke *et al.*, 2006).

### 16.6.6. Bioregulators

It has been reported that paclobutrazol (cultar) has good potential both as foliar spray and soil application in overcoming biennial bearing in mango. Soil application of paclobutrazol at 10g mixed in 5 litres of water in the ring of 15 cm depth dug 60 cm away around the tree trunk in August resulted in appreciable flowering in the following expected "off" year in Alphanso (Rao *et al.*, 1997, Srihari and Rao, 1996) (Table 16.67).

**Table 16.67.** Effect of paclobutrazol on the induction of flowering in the expected "off" year in mango cv. Alphanso

| Treatment | % flowering in expected "off" year | Yield in expected "off" year (No. of fruits/tree) | "off" year yield as % of the previous "on" year |
|---|---|---|---|
| Paclobutrazol 2000 ppm foliar spray | 67.3 | 271 | 55.0 |
| Paclobutrazol 3000 ppm foliar spray | 59.3 | 250 | 48.4 |
| TIBA 100 ppm foliar spray | 53.3 | 183 | 51.4 |
| Paclobutrazol 30 ml (7.5g a.i.) soil application | 51.0 | 237 | 76.0 |
| Paclobutrazol 40 ml (7.5g a.i.) soil application | 64.0 | 318 | 83.0 |
| Control | 6.0 | 10 | 2.5 |

### 16.6.7. Intercropping

Lentil was found as the best intercrop in young mango orchard among different legumes which also produced maximum yield of mango fruits (9.723 MT/ha) and lentil (0.758 MT/ha). This treatment also gave the maximum gross return of Rs. 1,05,564.60 per hectare. Similarly, the best intercrop among different vegetable crops was bottle gourd in *kharif* and radish in *rabi*. Intercropping in young orchard of mango cv. Langra produced maximum yield of mango fruits (9.290 MT/ha) with bottle gourd (0.762 MT/ha) resulting into maximum/gross return of Rs. 94,447.32 per hectare (Table 16.68).

**Table 16.68.** Economics of comparative performance of various horticultural crop combinations

| Crop combination | Yield (MT/ha) | | Net returns (Rs/ha) |
|---|---|---|---|
| | Main crop | Inter crop | |
| Mango | 2.60 | --- | 24,000 |
| Mango + Cowpea | 2.55 | 5.537 | 37,111 |
| Mango + Okra | 2.56 | 7.625 | 43,475 |
| Mango + Dolichos | 2.59 | 2.375 | 28,025 |
| Mango + Cluster bean | 2.66 | 2.312 | 21,600 |
| Maize + Pigeon pea (Traditional system) | 0.81 | 0.320 | 6,200 |

Among several intercrops (bottle gourd, cowpea, French bean, okra and black gram) tried during rainy season (1997 and 1998) for the grown up mango orchard at Ranchi, Jharkhand; cowpea produced significantly higher additional mango-equivalent yield in both the years (1.713 MT/ha and 3.088 MT/ha during first and second year, respectively). The intercrops did not affect the growth and production of mango base crop. Hence, the intercropping of cowpea in rainy season is recommended in mango orchard under agro-climatic conditions of Chotanagpur region (Singh *et al.*, 1999).

### 16.6.8. Mulching

Maximum soil moisture retention was recorded during September-November months under black polythene mulch which conserved 46-50% higher moisture at 15 and 30 cm depths as compared to control. The above treatment also gave highest increases in plant height (85%) and stem girth (41%). Black polythene mulch increased the marketable yield by 80% over control while organic mulches by 58-66% only.

Mulching of soil with black plastic (100$\mu$ thick) during the season of flower bud differentiation stage (October-November) recorded maximum fruit yield (43.33 kg/plant) compared to control (28.90 kg/tree) (Singh *et al.*, 2006).

### 16.6.9. Pruning

Pruning treatment T2 (heading back up to secondary branches) gave maximum fruit yield of mango cv. Chausa (317.50 kg/tree) followed by pruning treatment T3 (centre opening) (300.45 kg/tree) (Lal and Mishra, 2006).

Highest fruit production in mango cv. Mallika (80.14 kg/tree) was recorded from trees which received crown thinning treatment followed by 74.85 kg/tree in crown reduction treatment, while only 52.14 kg/tree was recorded in un-pruned trees. Trees subjected to crown thinning treatment showed 155.75% increase in available solar radiation within the canopy followed by 48.75% increase in available solar radiation within the canopy in crown reduction treatment compared to control (Mishra *et al.*, 2006).

### 16.6.10. Integrated pest management

#### i) Hoppers and thrips

Foliar spray of 0.005% imidacloprid provided maximum protection against hoppers and thrips and gave highest fruit yield (184.96 fruits/tree and 45.37 kg/tree) and benefit cost ratio (22.22). Imidacloprid treated trees also recorded appreciably higher population of natural enemies viz; coccinellids and predatory spiders as compared to the conventional insecticides (Kumar and Patel, 2006a).

The incidence of hoppers in treated plots consisting of three sprays viz; imidacloprid (0.005%) at panicle emergence stage, endosulfan (0.07%) at 21 days after first spray and its subsequent spray at 21 days after the second spray not only recorded significantly lower hopper population over untreated trees (2.77 v/s 8.80 hoppers/panicle) but also increased the fruit yield (6.397 MT/ha v/s 2.211 MT/ha in control) (Kumar and Patel, 2006b).

#### Validated IPM technology in mango (Uttarakhand)

- Spraying of 0.3% copper oxychloride for control of die-back, anthr-acnose and red rust diseases wherever they appeared during September-October..
- Ploughing of orchard in November-December to expose pupae of fruit flies, midges, leaf hoppers and eggs of mealy bugs to natural enemies.
- Polythene banding of tree trunk in December-January and application of 5% NSKE and *Beauveria bassiana* in January.
- Spraying of 0.2% sulfex for the control of powdery mildew disease.
- Spraying of imidacloprid at 0.5 ml/l during April to control hoppers and adult female mealy bugs.
- Spraying of *Verticillium lecanii* in orchards for control of hoppers.
- Fixing methyl eugenol traps (wooden blocks impregnated with methyl eugenol) to control fruit flies from April to August.
- Mechanical removal of mango leaf webber larvae and webs by leaf web removing device (developed by the Central Institute of Sub-tropical Horticulture, Lucknow) from April to September-October.

The IPM package was successfully validated in 16.8 ha of mango orchards in Gulabkhera, Habibpur, Budhadia, Pathakganj, Rehmankhera and Kanar villages in Malihabad and Kakori belt of mango near Lucknow on Dashehari variety during 2000-04. As a result of adoption of IPM, yield of mango increased to 6.0-9.0 MT/ha as it was 3.5-7.0 MT/ha earlier in non-IPM orchards. By adopting IPM, the mango growers in that area earned a profit of

Rs. 30,000/- to Rs.55,000/- while the farmers who did not adopt IPM, earned a profit of Rs. 17,000/- to 35,000/- per ha only (Table 16.69) (Amerika Singh *et al.*, 2004).

**Table 16.69.** Economics of IPM in mango

| Parameters | IPM plots | Non- IPM plots | % increase |
|---|---|---|---|
| Yield MT/ha | 6.0-9.0 | 3.5-7.0 | 28.57-71.43 |
| Net profit (Rs/ha) | 30,000-55,000 | 17,000-35,000 | 57.14-76.47 |

## 16.7. AVOCADO, *PERSEA AMERICANA*

### 16.7.1. High yielding variety

**Table 16.70.** High yielding variety of avocado

| High yielding variety | Fruit yield (MT/ha) | Developed at |
|---|---|---|
| TKD-1 | 26.4 (264 kg/tree) | TNAU, Coimbatore |

### 16.7.2. High density planting

The total combined production over the 12-year period in the mixed 6 acres is 1,391,275 lbs. and the regular density planting (15 x 15 ft.) is 820,700 lbs. This is 570,575 lbs. more, a 70% increase in production. 'Reed' trees planted in a conventional planting would take an additional 5.78 years to equal the production of 12 years in a high density planting. At year 17, the conventional planting is becoming less productive due to shading whereas, the high density planting is back in peak production and beginning of the third cycle of replanting. The cost of establishment for this type of planting is primarily in the cost of trees and the increased labour for planting. As for the infrastructure setup, the additional cost for increased irrigation emitters for a high density planting is not significant.

## 16.8. GRAPEVINE, *VITIS VINIFERA*

### 16.8.1. High yielding varieties

**Table 16.71.** High yielding varieties of grapes

| High yielding varieties | Yield (MT/ha) | Developed by |
|---|---|---|
| Arka Chitra | 38 | IIHR, Bangalore |
| Arka Krishna | 33 | IIHR, Bangalore |
| Arka Majestic | 38 | IIHR, Bangalore |
| Arka Neelmani | 28 | IIHR, Bangalore |
| Arka Soma | 40 | IIHR, Bangalore |
| Arka Trishna | 26 | IIHR, Bangalore |
| Sweta Seedless | 31 | |
| Bangalore Blue | 20-25 | |

| | |
|---|---|
| Anab-e-Shahi | 30-37 |
| Thompson Seedless | 15-18 |

Other high yielders include Pusa Seedless, Pusa Navrang, Pusa Urvashi (IARI, New Delhi); Cheema Sahebi (College of Agri., Pune); Madhu Angoor and Kishmish Chernyi (ANGRAU, Hyderabad).

### 16.8.2. Fertigation

With the advent of drip irrigation system and availability of liquid fertilizers, the nutrient application technology is considered to be the most efficient. Results available in grape indicate that 20% saving of nutrients without adversely affecting the yield and leaf nutrient status of plants (Table 16.72).

**Table 16.72.** Effect of the fertilizer application on the yield of Bangalore Blue grapes

| Treatment | Fruit yield (kg/vine) | % increase over control |
|---|---|---|
| Soil application | | |
| Normal fertilizers 100% + basin irrigation (control) | 10.44 | --- |
| Normal fertilizers 100% + drip irrigation | 14.51 | 39.00 |
| Water soluble fertilizers 100% + drip irrigation | 16.86 | 61.49 |
| Drip fertigation | | |
| Water soluble fertilizers 80% | 20.18 | 93.29 |
| Water soluble fertilizers 100% | 17.27 | 65.41 |
| Normal fertilizers 80% | 16.77 | 60.63 |
| Normal fertilizers 100% | 16.63 | 59.29 |

Shikamany (2001) reported higher yield through fertigation in grapes as compared to soil application of fertilizers (Table 16.73).

**Table 16.73.** Comparison of soil application of fertilizers and fertigation in grapes

| Parameter | Soil application | Fertigation |
|---|---|---|
| Yield (MT/ha) | 30 | 32 |
| TSS (°Brix) | 20 | 22 |
| Exportable grape yield (MT/ha) | 10 | 12 |
| Total returns on table grapes (Rs.) | 540000 | 660000 |
| Raisin yield (MT/ha) | 7.2 | 8.0 |
| Returns on raisins (Rs.) | 540000 | 600000 |
| Cost of manuring | 62725 | 74063 |

Studies on the influence of source and level of nitrogen on yield of grapes in Anab-e-Shahi variety revealed that the yield was highest (27.2 kg/plant) with ammonium nitrate fertigated at 40% of the recommended dose (Table 16.74). The yield increase with ammonium nitrate fertigated at 40% of the recommended dose was nearly 35%.

**Table 16.74.** Yield and quality of grapes (Anab-e-Shahi) as influenced by source and level of nitrogen

| Nitrogen source and level | Yield (kg/plant) |
|---|---|
| Urea – 60% RDF | 17.76 |
| Urea – 40% RDF | 25.20 |
| Ammonium nitrate – 60% RDF | 23.00 |
| Ammonium nitrate – 40% RDF | 27.20 |
| CD (P = 0.05) | 8.28 |

Application of 350: 250: 600 g NPK/vine in the form of water soluble fertilizers from the second year onwards through fertigation at fortnightly intervals is recommended for Bangalore Blue grapes for getting maximum productivity of improved quality and high returns (Khan *et al.*, 2001) (Table 16.75).

**Table 16.75.** Productivity and economics of grape cv. Bangalore Blue as influenced by sources and level of fertilizers

| Treatment | Fruit yield (MT/ha) | Net income (Rs.) | Benefit: cost ratio |
|---|---|---|---|
| Normal fertilizers 100% + Basin irrigation | 11.59 | 32,692 | 0.55 |
| Normal fertilizers 100% + Drip irrigation | 16.11 | 84,595 | 1.91 |
| Water soluble fertilizers 100% + Drip irrigation | 19.00 | 74,847 | 0.99 |
| Water soluble fertilizers 80% Fertigation | 22.00 | 111,918 | 1.79 |
| Water soluble fertilizers 100% Fertigation | 19.22 | 82,239 | 1.14 |
| Normal fertilizers 80% Fertigation | 18.63 | 111,205 | 2.93 |
| Normal fertilizers 100% Fertigation | 18.40 | 106,947 | 2.62 |

### 16.8.3. Integrated nutrient management

As compared with the NP treatment, the application of three different forms of potash fertilizers, i.e. potassium chloride, potassium sulphate and kali magnesia (potassium magnesium sulphate), at 200 kg $K_2O$/ha increased grapes yield by 22%, 35% and 41%, respectively.

### 16.8.4. Bioregulators

Use of $GA_3$ has increased the productivity of Thompson seedless variety of grapes in Maharashtra and North Karnataka remarkably. Application of

$GA_3$ has caused multifold increase in berry size and consequently the productivity has been raised from 6- 8 MT/ha to 15-18 MT/ha.

### 16.8.5. Integrated disease management

Grape cv. Madhu Angoor developed by Acharya N.G. Ranga Agricultural University, Hyderabad, is tolerant to downy mildew disease.

### 16.8.6. Integrated nematode management

#### i) Root-knot nematode, *Meloidogyne* spp.

Soil application of talc formulation of *P. fluorescens* containing 15 x $10^8$ cfu/g at 4 g/vine around root-knot infested grapevine at 15 cm depth in the basin at the time of pruning (during July) significantly reduced root galling due to *M. incognita* (39%), number of egg masses (250%) and increased fruit yield (166%) (Shanthi *et al.*, 1998) (Table 16.76).

**Table 16.76.** Field evaluation of biocontrol potential of *Pseudomonas fluorescens* against *Meloidogyne incognita* infecting grapevine

| Treatment | No. of galls/ 5 g roots | No. of egg masses/5 g roots | Root colonization ($10^8$ cfu/g) | Yield (MT/ha) |
|---|---|---|---|---|
| *P. fluorescens* – 1 g/vine | 428c | 94c | 24 | 12.07b |
| *P. fluorescens* – 2 g/vine | 390b | 85b | 36 | 15.41c |
| *P. fluorescens* – 4 g/vine | 326a | 72a | 58 | 22.07d |
| Carbofuran 3G – 60 g/vine | 300a | 67a | --- | 31.65e |
| Control (Untreated) | 535d | 180d | --- | 8.33a |

Figures with different letters are significantly different from each other at 5% level by analysis of variance test.

The combined application of two bioagents *P. lilacinus* and *P. penetrans* was most effective in increasing the fruit yield of grapevine besides checking *M. incognita* with cost: benefit ratio of 1: 3.3 (Sivakumar and Vadivelu, 1999).

### 16.8.7. Integrated management of disease complex

#### i) Root-knot nematode, *Meloidogyne incognita* and wilt, *Fusarium moniliforme* disease complex

Treatment of grapevines with *P. fluorescens* at 100g/vine + FYM at 20kg/ vine gave highest reduction in final soil nematode population of *M. incognita* (56.9%), least root gall index (1.8) and least wilt disease (*Fusarium moniliforme*) incidence (24.8%). The bunch weight of grapevine increased by 155.4% compared to untreated control (Senthil Kumar and Rajendran, 2004).

## 16.9. GAUVA, PSIDIUM GUAJAVA

### 16.9.1. High yielding varieties

**Table 16.77.** High yielding varieties of guava

| High yielding varieties | Yield (kg/tree) | Developed by |
|---|---|---|
| TRY (G)-1 | 40.52 (163 MT/ha) | TNAU, Coimbatore |
| Shweta | High-yielding | CISH, Lucknow |
| Arka Amulya | Heavy yield | IIHR, Bangalore |
| Hybrid-1 | High yield | BAC, Sabour |
| Hybrid-2 | High yield | BAC, Sabour |

**Table 16.78.** Guava varieties resistant to pests/diseases

| High yielding varieties | Yield (kg/tree) | Resistance to | Developed by |
|---|---|---|---|
| TRY (G)-1 | 40.52 (163 MT/ha) | Fruit fly, mealy bug (T), scale (T), mite (T), wilt (T) | TNAU, Coimbatore |

### 16.9.2. High density planting

By grafting on dwarfing rootstock Pusa Srijan (Aneuploid No. 82), high density planting can be achieved in guava. A full bearing high density guava orchard on Aneuploid No. 82 rootstock may yield 18 to 20 MT/ha compared to 8-10 MT/ha in low density planting. Planting distance varies from 5.4 to 7.5m. The trials conducted under All-India Co-ordinated Fruit Improvement Project at different centres, viz. Faizabad, Ranchi and Basti indicated that double-hedge-row system of planting with 800-900 plants/ha is beneficial for achieving high yield per unit area in Allahabad Safeda. At the Central Institute for Subtropical Horticulture, Lucknow; planting was tried at four densities e.g. 1.5 x 3.0m (2,222 plants/ha), 3.0 x 3.0m (1,111 plants/ha), 3.0 x 6.0m (555 plants/ha) and 6.0 x 6.0m (277 plants/ha). Average yield of 73, 41, 26 and 13 kg/plant from pruned trees in comparison to 57, 29, 22 and 10 kg/plant under unpruned trees from above densities was obtained.

Number of guava trees under high density orcharding vary from 833 to 2700/ha with fruit yield of 18-58.3 MT/ha (Table 16.79). Meadow orchard of Genteno Prolific guava cultivar has yielded 30.64 MT/ha in second year in Barbados, West Indies.

**Table 16.79.** Yield of guava under high density orcharding

| Guava variety | Rootstock | Spacing (m) | Density (plants/ha) | Yield (MT/ha) | Tree age (years) | Reference |
|---|---|---|---|---|---|---|
| Traditional system | --- | 6 x 6 | 277 | 15.7 | --- | Chundawat *et al.*, 1992 |
| Sardar | Seedling | 2 x 2 | 2500 | 58.3 | 5 | Lal *et al.*, 1996 |

| | | | | | | |
|---|---|---|---|---|---|---|
| Allahabad Safeda | L-49 seedling | 2 x 2 | 2500 | 18.0 | 4 | Kalra *et al.*, 1994 |
| Allahabad Safeda | L-49 seedling | 4 x 2 | 1250 | 47.2 | 4 | Kalra *et al.*, 1994 |
| Allahabad Safeda | L-49 seedling | 6 x 2 | 833 | 25.4 | 4 | Kalra *et al.*, 1994 |
| Allahabad Safeda | Aneuploid No. 82 | 3 x 2 | 1111 | 20.4 | - | Sharma *et al.*, 1992 |
| Allahabad Safeda | --- | 6 x 2 | 833 | 21.9 | 5 | Chundawat *et al.*, 1992 |
| Genteno Prolific | --- | 0.6 x 0.6 | 2700 | 30.6 | 2 | Mohammed *et al.*, 1984 |
| Lucknow-49 | --- | 2 x 2 | 2500 | 30.6 | 4 | Chandra & Govind, 1995 |

Among different planting systems, maximum fruit yield (18.685 MT/ha) was recorded in double hedgerow system (Table 16.80) (Mahajan *et al.*, 2005).

**Table 16.80.** Effect of planting systems on fruit yield

| Planting system | Yield (MT/ha) | | Total yield (MT/ha) |
|---|---|---|---|
| | Rainy season | Winter season | |
| Square system (9 plants/441 $m^2$) | 12.211 | 0.697 | 12.908 |
| Hedgerow system (15 plants/441 $m^2$) | 16.011 | 0.669 | 16.679 |
| Double hedgerow system (20 plants/441 $m^2$) | 17.861 | 0.828 | 18.685 |
| Paired row system (12 plants/441 $m^2$) | 15.454 | 0.685 | 16.139 |
| Cluster system (16 plants/441 $m^2$) | 16.655 | 0.829 | 17.494 |

In meadow orchard, plants spaced at 2.0 x 1.0m (plant density of 5000 plants/ha) gave fruit yield of 12.5 MT/ha in the first year of planting itself. The yield increased up to 48.0 MT/ha after five years of planting in a multi-shoot meadow orchard system (Singh, 2006).

### 16.9.3. Integrated nutrient management

Application of FYM and the bio-fertilizer, *Azotobacter* on 3-year-old guava cv. Allahabad Safeda, gave maximum fruit yield (92 fruits/plant and 13.69 kg/plant) (Table 16.81).

**Table 16.81.** Effect of different organic/biodynamic treatments on yield of guava.

| Treatment ( Dose/tree) | No. of fruits/tree | Fruit yield (kg/tree) |
|---|---|---|
| Vermicompost – 5 kg | 72 | 11.45 |
| Vermiwash + 10 kg FYM | 44 | 5.90 |
| BD 500 + 10 kg FYM | 58 | 8.70 |
| CPP – 500 g | 50 | 8.61 |

| | | |
|---|---|---|
| 20 kg FYM + 250 g *Azospirillum* | 63 | 7.88 |
| 20 kg FYM + 250 g *Azotobacter* | 92 | 13.69 |
| 10 kg FYM + 5 kg Celrich | 61 | 9.66 |
| 350 g N + 200 g $P_2O_5$ + 350 g $K_2O$ | 48 | 8.09 |
| CD (P = 0.0 5) | 24.81 | 5.18 |

The treatment *in situ* vermiculturing at 50 worms per plant recorded maximum guava (cv. Sardar) fruit yield (29.10 kg/plant and 8.10 MT/ha) which was on par with 75% RDF + 10 kg vermicompost/plant, and 100% RDF (Athani *et al.*, 2006a).

### 16.9.4. Bioregulators

Foliar spraying of 2,4-D at 5 and 10 ppm and ethephon at 300 ppm gave maximum fruit yield in rainy season guava crop at Sabour, while NAD at 50 and 75 ppm and NAA at 200 ppm gave higher fruit yield in winter season (Kumar and Hooda, 1977) (Table 16.82).

**Table 16.82.** Effect of growth regulators on fruit yield of guava cv. Allahabad Safeda

| Growth regulators | Fruit yield (kg/tree) (Pooled for 1984 and 1985) | |
|---|---|---|
| | Monsoon season | Winter season |
| NAA 100 ppm | 18.58 | 26.83 |
| NAA 200 ppm | 12.71 | 33.38 |
| 2,4-D 5 ppm | 26.52 | 16.98 |
| 2,4-D 10 ppm | 23.05 | 21.28 |
| NAD 50 ppm | 9.85 | 38.90 |
| NAD 75 ppm | 10.79 | 36.50 |
| Etephon 300 ppm | 20.61 | 24.27 |
| Control | 3.88 | 3.82 |
| CD (P = 0.05) | 1.779 | 1.616 |

Foliar spraying of Gibberellic acid at 150 ppm significantly increased fruit yield (78.90 kg/tree) followed by 0.4% $ZnSO_4$ (64.45 kg/tree) (Kumar and Shukla, 2009).

**Table 16.83.** Intercropping in guava

| Treatment | Crop yield (MT /ha) | Net return (Rs./ha) |
|---|---|---|
| Guava + groundnut | 1.44 | Medium |
| Guava+ soybean | 1.45 | Low |
| Guava+ rice | 1.95 | Low |
| Guava + ginger | 9.28 | High |
| Guava+ turmeric | 10.69 | Medium |

### 16.9.5. Intercropping

The results of an investigation conducted in eastern parts of Himalayas, indicated that the high net return was obtained when guava is grown with ginger. No adverse effect of intercrops on the growth of guava was observed (Tables 16.83 and 16.84).

**Table 16.84.** Economics of comparative performance of various horticultural crop combinations

| Crop combination | Yield (MT/ha) | | Net returns (Rs/ha) |
|---|---|---|---|
| | Main crop | Inter crop | |
| Guava | 4.60 | --- | 21,000 |
| Guava + Cowpea | 4.60 | 5.066 | 33,198 |
| Guava + Okra | 4.50 | 6.328 | 36,484 |
| Guava + Dolichos | 4.50 | 2.583 | 25,249 |
| Guava + Cluster bean | 4.60 | 2.416 | 25,248 |
| Maize + Pigeon pea (Traditional system) | 0.81 | 0.320 | 6,200 |

### 16.9.6. Mulching

Mulching of guava with banana leaf gave highest fruit yield (11.8 kg/plant) followed by black polyethylene film (10.0 kg/plant). The average soil moisture was highest in black polyethylene film (74.3%) followed by paddy straw (71%) and banana leaf (70%) (Garg *et al.*, 2006).

## 16.10. PASSION FRUIT, *PASSIFLORA EDULIS*

### 16.10.1. High yielding varieties

Passion fruit cv. Kaveri (IIHR, Bangalore) is high yielding.

### 16.10.2. Integrated nutrient management

The treatment consisting of 20 kg FYM + *Azospirillum* at 5 g/plant gave the maximum number of fruits/vine (142) and the maximum yield (4 MT/ha). The above treatment also resulted in significant improvement in fruit quality in terms of fruit weight, fruit volume, TSS, reducing sugar and total sugar (Kishore *et al.*, 2006).

### 16.10.3. Integrated disease management

Passion fruit cv. Kaveri (IIHR, Bangalore) is tolerant to collar rot, wilt, brown leaf spot and root-knot nematode.

## 16.11. APPLE, *MALUS PUMILA*

### 16.11.1. High yielding varieties

**Table 16.85.** High yielding varieties of apple

| High yielding varieties | Fruit yield (MT/ha) | Developed at |
|---|---|---|
| Firdous | 7.2 | SKUAST, Srinagar |
| KKL-1 | 22.0 | HRS, TNAU, Kodaikanal |

### 16.11.2. High density planting

There are four categories of high density planting based on plant densities used in apple, i.e. low (less than 250 plants/ha), moderate (250 to 500 plants/ha), high (500 to 1,250 plants/ha) and ultra-high (more than 1250 plants/ha) density (Fig. 16.3). Apart from these, super high density plantings or meadow orcharding (20,000 to 70,000 plants/ha) have been demonstrated in some European countries. The optimum density ranges from over 3,000 trees/ha in the Netherlands to less than 2,000 trees/ha in USA under single and double row system of planting. Use of pruning, rootstocks and spur type scion varieties is done to establish HDP. Other techniques like training and pruning methods or bending of branches and use of growth retardant are also being used.

**Fig 16.3.** High density planting in apple.

The most common dwarfing/semi-dwarfing rootstocks for HDP are M9, M26, M4, M7 and MM106, whereas M27 has been used for raising meadow orchards. In Himachal Pradesh, MM106 and MM109 for spur types and M7 and MM111 for non-spur cultivars are being popularized. Of late, MM 106 has been recommended for establishing HDP in Jammu and Kashmir. At present, non-availability of clonal rootstocks is the major problem in area expansion. For this purpose, stooling and micropropagation techniques are required to be adopted on a large scale to meet such demand. The spur-type varieties are genetically dwarf. Some of the promising spur varieties which need popularization are Starkrimson, Red Spur, Gold Spur, Well Spur, Oregon Spur, Silver Spur, Red Chief and Hardispur. Besides, many other varieties such as Bright 'N' Early, Royal Gala and Red Fuji are promising columnar varieties suited for HDP. Training systems, such as Spindle Bush, Dwarf Pyramid, Espalier, Cordon, Palmette and Noored Holland Spindle Bush are recommended for intensive orchards. On dwarfing rootstocks, plantings are trained on Spindle Bush, Dwarf Pyramid and Cordon in Himachal Pradesh.

Data in Table 16.86 show recommendation of rootstocks and cultivars for specific planting densities. More than 50 MT/ha fruit yield can be obtained with Golden Delicious apple from high density orchards.

**Table 16.86.** Yield of apple under high density planting

| Cultivar | Rootstock | Planting distance (m) | Density (plants/ha) | Yield (MT/ha) | Reference |
|---|---|---|---|---|---|
| Traditional system | --- | 5-8 x 5-8 | 400-227 | 30-50 | Singh, 1995 |
| Jonathan | MM-106 | 3.0 x 7.5 | 2222 | 68.75 | Gher Ghi *et al.*, 1994 |
| Jonathan | M-9 | 2.0 x 1.0 | 5000 | 49.00 | Suta *et al.*, 1986 |
| Goldspur | M-9 | 2.0 x 1.0 | 5000 | 48.00 | Suta *et al.*, 1986 |
| Goldenspur | --- | 3.0 x 2.0 | 1666 | 15.97 | Meng *et al.*, 1997 |
| Golden Delicious | M-4 | 4.0 x 5.0 | 833 | 27.00 | Khan *et al.*, 1998 |
| Golden Delicious | --- | 4.0 x 2.0 | 1250 | 44.10 | Senin & Kovaleva, 1992 |
| Golden Delicious | --- | 4.0 x 1.5 | 1656 | 46.60 | Senin & Kovaleva, 1992 |
| Golden Delicious | --- | 4.0 x 1.0 | 2500 | 53.50 | Senin & Kovaleva, 1992 |
| Golden Delicious | M-27 | 3.5 x 1.75 | 1623 | 47.18 | Antognozzi *et al.*, 1988 |
| Golden Delicious | M-9 | 5.0 x 3.5 | 571 | 52.30 | Ceausescu & Isac, 1986 |
| Oregen Delicious | M-106 | 1.0 x 0.5 | 20000 | 41.00 | Khan *et al.*, 1998 |
| Vance Delicious | MM-106 | 3.0 x 7.5 | 2222 | 76.00 | Awsthi & Mehta, 2000 |
| Fuji | *M. prunifolia* | 5.0 x 3.5 | 571 | 57.00 | Shiozaki *et al.*, 1988 |
| Fuji | M-26 | 5.0 x 3.5 | 571 | 58.00 | Shiozaki *et al.*, 1988 |
| Mutsu | M-26 | 4.0 x 3.5 | 714 | 46.00 | Shiozaki *et al.*, 1988 |
| Antonovka | *M. silvestris* | 4.0 x 3.0 | 832 | 48.20 | Devyatov, 1991 |

| | | | | | |
|---|---|---|---|---|---|
| Starkimson | --- | 4.0 x 2.0 | 1250 | 30.5 | Senin & Kovaleva, 1992 |
| Starkimson | --- | 4.0 x 1.5 | 1656 | 35.50 | Senin & Kovaleva, 1992 |
| Starkimson | --- | 4.0 x 1.0 | 2500 | 42.90 | Senin & Kovaleva, 1992 |

A yield as high as 76 MT/ha has been obtained with a planting density of 2222 trees/ha in Vance Delicious apple on MM 106 rootstock.

### 16.11.3. Fertigation

Klien *et al.* (1989) reported that application of 150 kg N/ha proved best for producing higher fruit yield (54.27 MT/ha), whereas Neilsen *et al.* (2004) stressed the importance of supplying potash via fertigation at 15 g K/tree which increased fruit yield (Table 16.87).

**Table 16.87.** Fruit yield in apple as affected by N and K fertigation

| Treatment | Fruit yield (MT/ha) |
|---|---|
| N fertigation | |
| 50 kg N/ha | 47.72 |
| 100 kg N/ha | 54.27 |
| 150 kg N/ha | 45.06 |
| 200 kg N/ha | 47.13 |
| K fertigation | Fruit yield (kg/tree) |
| 0 kg P/ha | 4.53 |
| 15 kg P/ha | 5.51 |

In the fertigation experiment on young apple trees, 30% of total N, 70% of total P and 40% of total K were applied as a base fertilizer and the rest of NPK was fertigated. The application of the highest rate of K (540 kg $K_2O$/ha) produced the highest cumulative yield of apples for two years compared to the medium (405 kg $K_2O$/ha) and lowest K rates (270 kg $K_2O$/ha), with yield increment by 45% and 63%, respectively. Such apple quality parameters as size and firmness of fruits were also enhanced when 540 and 405 kg $K_2O$/ha were applied as compared with the lowest rate of 270 kg $K_2O$/ha.

### 16.11.4. Integrated pest management

#### Validated IPM technology in apple (Himachal Pradesh)

#### Winter module (October-November)

- Use of 5% urea at leaf shedding stage for early decomposition of the infested leaves and to encourage the population of antagonists in the plant rhizosphere.
- Use of Bordeaux paint during autumn on the naked plant stem to overcome the direct effect of UV rays on the plant skin to reduce sun burn and cankers disease complex.

- Overwinter spray of Bordeaux mixture as eradicative action to pathogens and total disinfection of plant surface.

**Spring module (April-June)**

- Use of Neemarin at pink bud stage i.e. pre-bloom stage to manage blossom thrips population.

**Summer module (July-September)**

- Use of *Bacillus thurigiensis* at fruit development stage for the management of fruit scrapper insect pests.
- Use of *Trichoderma viride* (Bioderma) for the control of root rot fungus.
- Use of Bordeaux mixture for the control of root and collar rots.

During the period 2001-04, IPM for apple crop was validated and promoted in 30 ha of orchards in Kotkhai, Jhubbal, Theneder and Rohroo villages of Himachal Pradesh. By adopting IPM package, farmers were able to harvest 580 boxes (4.99 MT/ha) of apple in IPM plots as compared to 380 boxes (4.12 MT/ha) in case of non-IPM farmers. Apple growers who adopted IPM earned a profit of Rs. 145733/- while the farmers who did not adopt IPM earned a profit of Rs. 110889/-. Average benefit cost ratio of IPM to non-IPM was 4.07 to 3.01 (Table 16.88) (Amerika Singh *et al.*, 2004).

**Table 16.88.** Economics of apple IPM

| Parameters | IPM plots | Non-IPM plots | % increase |
|---|---|---|---|
| Yield (MT/ha) | 4.99 | 4.12 | 21.11 |
| Net profit (Rs/ha) | 145733 | 110889 | 31.42 |
| Benefit: cost ratio | 4.07 | 3.01 | 35.21 |

### 16.11.5. Integrated disease management

Apple cvs. Firdous and Shireen (SKUAST, Srinagar) are reported to be resistant to scab, powdery mildew and *Alternaria* leaf spot; while Akber (SKUAST, Srinagar) to scab and powdery mildew diseases. Apple cvs. Ambstarking and Ambrich (Dr. YSPUHF, Solan) are reported to be tolerant to scab.

## 16.12. PEACH, PRUNUS PERSICA

### 16.12.1. High density planting

Peach bears fruits on one-year-old shoots, which require regular annual pruning to make them suitable for high density orcharding. High density in peach has been tried with planting density varying from 930 to 2,050 trees per hectare yielding 13 to 34 MT/ha following different training systems viz. hedge row, Y shape trellis, spindle, modified free spindle and modified leader. A plant population of 2,000 trees/ha trained on Tatura trellis system gave fruit yield of 31.5 MT/ha (Table 16.89). Increase in plant density from 400 to 5000 trees/ha increased the yield proportionately.

**Table 16.89.** Yield of peach under high density orcharding

| Peach variety | Rootstock | Spacing (m) | Density (plants/ha) | Yield (MT/ha) | Tree age (years) | Reference |
|---|---|---|---|---|---|---|
| Traditional system | --- | 6.0 x 6.0 | 277 | 7.00-10.00 | | Pathak & Pathak, 1991 |
| Yanco green | --- | 3.0 x 3.0 | 1111 | 34.65 | --- | Hutton *et al.*, 1987 |
| Yanco green | --- | 3.1 x 2.3 Hedge row | 1396 | 23.30 | 5 | Reader *et al.*, 1980 |
| Suncrest | Seedling | --- | 2050 | 30.10 | First year | Guerriero *et al.*, 1981 |
| Benishimizu | --- | 4.0 x 2.0 | 1250 | 28.72 | 2-4 | Furukawa *et al.*, 1992 |
| Florida prince | GF-667 | 4.5 x 1.2 Y-shape | 1850 | 32.20 | 3 | Caruso *et al.*, 1999 |
| Florida prince | GF-667 | 4.5 x 2.4 Modified free spindle | 930 | 20.60 | 3 | Caruso *et al.*, 1999 |
| Shan-e-Punjab | --- | 3.0 x 3.0 Modified Leader system | 1111 | 13.60 | 5 | Kanwar *et al.*, 2000 |
| Shan-e-Punjab | --- | 3.0 x 2.0 Modified Leader system | 1667 | 14.20 | 6 | Kanwar *et al.*, 2000 |
| Shan-e-Punjab | --- | 2.5 x 2.0 Tatura trellis system | 2000 | 31.50 | --- | Awasthi & Mehta, 2000 |
| Shan-e-Punjab | --- | 2.5 x 2.0 Bed system | 2000 | 32.50 | --- | Awasthi & Mehta, 2000 |

Peach planting at a spacing of 5 x 1m under Tatura Trellis system of training gave 3 times more yield of better quality fruits in comparison to traditional system of planting.

## 16.13. PLUM, *PRUNUS SALICINA*

### 16.13.1. High density planting

Plum requires annual pruning for growth and fruiting which makes it suitable for high density planting. Fruit yields of 14-34 MT/ha were obtained in plantings having 825-1250 trees/ha (Table 16.90).

**Table 16.90.** Yield of plums under high density planting

| Cultivar | Rootstock | Density (plants/ha) | Yield (MT/ha) | Tree age (years) | Reference |
|---|---|---|---|---|---|
| Sanctus Hubertus | St. Julien | 1250 | 34 | 2-4 | Wustenbergh & Keulemans, 1997 |
| Sanctus Hubertus | St. Julien | 825 | 26 | 2-4 | Wustenbergh & Keulemans, 1997 |
| Czar | St. Julien | 1250 | 17 | 2-4 | Wustenbergh & Keulemans, 1997 |
| Czar | St. Julien | 825 | 14 | 2-4 | Wustenbergh & Keulemans, 1997 |
| Premier | St. Julien | 5000 | 15 | 3 | Wustenbergh & Keulemans, 1997 |
| Methley | St. Julien | 5000 | 15 | 3 | Gosh, 1986 |
| Kabullireengnage | St. Julien | 864 | --- | 4 | Balakrishna *et al.*, 1998 |

## 16.14. PEAR, *PYRUS COMMUNIS*

### 16.14.1. High yielding varieties

**Table 16.91.** High yielding varieties of pear

| High yielding varieties | Fruit yield (kg/tree) | Developed at |
|---|---|---|
| Punjab Nector | 80 | PAU, Ludhiana |
| Punjab Gold | 80 | PAU, Ludhiana |
| Punjab Soft | 86 | PAU, Ludhiana |

### 16.14.2. High density planting

In pear, hedge row system gives a plant density of 888 plants/ha and causes significant increment in yields in Bagugosha pear.

## 16.15. APRICOT

### 16.15.1. High density planting

High density orcharding in apricot has great promise giving fruit yield of 20.6 to 120 MT/ha when planted at 1250-2367 trees/ha (Table 16.92).

**Table 16.92.** Yield of apricot under high density planting

| Cultivar | Rootstock | Planting distance (m) | Density (plants/ha) | Yield (MT/ha) | Reference |
|---|---|---|---|---|---|
| Traditional system | --- | 6.0 x 6.0 | 277 | 13-22 | Rathore, 1991 |
| Traditional system | --- | 4.0 x 2.0 | 1250 | 28.3 | Bunea *et al.*, 1983 |

| | | | | | |
|---|---|---|---|---|---|
| Rougede Raussibon | PG 88-2 | 3.0 x 2.0 | 1666 | 20.6 | Lichou *et al.*, 1989 |
| Rougede Raussibon | PG 88-2 | 3.0 x 1.5 | 222 | 27.2 | Lichou *et al.*, 1989 |
| Regina | GM-61 | 4.0 x 2.0 | 1250 | 10.0 | Ghiesen, 1990 |
| San Castrese | Myrobalon | --- | 2222 | 84.0 | Guerriore & Scalabrelli, 1981 |
| San Castrese | Myrobalon | --- | 2367 | 120.0 | Monastra *et al.*, 1999 |
| Tyrinthos | Peach seedling | 4.5 x 1.5 | 1481 | 64.4 | Monastra *et al.*, 1999 |
| Tyrinthos | Peach seedling | 4.5 x 1.5 | 1481 | 36.2 | Mcland & Ystaas, 1998 |

## 16.16. LITCHI, *LITCHI CHINENSIS*

### 16.16.1. High yielding varieties

**Table 16.93.** Increased productivity in high yielding varieties of litchi

| High yielding variety | Productivity (kg/tree) | Developed at |
|---|---|---|
| *Sabour Madhu* | 70 | RFRS, Sabour |
| *Rajendra Hybrid* | 80 – 100 | RAU, Pusa |
| *Rajendra Sabour Bedana* | 80 – 100 | RAU, Pusa |

### 16.16.2. High density planting

Double hedge row planting with 222 litchi trees/ha produced the highest yield of 20.3 MT/ha as against 11.5 MT/ha with 100 tress/ha under square system of planting.

### 16.16.3. Micro irrigation

Hasan and Chattopadhaya (1990) obtained the maximum fruit yield of 61.6 kg/plant in litchi cv. Bambai by irrigation at 30% depletion of available soil moisture (ASM) giving 6.22 cm water at each irrigation followed by 59.7 kg/plant by irrigation at 45% ASM giving 9.33 cm water at each irrigation (Table 16.94).

**Table 16.94.** Effect of irrigation levels on fruit yield in litchi

| % depletion of available soil moisture | Fruit yield (kg/tree) | % increase |
|---|---|---|
| 30 | 61.6 | 71.86 |
| 45 | 59.7 | 66.29 |
| 60 | 46.6 | 27.02 |
| No irrigation | 35.9 | --- |

Micro sprinkler + drip irrigation gave maximum litchi fruit yield of 3.74 MT/ha as compared to 2.71 MT/ha in surface irrigation. The significant increase in yield in the above treatment might be due to frequent irrigation, uniform moisture distribution in the soil profile and better utilization of nutrients by the crop for its full growth (Bansod and Shukla, 2006b).

### 16.16.4. Integrated nutrient management

Litchi plants treated with *Azotobacter* + AMF each at 150 g/plant gave maximum fruit yield (98.42 kg/plant), TSS (20.8° Brix), total sugars (14.80% fresh weight), ascorbic acid (37.42 mg/100 g) and anthocyanin (23.70 mg/100 g peel) (Dutta *et al.*, 2009).

### 16.16.5. Bioregulators

Foliar spray of NAA at 40 mg/L helped in maximum retention of fruits (47.12%), gave highest fruit yield (110.53 kg/plant) and highest benefit-cost ratio (5.64) as compared to control (2.78) in litchi cv. Bombai (Sarkar and Ghosh, 2009).

## 16.17. STRAWBERRY, *FRAGARIA VESCA*

### 16.17.1. Fertigation

Among the N levels, the highest fruit yield was obtained with 7.2 m mol N/litre owing to increases in both fruit weight/plant (465 g) and number of fruits/plant (35.5). Among the N sources, the highest yield was obtained with urea owing to better fruit setting (Table 16.95) (Papadopoulos, 1987).

**Table 16.95.** Nitrogen fertigation of strawberries under greenhouse

| Treatment | Fruit yield/plant (g) |
|---|---|
| Level of N (m mol/L) | |
| 3.6 | 405 |
| 7.2 | 465 |
| 10.8 | 376 |
| Form of N ($NH_4/NO_3$) (m mol/L) | |
| 7/0 | 650 |
| 3.5/3.5 | 547 |
| 0/7 | 568 |

### 16.17.2. Mulching

Studies conducted at the Y.S. Parmar University of Horticulture and Forestry, Nauni, Solan (Himachal Pradesh) revealed that black polythene mulching gave 55% increase in strawberry yield as compared to unmulched plot.

In Himachal Pradesh, mulching with polythene and pine needles at 10 MT/ha increased the fruit yield by 68 and 33%, respectively over that in control in the strawberry cv. Tioga (Badiyala and Aggarwal, 1981). It also induced early ripening and improved fruit quality.

### 16.17.3. Integrated weed management

Studies conducted at the Y.S. Parmar University of Horticulture and Forestry, Nauni, Solan (Himachal Pradesh) revealed that black polythene mulching gave 85-90% weed control in strawberry.

### 16.17.4. Integrated disease management

#### i) Fruit rot, *Phytophthora cactorum*

Maximum disease control was achieved in plots which were mulched with polythene and sprayed with ridomil MZ (74.47%) followed by paddy straw mulch and ridomil MZ sprays (64.92%). The fruit yield of strawberry also increased significantly in combined treatment of polythene mulch and ridomil MZ (97.57%) followed by paddy straw mulch and ridomil MZ (90.04%) (Table 16.96) (Bharadwaj and Sharma, 2000).

**Table 16.96.** Effect of fungicide alone and in combination with mulch on incidence of fruit rot and yield of strawberry

| Treatment | Disease incidence (%) | Fruit rot control (%) | Fruit yield (kg/plot) | % increase in fruit yield |
|---|---|---|---|---|
| Polythene mulch | 18.43 | 54.36 | 1.600 | 71.99 |
| Ridomil MZ | 24.67 | 38.92 | 1.258 | 35.23 |
| Paddy straw mulch | 28.02 | 30.63 | 1.056 | 13.51 |
| Polythene mulch + Ridomil MZ | 10.31 | 74.47 | 1.838 | 97.57 |
| Paddy straw mulch + Ridomil MZ | 14.17 | 64.92 | 1.768 | 90.04 |
| Control | 40.39 | --- | 0.930 | --- |
| CD (P = 0.05) | 5.025 | 14.263 | --- | --- |

## 16.18. SWEET CHERRY, *PRUNUS AVIUM, P. CERASUS*

### 16.18.1. High density planting

Studies on high density orcharding in sweet cherry revealed that as high as 16 MT/ha fruit yield could be obtained which is about double to the traditional planting (Table 16.97).

**Table 16.97.** Yield of sweet cherry under high density planting

| Cultivar | Rootstock | Spacing (m) | Density (plants/ha) | Yield (MT/ha) | Reference |
|---|---|---|---|---|---|
| Traditional system | --- | 8 x 5 | 250 | 6.3-9.0 | Teskey & Shoemaker, 1972 |

| | | | | | |
|---|---|---|---|---|---|
| Van | Damil | --- | 5000 | 14.50 | Giesen, 1990 |
| Van | Damil | 4 x 2 | 1250 | 16.00 | Jacyna *et al.*, 1992 |
| Regina | GM-61 | 4 x 2 | 1250 | 12.40 | Jacyna *et al.*, 1992 |
| Regina | GM-61 | 4 x 2 | 1250 | 11.25 | Jacyna *et al.*, 1992 |
| Bing | Mahaleb | --- | 1333 | 4.90 | Jacyna *et al.*, 1992 |

## 16.19. ALMOND

### 16.19.1. High yielding varieties

**Table 16.98.** High yielding varieties of almond

| High yielding varieties | Yield (kg/ha) | Developed at |
|---|---|---|
| Makhdoom | 500 | SKUAST, Srinagar |
| Parbat | 500 | SKUAST, Srinagar |
| Waris | 600-700 | SKUAST, Srinagar |
| Shalimar | 500 | SKUAST, Srinagar |

## 16.20. POMEGRANATE, *PUNICA GRANATUM*

### 16.20.1. High yielding varieties

**Table 16.99.** High yielding varieties of pomegranate

| High yielding varieties | Fruit yield (kg/tree) | Developed by |
|---|---|---|
| Ganesh | 12.21 | MPKV, Rahuri |
| G-137 | 11.97 | MPKV, Rahuri |
| Mridula | 14.36 | MPKV, Rahuri |
| Phule Arakta | 29.83 (30-35) | MPKV, Rahuri |
| Bhagwa | 30.38 (30-40) | MPKV, Rahuri |
| Ruby | 16-18 MT/ha | IIHR, Bangalore |
| CO 1 | 17 | TNAU, Coimbaore |
| YCD 1 | 25.9 | HRS, TNAU, Yercaud |

### 16.20.2. High density planting

Normally pomegranate is planted at 5 x 5 m or 6 m x 4 m spacing. High density planting has been reported suitable in pomegranate for high yield when planting is done at 5 m x 2 m spacing in cv. Ganesh. High density plantation at 5 x 2 m spacing with higher water and fertilizer input (22 irrigations/year and 625: 250: 250 g NPK/plant) under semi-arid agroclimate receiving 800 mm rainfall gave nearly 2.5 times fruit yield of that obtained at the normal spacing (Table 16.100).

**Table 16.100.** Effect of high density planting on fruit yield of pomegranate cv. Ganesh

| Spacing (m) | Fruit yield (MT/ha) |
|---|---|
| 5 x 2 | 8.57 |
| 5 x 3 | 6.20 |
| 5 x 4 | 3.84 |
| 5 x 5 | 3.66 |
| C.D (P = 0.05) | 2.53 |

### 16.20.3. Micro irrigation

The fruit yield and water use efficiency were considerably increased by trickle irrigation in Ganesh pomegranate at Rahuri in Maharashtra state (Anon, 1987) (Table 16.101).

**Table 16.101.** Effect of method of irrigation on yield of Ganesh pomegranate

| Method of irrigation | Fruit yield (MT/ha) | Water used (litres) Mar – June | Water use efficiency (q/ha cm) |
|---|---|---|---|
| Check basin – 0.6 IW/CPE | 4.06 | 4800 | 0.82 |
| Check basin – 0.8 IW/CPE | 4.75 | 6600 | 0.69 |
| Trickle – daily, 20% wetted area | 5.25 | 3580 | 1.25 |
| Trickle – daily, 30% wetted area | 4.49 | 5322 | 0.72 |
| Trickle – alternate day, 20% wetted area | 4.86 | 3580 | 1.16 |
| Trickle – alternate day, 30% wetted area | 5.73 | 5322 | 0.92 |

The maximum fruit yield (4.25 MT/ha) was recorded in 0.75 CPE through drip system which was on par with 1.00 CPE through micro sprinkler and drip system. Considering the water used and yield of the crop, the maximum water efficiency (3.02 q/ha-cm) was estimated under drip system which was higher by 12, 23 and 25% over micro sprinkler, bubbler and open channel irrigation system, respectively. Irrigation through drip system had better water use efficiency, saved 25% water and produced 16, 45 and 62% higher fruit yield over micro sprinkler, bubbler and open channel irrigation system, respectively (Sharma *et al.*, 2006a).

Irrigation at regular turn + mulching with sugarcane trash significantly increased fruit yield in Ganesh pomegranate at Rahuri in Maharashtra (Anon, 1987) (Table 16.102).

**Table 16.102.** Effect of irrigation interval, mulching and anti-transpirant spray on the productivity of Ganesh pomegranate

| Treatment | Fruit yield (kg/tree) |
|---|---|
| Irrigation at regular turn | 8.32 |
| Irrigating half basin at one time and the other half at next | 6.68 |

| | |
|---|---|
| turn | |
| Irrigation at regular turn + mulching with sugarcane trash | 9.39 |
| Irrigation at alternate turn | 5.33 |
| Irrigation at alternate turn + mulching with sugarcane trash | 5.80 |
| Irrigation at alternate turn + 5% kaolin spray | 8.49 |
| CD (P = 0.05) | 1.98 |

### 16.20.4. Fertigation

The yield obtained with 100% NPK (solid soluble fertilizers) + drip irrigation (11.88MT/ha) was significantly higher than 100% NPK (conventional fertilizers) + drip irrigation (9.54MT/ha) and was at par with 70% NPK (solid soluble fertilizers) + drip irrigation (10.81MT/ha). Hence fertigation with 70% NPK (solid soluble fertilizers) + drip irrigation resulted in 36% more fruit yield and 30% fertilizer savings over control (Table 16.103) (Firake and Kumbhar, 2002).

**Table 16.103.** Fertigation in pomegranate

| Treatment | Fruit yield (MT/ha) |
|---|---|
| Conventional fertilizers | |
| 100% RDF + Surface irrigation | 8.71 |
| 100% RDF + Drip irrigation | 9.54 |
| 100% RDF + N drip | 10.36 |
| Solid soluble fertilizers | |
| 100% NPK + Drip irrigation | 11.88 |
| 70% N, 80% P&K + Drip irrigation | 11.27 |
| 70% NPK + Drip irrigation | 10.81 |
| 50% N, 80% P&K + Drip irrigation | 9.14 |
| 50% N, 70% P&K + Drip irrigation | 8.90 |

Firake and Deolanker (2000) observed maximum number of fruits/plant (45.33) and fruit yield (4.95 MT/ha) in the treatment with 100% recommended dose of solid soluble fertilizer + microsol (micronutrients), as compared to control (38.3 fruits/plant and 3.85 MT/ha). They concluded that the fertigation to pomegranate at 75% of recommended dose (625 N, 250 $P_2O_5$ and 250 $K_2O$ g/plant) of NPK through soluble fertilizers was as good as 100% recommended dose of conventional fertilizers. Application of microsol increased the yield by 9.65% when applied with the best treatment.

### 16.20.5. Integrated nutrient management

Application of organic N through FYM (50% N) + castor cake (50% N) gave maximum fruit retention (47.75 fruits/plant) and fruit yield (8.84 kg/plant) compared to 100% N applied through urea (fruit retention of 32.75 fruits/plant and fruit yield of 5.80 kg/plant) (Table 16.104).

**Table 16.104.** Effect of organic and inorganic N application on yield of pomegranate.

| Treatment | No. of fruits retained/plant | Fruit yield (kg/plant) |
|---|---|---|
| FYM (100%) | 46.00 | 8.23 |
| Castor cake (100%) | 46.00 | 8.20 |
| FYM (50%) + Castor cake (50%) | 47.75 | 8.84 |
| Urea (100%) | 32.75 | 5.80 |
| CD at 5% | 12.05 | 2.19 |

Application of boron at 0.2% produced maximum fruit yield of 25.3 kg/tree in 2004 and 31.4 kg/tree in 2005 and improved the quality in terms of TSS, acidity and colour of fruits and arils. Minimum cracking was also observed when boron was applied at 0.2% concentration on early *Mrig bahar* flowering (Singh and Kingsly, 2006).

## 16.21. BER, *ZIZIPHUS MAURITIANA*

### 16.21.1. High yielding varieties

**Table 16.105.** Increased productivity in high yielding varieties of ber

| High yielding variety | Yield (kg/tree) |
|---|---|
| Gola | 100-115 |
| Seb | 100 |
| Kaithli | 130 |
| Umran | 150-200 |
| Sanaur-2 | 200-250 |
| Sanaur-5 | 100-125 |
| Ilachi | 115 |
| ZG-2 | 150-200 |

### 16.21.2. Rootstocks

The maximum fruit yield in ber cv. Umran was observed in *Z. mauritiana* var. Tikadi rootstock, while Gola gave maximum fruit yield on *Z. rotundifolia* rootstock (Table 16.106).

**Table 16.106.** Effect of rootstocks on yield of ber

| Rootstock | Fruit yield (kg/tree) | |
|---|---|---|
| | Umran | Gola |
| *Z. mummularia* | 10.80 | 25.70 |
| *Z. rotundifolia* | 20.50 | 44.50 |
| *Z. mauritiana* var. Sukhawani | 12.30 | 28.40 |

| | | |
|---|---|---|
| *Z. mauritiana* var. Tikadi | 27.50 | 30.90 |
| Mean | 17.77 | 32.30 |

### 16.21.3. Fertigation

The maximum fruit yield (5.85 MT/ha) was recorded in 75% of recommended dose of nitrogen with 1.00 CPE irrigation through drip which was on par with 75% of recommended dose of nitrogen with 0.75 CPE level of drip irrigation (5.55 MT/ha). Considering the fruit yield and N applied, maximum N use efficiency (3.58 q/ha-cm) was estimated in 75% of recommended dose of N with 0.75 CPE level of drip irrigation which was 35% and 50% higher over 50% of recommended dose of N and control (no N), respectively (Sharma *et al.*, 2006b).

### 16.21.4. Intercropping

Intercropping ber with okra did not affect the fruit yield of ber and gave maximum net returns (Table 16.107).

**Table 16.107.** Economics of comparative performance of various horticultural crop combinations

| Crop combination | Yield (MT/ha) | | Net returns (Rs/ha) |
|---|---|---|---|
| | Main crop | Inter crop | |
| Ber | 10.85 | --- | 51,250 |
| Ber + Cowpea | 10.88 | 1.735 | 53,605 |
| Ber + Okra | 10.79 | 4.810 | 62,380 |
| Ber + Dolichos | 10.81 | 1.373 | 52,169 |
| Ber + Cluster bean | 11.02 | 1.435 | 53,405 |
| Maize + Pigeon pea (Traditional system) | 0.81 | 0.320 | 6,200 |

### 16.21.5. Mulching

In low rainfall areas of Gujarat, black polythene mulch gave the highest fruit yield of 63.1 kg/tree in Umran ber (compared to 52.6 kg/tree in the control) as a result of increased soil moisture conservation (Anon, 1989).

### 16.21.6. Integrated pest and disease management

Two sprays of dimethoate (0.05%) at monthly interval commencing from 21 days after pruning followed by an application of endosulfan (0.07%) at triweekly interval during flowering and subsequent sprays of fenvalerate (0.005%) given along with carbendazim (0.1%) followed by sulfex (0.25%) at 10 days interval during fruiting time effectively reduced the incidence of fruit fly (*Carpomyia vesuviana*), fruit borer (*Meridarchis scyrodes*) and powdery mildew infesting ber cv. Umran. A significant increase to the tune of 13% in the marketable yield of ber was also recorded in treated plants (Bagle *et al.*, 2006).

## 16.22. AMLA, *PHYLLANTHUS EMBLICA*

### 16.22.1. High yielding varieties

**Table 16.108.** High yielding varieties of amla

| High yielding varieties | Fruit yield (kg/tree) | Developed at |
|---|---|---|
| Chakaiya | 31.60 | |
| Krishna | 41.80 | NDUAT, Faizabad |
| Kanchan | 45.75 | NDUAT, Faizabad |
| NA-6 | 36.20 | NDUAT, Faizabad |
| NA-7 | 56.70 | NDUAT, Faizabad |
| NA-10 | 38.90 | NDUAT, Faizabad |
| BSR 1 | 155 | ARS, TNAU, Bhavanisagar |

### 16.22.2. High density planting

Amla is generally planted at a distance of 6-7 m. However, by systematic pruning, the grafted trees of high yielding varieties can be maintained under HDP. High yielding varieties such as NA 7, Krishna and Chakia can successfully be grown under HDP concept at a spacing of 3m x 3m under hexagonal or triangular systems accommodating over 1,200 plants per ha.

### 16.22.3. Intercropping

Intercropping in fruit orchards increased the income substantially. Okra was found as the most suitable *kharif* intercrop for income generation. The amla + okra crop combination gave the highest net return of Rs.72,502/ha (Raturi and Hiwale, 1999) (Table 16.109).

**Table 16.109.** Economics of comparative performance of various horticultural crop combinations

| Crop combination | Yield (MT/ha) | | Net returns (Rs/ha) |
|---|---|---|---|
| | Main crop | Inter crop | |
| Amla | 5.57 | --- | 53,700 |
| Amla + Cowpea | 5.52 | 5.242 | 65,926 |
| Amla + Okra | 5.64 | 7.034 | 72,502 |
| Amla + Dolichos | 5.48 | 2.460 | 57,180 |
| Amla + Cluster bean | 5.59 | 2.243 | 57,629 |
| Maize + Pigeon pea (Traditional system) | 0.81 | 0.320 | 6,200 |

### 16.22.4. Mulching

Studies conducted in Orissa University Agriculture & Technology, Bhubaneshwar (Orissa) revealed that mulching with black polythene gave 49% increase in yield as compared to control and 32% weed control.

## 16.23. CUSTARD APPLE, *ANNONA* SPP.

### 16.23.1. High yielding varieties

**Table 16.110.** High yielding varieties of custard apple

| High yielding varieties | Fruit yield (MT/ha) | Developed at |
|---|---|---|
| Arka Sahan | 12.0 | IIHR, Bangalore |
| APK(Ca) 1 | 7.3 | TNAU, Coimbatore |

## 16.24. FIG, *FICUS CARICA*

### 16.24.1. High yielding varieties

**Table 16.111.** High yielding varieties of fig

| High yielding varieties | Fruit yield (fruits/tree) | Developed at |
|---|---|---|
| YCD-1 Timla Fig | 4,000 | HRS, TNAU, Yercaud |

## 16.25. TAMARIND, *TAMARINDUS INDICUS*

### 16.25.1. High yielding varieties

**Table 16.112.** High yielding varieties of tamarind

| High yielding varieties | Fruit yield (kg/tree) | Developed at |
|---|---|---|
| PKM (MT)-1 | 263.3 (11.85 MT/ha) | HC&RI, TNAU, Periyakulam |

## 16.26. DATE PALM, *PHOENIX DACTYLIFERA*

### 16.26.1. High yielding varieties

**Table 16.113.** High yielding varieties of date palm

| High yielding varieties | Fruit yield (kg/tree) |
|---|---|
| Halaway | 80-100 |
| Barhee | 80-120 |
| Medjool | 80-100 |
| Zahidi | 125 |
| Shamran | 100 |

## 16.27. JACK FRUIT, *ARTOCARPUS HETEROPHYLLUS*

### 16.27.1. High yielding varieties

**Table 16.114.** High yielding varieties of jack fruit

| High yielding varieties | Yield (kg/tree) | Developed by |
|---|---|---|
| PLR 1 | 900 (80 fruits/tree) | TNAU, Coimbatore |
| PPI 1 | 1818 (107 fruits/tree) | TNAU, Coimbatore |
| PLR(J) 2 | 1600-1950 (95-110 fruits/tree) | TNAU, Coimbatore |

Chapter 17

# VEGETABLE CROPS

China is the largest producer of vegetables contributing 47.0% of the total world production (Table 17.1). Vegetables occupy an area of 23.936 million ha in China with a production of 448.983 million MT. Productivity of vegetable crops is highest (33.0 MT/ha) in Spain.

**Table 17.1.** Major vegetable producing countries of the world (2007-08) (Bijay Kumar, 2009)

| Country | Area (in '000 ha) | Production (in '000 MT) | Productivity (in MT/ha) | % share of world production |
|---|---|---|---|---|
| China | 23936 | 448983 | 19 | 47 |
| India | 7803 | 125887 | 16 | 13 |
| USA | 1333 | 38075 | 29 | 4 |
| Turkey | 996 | 24454 | 25 | 3 |
| Russia | 970 | 16516 | 17 | 2 |
| Egypt | 598 | 16041 | 27 | 2 |
| Iran | 641 | 15993 | 25 | 2 |
| Italy | 528 | 13587 | 26 | 1 |
| Spain | 379 | 12676 | 33 | 1 |
| Japan | 433 | 11938 | 28 | 1 |
| Others | 16957 | 222625 | 13 | 24 |
| Total | 54573 | 946774 | | |

*Source:* FAO (except Indian data)
(Indian data source: Indian Horticulture Database 2008)

India is the second largest producer of vegetables next only to China contributing 13.0% of the total world production. Vegetables occupy an area of 7.803 million ha with a production of 125.887 million MT (Bijay Kumar, 2009) (Tables 17.2 and 17.3). Vegetable crops contribute 58.93% share to the total production of horticultural crops. India produces 18.0% of the world onions. In vegetables, India occupies the first position in peas; second in cabbage, cauliflower, onion, brinjal; third position in potato and tomato. The per capita availability of vegetables has increased from 76 g in 1951 to 256 g in 2001 as against 280 g/day recommended by ICMR for a balanced diet. Increase in production is mainly due to increase in productivity. Productivity is still lower in India than the world average productivity. There is scope for increasing the productivity of vegetables in India further. The vegetable production target for 2020 has been fixed at 150 million MT.

**Table 17.2.** Crop-wise area, production and productivity of vegetable crops in India (2007-08) (Bijay Kumar, 2009)

| Vegetable crops | Area ('000 ha) | Production ('000 MT) | Productivity (MT/ha) | % share of India production |
|---|---|---|---|---|
| Potato | 1786 | 34463 | 19.3 | 27.4 |
| Onion | 805 | 12157 | 15.1 | 9.7 |
| Tomato | 572 | 10261 | 17.9 | 8.2 |
| Brinjal | 566 | 9596 | 17.0 | 7.6 |
| Tapioca | 270 | 9054 | 33.5 | 7.2 |
| Cabbage | 265 | 5888 | 22.2 | 4.7 |
| Cauliflower | 321 | 5797 | 18.1 | 4.6 |
| Okra | 409 | 4193 | 10.3 | 3.3 |
| Peas | 314 | 2560 | 8.2 | 2.0 |
| Sweet Potato | 126 | 1146 | 9.1 | 0.9 |
| Others | 2370 | 30774 | 13.0 | 24.4 |
| Total | 7803 | 125887 | 16.1 | |

(*Source:* Indian Horticulture Database 2008)

**Table 17.3.** State-wise area, production and productivity of vegetable crops in India (2007-08) (Bijay Kumar, 2009)

| State | Area ('000 ha) | Production ('000 MT) | Productivity (MT/ha) | % share of India production |
|---|---|---|---|---|
| West Bengal | 1313.2 | 22456.8 | 17.1 | 17 |
| Uttar Pradesh | 960.9 | 19790.3 | 20.6 | 15 |
| Bihar | 823.7 | 14067.7 | 17.1 | 11 |
| Orissa | 660.8 | 1214.6 | 12.4 | 7 |
| Tamil Nadu | 262.7 | 7975.7 | 30.4 | 6 |
| Gujarat | 411.7 | 7402.9 | 18.0 | 6 |
| Maharashtra | 455.3 | 6454.8 | 14.2 | 5 |
| Karnataka | 406.0 | 5030.9 | 12.4 | 4 |
| Andhra Pradesh | 284.1 | 4769.6 | 16.8 | 4 |
| Assam | 328.9 | 4474.2 | 13.6 | 4 |
| Jharkhand | 238.9 | 3639.7 | 15.2 | 3 |

| | | | | |
|---|---|---|---|---|
| Kerala | 166.9 | 3479.7 | 20.8 | 3 |
| Haryana | 274.7 | 3277.1 | 11.9 | 3 |
| Others | 915.2 | 2185.3 | 2.3 | 12 |
| Total | 7803 | 125887 | 16.1 | |

Potato, tomato, brinjal, chilli, cabbage, cauliflower, onion, peas, gourds, okra and melons are now grown on commercial scale with increasing demand, better output, change in production system and development of suitable cvs. These crops have high potential for foreign exchange earning. Today the emphasis on vegetable breeding is on value addition, disease resistance and better quality.

The important vegetables exported include onions-7.1%, cabbage-4%, cauliflower-4%, okra-3% and peas-3%. Traditional vegetables like onion, potato, okra, bottle gourd and chilli are exported to Malaysia, Singapore, Gulf, Sri Lanka, Bangladesh, Pakistan and Nepal. Non-traditional vegetables like asparagus, cabbage, capsicum, sweet corn, baby corn, green peas, French bean, cucumber, gherkin, cherry tomato, Brussels sprout, broccoli, processed and frozen French bean and lima bean are exported to European countries, Australia, Gulf and S.E. Asian countries. French bean, lima bean, sweet corn and capsicum are exported to Gulf. Some of the new products introduced in export market are tomato paste in bulk aseptic packs, processed button mushrooms, freeze dried and instant frozen vegetables. Dehydrated onion and tomato products have great potential as export commodities.

## 17.1. POTATO, *SOLANUM TUBEROSUM*

### 17.1.1. High yielding varieties

**Table 17.4.** High yielding varieties of potato

| High yielding varieties | Tuber yield (MT/ha) | Developed by |
|---|---|---|
| CO Shimla potato | 12 | TNAU, Coimbatore |
| CO 1 | 25-30 | TNAU, Coimbatore |
| CO 2 | 22.65 | TNAU, Coimbatore |

**Table 17.5.** High yielding varieties resistant to pests/diseases developed at Central Potato Research Institute, Shimla

| High yielding varieties | Resistance to |
|---|---|
| Kufri Kuber | PLRV, PVY |
| Kufri Kumar | Wart, Charcoal rot |
| Kufri Kundan | Charcoal rot |
| Kufri Neela | Cyst nematodes |
| Kufri Sindhuri | Early blight (MR), PLRV (T) |
| Kufri Alankar, Kufri Naveen, Kufri Neelmani, Kufri Arun, Kufri Shailja, | Late blight (MR) |

| | |
|---|---|
| Kufri Himalini, Kufri Khyati, Kufri Jeevan | |
| Kufri Chamatkar | Early blight, Charcoal rot, Wart |
| Kufri Khasigaro | Late blight (MR), Early blight (MR) |
| Kufri Sheetman | Charcoal rot (MR) |
| Kufri Muthu, Kufri Jawahar, Kufri Sutlez, Kufri Pukhraj, Kufri Anand, Kufri Kanchan, Kufri Jyoti | Late blight (MR), Wart |
| Kufri Badshah | Late blight (MR), Early blight (MR), PVX |
| Kufri Bahar | Wart |
| Kufri Lalima | Early blight (MR), PVX |
| Kufri Sherpa | Late blight (MR), Early blight (MR), Wart |
| Kufri Swarna | Late blight (MR), Early blight (MR), Wart, Cyst nematodes |
| Kufri Megha, Kufri Chipsona 2, Kufri Himsona, Kufri Sadabahar, Kufri Girdhari | Late blight |
| Kufri Pushkar | Late blight, Wart |
| Kufri Surya | Hopper burn, Wart |
| Kufri Jyoti, Kufri Kuber, Kufri Lalima, Kufri Sindhuri, Kufri Ashoka | PVX, PVY, PLRV |

### 17.1.2. Micro irrigation

Irrigation at weekly intervals gave potato tuber yield of 31.51 MT/ha which was significantly higher than that of one irrigation after earthing up (23.65 MT/ha) and irrigation at 30 days interval (22.63 MT/ha). The above treatment was on par with three irrigations after earthing up and five irrigations after earthing up (Singh *et al.*, 2006).

### 17.1.3. Fertigation

Application of 100 : 100 : 112 kg NPK/ha in the form of water soluble fertilizers through fertigation at fortnightly intervals is recommended for getting maximum marketable yield of potato and high returns (Khan *et al.*, 2001) (Table 17.6).

**Table 17.6.** Productivity and economics of potato cv. Kufri Jyothi as influenced by sources and level of fertilizers

| Treatment | Fruit yield (MT/ha) | Net income (Rs.) | Benefit: cost ratio |
|---|---|---|---|
| Normal fertilizers 100% + Furrow irrigation | 18.94 | 32,583 | 2.06 |
| Normal fertilizers 100% + Drip irrigation | 19.42 | 33,604 | 2.11 |
| Normal fertilizers 60% Fertigation | 17.85 | 29,757 | 2.04 |

| | | | |
|---|---|---|---|
| Normal fertilizers 80% Fertigation | 19.16 | 33.232 | 2.14 |
| Normal fertilizers 100% Fertigation | 20.04 | 36,950 | 2.24 |
| Water soluble fertilizers 60% Fertigation | 20.35 | 32,003 | 1.95 |
| Water soluble fertilizers 80% Fertigation | 22.55 | 38,256 | 2.06 |
| Water soluble fertilizers 100% Fertigation | 23.40 | 38,742 | 2.01 |

The highest yield of fresh tubers of 30.13 MT/ha was obtained with trickle fertigated crop (Table 17.7) (Chawla and Narda, 2000).

**Table 17.7.** Fertigation in potato

| Treatment | Tuber yield (MT/ha) | % increase |
|---|---|---|
| Trickle fertigation | 30.12 | 40 |
| Furrow irrigation | 21.50 | --- |

### 17.1.4. Integrated nutrient management

Sharma *et al.* (1980) showed an increase of 2.8 and 12.6% in potato tuber yield with 40 and 80 kg N/ha, respectively, through neem cake coated urea as compared to uncoated urea. Besides, neem cake coated urea also economized the dose of N by 25%. Likewise, Sharma *et al.* (1986) at Palampur found that neem coated urea increased potato yield by 16% (Table 17.8).

**Table 17.8.** Relative efficiency of urea and neem coated urea on productivity of potato

| Location | N (kg/ha) | Yield (MT/ha) | | % increase |
|---|---|---|---|---|
| | | Urea | Neem coated urea | |
| CPRI, Shimla | 40 | 19.79 | 20.35 | 2.8 |
| | 80 | 22.92 | 25.81 | 12.6 |
| HPKVV, Palampur | 120 | 15.94 | 18.49 | 16.0 |

Mineral phosphate solubilizing organisms such as *Aspergillus awamorii, Pseudomonas striata* and *Bacillus polymyxa* convert the fixed forms of phosphorous into ionic available forms that can be taken up by potato and increase tuber yields (Table 17.9).

**Table 17.9.** Influence of mineral phosphate solubilizing organisms on P uptake and yield of potato

| Organism | % increase in P uptake | % increase in tuber yield | Reference |
|---|---|---|---|
| *Aspergillus awamorii* | Increased P uptake | 16.0 | Gaur, 1990 |
| *Pseudomonas striata* | 20.4 | 60.3 | Gaur, 1990 |
| *Bacillus polymyxa* | 20.0 | 52.2 | Gaur, 1990 |

Mahendran and Kumar (1998) reported that application of 100 per cent recommended dose of NPK in combination with *Azospirillum* and

*Phosphobacteria* increased the starch content whereas the highest tuber yield was obtained when 100 per cent recommended dose of NPK was applied in two equal splits along with *Azospirillum* and *Phosphobacteria* (Table 17.10).

**Table 17.10.** Effect of biofertilizer on tuber yield and quality in potato

| Treatment | Tuber yield (MT/ha) |
|---|---|
| 100 % recommended dose* of NPK in combination with *Azospirillum* and *Phosphobacteria* | 21.07 |
| Two equal splits of 100 % recommended dose of NPK + *Azospirillum* + *Phosphobacteria* | 21.27 |

* Recommended dose – 120: 240: 120 kg NPK/ha

Application of N at 200 kg/ha gave potato tuber yield of 30.66 MT/ha which was significantly higher than N at 50 kg/ha (21.91 MT/ha). The above treatment was on par with N at 100 kg/ha and N at 150 kg/ha. Three irrigations after earthing up with 100 kg N/ha was the best combination treatment for optimum potato tuber yield (Singh *et al.*, 2006).

Application of 150 kg $K_2O$/ha increased tuber yield over the zero K treatment by 35%, 54% and 61% at 80, 160 and 240 kg N/ha, respectively. K and N application also improved tuber size by increasing the yield of medium and large sized tubers and decreasing the yield of medium and small sized tubers. In general, K application increased the percentage of large tubers from 29% (0 kg $K_2O$/ha) up to 40% (75 kg $K_2O$/ha) and further up to 44% (150 kg $K_2O$/ha).

Application of 100% NPK + FYM at 20 MT/ha + borax at 10 kg/ha + Zn $SO_4$, Mn $SO_4$ and Fe $SO_4$ each at 2 kg/ha and 100% NPK + FYM at 20 MT/ha + borax at 20 kg/ha + Zn $SO_4$, Mn $SO_4$ and Fe $SO_4$ each at 2 kg/ha resulted in significantly higher seed tuber yield and net return and both treatments were at par. The above fertility levels increased tuber yield and net return by 57.1 and 75.51% and 61.31 and 79.23%, respectively over the farmers practice (Chadha *et al.*, 2006).

Highest increase in tuber yield of large (4.95%), medium (10.46%) and total tubers (17.32%) were recorded with integrated application of FYM at 15 MT/ha + 75% RDF (Singh and Lal, 2006).

### 17.1.5. Intercropping

Potato was found the best companion crop for autumn planted sugarcane. Intercropping of sugarcane with potato produced 70 MT/ha of sugarcane and 15 MT/ha of potato tubers (Singh, 1974). The yield of sugarcane was not affected due to intercropping of potato in the centre of the ridge. Smothering effect of potato as an intercrop resulted in an 80% efficiency of weed control (Nankare, 1990).

Sowing of wheat in the furrows after the first earthing up of potato protected potato crop from frost. The yield up to 2.3 and 14.0 MT/ha of wheat and potato, respectively was obtained (Khan *et al.*, 1965).

Highest yield of potato (12 MT/ha) and maize (4.19 MT/ha) were obtained when these were grown in alternate rows on ridges at a row distance of 37.5 cm (Anon, 1990).

Intercropping of potato with mustard fertilized with 150 kg N/ha was found highly productive than their sole cropping. Highest value of potato yield equivalent to 25.03 MT/ha was recorded with intercropping of mustard with potato which was 3.07 MT/ha higher than sole crop of mustard.

### 17.1.6. Transgenic potato

Scientists claim the controversial potato, known as the "protato", contains at least a third more protein than normal tubers, and that it holds "high-quality nutrients". It has been created by the addition of a gene called AmA1 taken from the amaranth plant, which is native to South America and sold widely in western health food stores. The extra gene is said to give ordinary potatoes 30-50% more protein, as well as substantial amounts of the amino acids lysine and methionine. The transgenic plant, developed by a team of scientists led by Asis Datta, who also leads the review committee on genetic modification in the Indian government's Biotechnology Department, is now in its third year of field trials and could be approved within six months.

The protato would be given free to millions of poor children in order to reduce malnutrition in the country. It is planned to incorporate the vegetable into the government's free midday meal programme for schools. There has been a serious concern that malnutrition is one of the reasons for the blindness, the vitamin A deficiency, the protein deficiency. So it is really a very important global concern, particularly in the developing world.

Leading Indian food analyst Devinder Sharma dismissed the GM potato as another magic bullet from the trash can of biotechnology industry. He argued that protein could be better provided by the pulses used traditionally in India. Pulses contain 20%-26% proteins, whereas this potato has only 2.5% protein.

Greenpeace campaigners dismissed the protato as an advertisement for biotechnology. Years were spent in a lab trying to lever protein into potatoes, while cheap, protein-rich pulses grow abundantly all over India.

### 17.1.7. Integrated weed management

Soil solarization helps in reducing the weed density. In solarized soil, number of weeds and fresh weight of weeds per plot of 1.5 $m^2$ were, 5 and 4.29 g, respectively. Whereas, corresponding figures for control were 265 and 416 g, respectively (Table 17.11). Reduction in weed population in solarized plots could be due to direct thermal killing of weed seeds which might have occurred because of increase in temperature below the tarp.

**Table 17.11.** Effect of soil solarization and organic mulch on seed potato tuber yield and weeds.

| Treatment | Tuber | | Weed ( 45 DAP) | |
|---|---|---|---|---|
| | Number/1.5 m$^2$ | Yield (MT/ha) | Number/1.5 m$^2$ | Weight (MT/ha) |
| Soil solarization | 370 | 33.83 | 5 | 0.43 |
| Organic mulch | 304 | 22.20 | 259 | 41.07 |
| Control | 263 | 20.33 | 265 | 41.61 |
| CD (P = 0.05) | 44.6 | 3.37 | 31.6 | 5.08 |

### 17.1.8. Integrated disease management

#### i) Late blight, *Phytophthora infestans*

Protective application of all fungicides tested gave highly significant control of late blight on potato cv. Kufri Chandramukhi. The disease severity was found from 0 to traces in different treatments as compared to 86.6% in control. Tuber infection was also effectively controlled by all the treatments. It varied from 0.1 to 0.5% in case of Indofil M-45 at 0.3 and 0.2%, respectively compared to 16.6% in control. There was significant increase in tuber yield in all the treatments compared to untreated control (Table 17.12) (Thind *et al.*, 2000).

**Table 17.12.** Efficacy of protective spray schedule of fungicides against late blight of potato and tuber yield in cv. Kufri Chandramukhi

| Fungicide | Conc. (%) | Disease severity (%) | Tuber infection (%) | Tuber yield (kg/plot)* |
|---|---|---|---|---|
| Indofil M-45 | 0.2 | Traces | 0.5 | 16.5 |
| | 0.3 | Traces | 0.1 | 17.0 |
| Antracol | 0.2 | Traces | 0.0 | 16.4 |
| | 0.3 | Traces | 0.0 | 17.4 |
| Kavach | 0.2 | Traces | 0.0 | 17.2 |
| | 0.3 | Traces | 0.0 | 17.1 |
| Patafol | 0.2 | Traces | 0.0 | 17.0 |
| Ridomil MZ | 0.2 | Traces | 0.0 | 17.5 |
| Control | --- | 86.6 | 16.6 | 5.4 |

* Plot size = 4.0 x 3.5 m

Curative application of fungicides like Ridomil MZ gave effective control of late blight (8.0% disease severity) followed by Patafol and Curzate M-8 where the disease severity varied from 14.0 to 14.5%, respectively compared to 86.6% in control. Tuber infection traces in case of Ridomil MZ (0.5%) followed by 3.8% in Patafol and 4.0% in Curzate M-8 compared to 16.6% in control. Ridomil MZ gave maximum yield (16.0 kg/plot) followed by Patafol

(15.0 kg/plot) and Curzate M-8 (13.8 kg/plot) as compared to 5.4 kg/plot in control (Table 17.13) (Thind *et al.*, 2000).

**Table 17.13.** Efficacy of curative application of fungicides against late blight of potato and tuber yield in cv. Kufri Chandramukhi

| Fungicide | Conc. (%) | Disease severity (%) | Tuber infection (%) | Tuber yield (kg/plot)* |
|---|---|---|---|---|
| Indofil M-45 | 0.30 | 38.5 | 8.5 | 10.4 |
| Antracol | 0.30 | 33.0 | 8.2 | 10.5 |
| Dodine | 0.15 | 50.0 | 10.0 | 8.8 |
| Akomine | 0.20 | 77.0 | 15.2 | 6.7 |
| Ridomil MZ | 0.25 | 8.0 | 0.5 | 16.0 |
| Patafol | 0.27 | 14.0 | 3.8 | 15.0 |
| Curzate M-8 | 0.20 | 14.5 | 4.0 | 13.8 |
| Curzate 50 WP | 0.04 | 20.0 | 6.2 | 11.7 |
| Ofurace | 0.09 | 32.0 | 8.0 | 10.4 |
| Control | --- | 86.6 | 16.6 | 5.4 |

* Plot size = 4.0 x 3.5 m

Application of higher dose of potassic fertilizer along with recommended doses of nitrogenous and phosphatic fertilizers to potato gave decreased late blight disease severity and increased yield (Table 17.14) (Ojha and Yadav, 2000).

**Table 17.14.** Effect of potassic fertilizer on late blight disease severity and yield of potato

| Potato cultivar | Level of potassium (kg/ha) | Disease severity (%) | Tuber yield (MT/ha) |
|---|---|---|---|
| 92-PT-27 | 180 | 2.13 | 219.90 |
| | 120 | 4.56 | 216.00 |
| | 60 | 9.66 | 211.36 |
| | 0 | 16.23 | 205.96 |
| Kufri Badshah | 180 | 20.80 | 195.56 |
| | 120 | 31.83 | 178.60 |
| | 60 | 41.63 | 169.46 |
| | 0 | 48.26 | 160.56 |

### ii) Black scurf, *Rhizoctonia solani*

Potato seed tuber treatment with *T. viride* at 3.5-4.5g/kg seed was very effective which was at par with boric acid or organomercurials (Arora and Somani, 2001) (Table 17.15) (Fig. 17.1). Combination of soil solarization of black scurf-infested fields and seed treatment with *T. viride* further improved disease control (Arora, 2000).

**Table 17.15.** Control of black scurf disease through seed treatment with *Trichoderma viride*

| Treatments | Germination (%) | Disease incidence (%) | Disease index | Yield (MT/ha) |
|---|---|---|---|---|
| Control (Untreated) | 85.6 | 71.0 | 1.18 | 29.99 |
| *T. viride* at 3.5g/kg | 84.1 | 41.6 | 0.83 | 30.24 |
| Emisan dip in 0.25% | 79.5 | 24.7 | 0.39 | 28.51 |
| Boric acid dip/spray – 3% | 80.3 | 21.3 | 0.33 | 29.21 |
| CD (P = 0.05) | NS | 8.55 | 0.195 | NS |

**Fig. 17.1.** Management of *Rhizoctonia solani* on potato by *Trichoderma* spp. Left – Treated, Right – Control.

### (iii) Bacterial wilt and brown rot, *Ralstonia solanacearum*

In Eastern Plains and Deccan Plateau, the bacterial wilt disease of potato can be reduced to considerable extent by following cultural practices like use of disease-free seeds, adjusting the date of planting, application of bleaching powder, straw burning in infested soil, blind earthing, hot weather cultivation and crop rotation (Table 17.16).

**Table 17.16.** Control of bacterial wilt and brown rot of potato.

| Treatment | % reduction | | % yield increase |
|---|---|---|---|
| | Wilt | Brown rot | |
| Healthy seed | 50 – 76 | 40 – 46 | 18 – 27 |
| Bleaching powder at 12 kg/ha + Blind earthing | 70 – 94 | 50 – 85 | 22 – 28 |
| Summer ploughing in Plains | 50 – 83 | 50 – 70 | 5 – 25 |
| Winter ploughing in Hills | 60 – 75 | 62 – 78 | 35 – 40 |
| Early planting in Hills | 77 – 80 | 75 – 85 | 44 – 68 |

Bacterial antagonists were responsible for 57.6-72.4% reduction in bacterial wilt and 299.4 to 465.2% increase in tuber yield at Kumaon hills (Shekhawat *et al.*, 2000) (Table 17.17).

**Table 17.17.** Effect of antagonists on bacterial wilt of potato in Kumaon hills

| Antagonist | Reduction in wilt (%) | Increase in yield (%) |
|---|---|---|
| Avirulent *Ralstonia solanacearum* | 59.4 | 391.2 |
| *Enterobacter cloacae* | 66.3 | 391.7 |
| *Bacillus subtilis* strain B5 | 72.4 | 465.2 |
| *Bacillus cereus* strain B4 | 57.6 | 299.4 |

An enhancement of tuber yield by 16.7 to 25.6% has also been recorded by seed treatment with *Bacillus cereus* strain B4 and *B. subtilis* strain B5 (Sunaina *et al.*, 2001).

### 17.1.9. Integrated nematode management

#### (i) Root-knot nematode, *Meloidogyne incognita*

*Paecilomyces lilacinus* significantly reduced the tuber and root galling due to root-knot nematodes on potatoes (Jatala, 1985). Potato plants grown in plots inoculated with the fungus had a significantly lower root galling index than did those grown in the nematicide-treated plots. Similarly, 86% of the egg masses collected from the roots of fungus-treated plots were infected with *P. lilacinus*, and over 54% of the eggs within these egg masses were destroyed. Yield of marketable potatoes (non-infected tubers) was significantly higher in the fungus-treated plots than in those of the control and the nematicide-treated plots (Table 17.18).

**Table 17.18.** Effect of *Paecilomyces lilacinus* and nematicides on *Meloidogyne incognita* infecting potato.

| Treatment | Root gall index | % Egg masses infected | % Eggs destroyed | Marketable yield/400 m$^2$ |
|---|---|---|---|---|
| *Paecilomyces lilacinus* | 2.93 | 86.0 | 54.7 | 302.59 |
| Aldicarb at 2.5 kg a.i./ha | 3.36 | --- | --- | 147.12 |
| Phenamiphos at 2.5 kg a.i./ha | 3.44 | --- | --- | --- |
| Carbofuran at 2.5 kg a.i./ha | 3.57 | --- | --- | --- |
| Organic matter at 10 MT/ha | 3.62 | --- | --- | --- |
| Control | 3.60 | --- | --- | 122.69 |
| CD (P = 0.05) | 0.14 | --- | --- | 85.10 |

#### (ii) Cyst nematode, *Globodera rostochiensis*

Experiments on control of *G. rostochiensis* using *P. lilacinus* revealed that the fungus-treated plots had higher yield than did the control plots and those treated with carbofuran (Table 17.19). Similarly, there were significantly lower numbers of cysts in the *P. lilacinus*-treated plots than in nematicide-treated or control plots (Jatala, 1985).

**Table 17.19.** Biological control of *Globodera rostochiensis* in potato by *Paecilomyces lilacinus*

| Treatment | Tuber yield (MT/ha) | | No. of cysts/ 50 ml soil |
|---|---|---|---|
| | Marketable yield | Total yield | |
| *P. lilacinus* | 20.58 | 26.64 | 58.88 |
| Carbofuran at 3 kg a.i./ha | 19.68 | 24.90 | 67.33 |
| Control | 15.72 | 20.52 | 70.35 |
| CD (P = 0.05) | NS | NS | 5.73 |

Soil application of neem cake at 5 MT/ha along with *Trichoderma viride* at 0.5 kg/ha recorded maximum tuber yield (23.14 MT/ha) and reduction in potato cyst nematode multiplication (Reproduction factor-1.09) (Umamaheswari *et al.*, 2010a).

*Pseudomonas fluorescens* at 10 kg/ha gave maximum reduction in cyst nematode multiplication (Reproduction factor-1.12) and improvement in tuber yield (19.40 MT/ha) (Umamaheswari *et al.*, 2010b).

## 17.2. TOMATO, *LYCOPERSICON ESCULENTUM*

### 17.2.1. High yielding varieties/hybrids

**Table 17.20.** Increased productivity in high yielding varieties/F1 hybrids of tomato

| HYV/F1 hybrid | Yield MT/ha | Developed at |
|---|---|---|
| Pusa Ruby | 28-30 | IARI, New Delhi |
| Pusa Sheetal | 35 | IARI, New Delhi |
| Pusa Gaurav | 40 | IARI, New Delhi |
| Pusa Uphaar | 37 | IARI, New Delhi |
| Pusa Sadabahar | 35 | IARI, New Delhi |
| Pusa Rohini | 41.2 | IARI, New Delhi |
| Pusa Hybrid-2* | 60.0-62.5 | IARI, New Delhi |
| Pusa Hybrid-4* | 55 | IARI, New Delhi |
| Pusa Divya* | 55 | IARI, New Delhi |
| Arka Vikas | 40.0-42.5 | IIHR, Bangalore |
| Arka Saurabh | 30-35 | IIHR, Bangalore |
| Arka Abha | 43 | IIHR, Bangalore |
| Arka Alok | 46 | IIHR, Bangalore |
| Arka Vishal* | 75 | IIHR, Bangalore |

| | | |
|---|---|---|
| Arka Vardan* | 75 | IIHR, Bangalore |
| Arka Ashish | 38 | IIHR, Bangalore |
| Arka Ahuti | 42 | IIHR, Bangalore |
| Arka Shreshta* | 76 | IIHR, Bangalore |
| Arka Abhijit* | 65 | IIHR, Bangalore |
| CO 1 | 35 | TNAU, Coimbatore |
| CO 2 | 41.0-42.5 | TNAU, Coimbatore |
| Marutham (CO-3) | 45-47 | TNAU, Coimbatore |
| PKM 1 | 36-38 | TNAU, Coimbatore |
| Paiyur 1 | 30 | TNAU, Coimbatore |
| COTH 1 | 95.9 | TNAU, Coimbatore |
| COLCRH 3 | 90.2 | TNAU, Coimbatore |
| Narendra Tomato-1 (NDT-5) | 40-45 | NDUAT, Faizabad |
| Narendra Tomato-2 (NDT-120) | 40-45 | NDUAT, Faizabad |
| Narendra Tomato-5 (NDT-96) | 40-45 | NDUAT, Faizabad |
| Kalyanpur Angoorlata | 48-50 | CSAUAT, Kanpur |
| Punjab NR-7 | 38-40 | PAU, Ludhiana |
| Punjab Selection No. 12 | 45-48 | PAU, Ludhiana |
| TH 802 | 63 | PAU, Ludhiana |
| TH 2312 | 63 | PAU, Ludhiana |
| Vasundhara | 45-47 | MPKV, Rahuri |
| Utkal Kumari | 40.0-41.5 | OUAT, Bhubaneshwar |
| Solan Vajra | 45.0-47.5 | YSPUHF, Solan |
| Solan Shagun* | 49-50 | YSPUHF, Solan |
| Parbhani Yashasri | 60.0-62.5 | MAU, Parbhani |
| Swarna Baibhav | 70.0-72.5 | HARP, Ranchi |
| Swarna Sampada* | 60-65 | HARP, Ranchi |
| Swarna Naveen | 58-60 | HARP, Ranchi |
| Swarna Lalima | 50-55 | HARP, Ranchi |
| Kashi Amrit | 62.0-63.5 | IIVR, Varanasi |
| Kashi Anupam | 57-60 | IIVR, Varanasi |
| Kashi Sharad | 48-50 | IIVR, Varanasi |
| Kashi Vishesh | 43-45 | IIVR, Varanasi |
| Kashi Hemant | 40.0-42.5 | IIVR, Varanasi |
| VRTH-1 | 110 | IIVR, Varanasi |
| VRTH-2 | 90 | IIVR, Varanasi |
| Hybrid Tomato (ARTH-3)* | 88-90 | IIVR, Varanasi |
| Hybrid Tomato (ARTH-4)* | 100-102 | IIVR, Varanasi |
| Mani Leima | 63-65 | ICARRC, Barapani |

| | | |
|---|---|---|
| Rajashree | 75 | MPKV, Rahuri |
| Mukthi | 43-45 | KAU, Vellanikkara |
| Sel-1-6-1-4 | 56-58 | |
| Gulmohar (MTH-6) | 93-95 | |
| Meenakshi * | 101-102 | |

* Hybrids

**Table 17.21.** Increased productivity in high yielding varieties/F1 hybrids of tomato with disease resistance (Swamy and Sadashiva, 2004)

| Improved cultivar/ hybrid | Yield (MT/ha) | Resistance to disease | Developed at |
|---|---|---|---|
| Arka Alok | 46 | Bacterial wilt | IIHR, Bangalore |
| Arka Asish | 38 | Powdery mildew (T) | IIHR, Bangalore |
| Arka Vardan (F1)* | 75 | Root-knot nematodes | IIHR, Bangalore |
| Arka Shreshta (F1)* | 76 | Bacterial wilt | IIHR, Bangalore |
| Arka Abhijit (F1)* | 65 | Bacterial wilt | IIHR, Bangalore |
| Pusa-120 | 30-32 | Root-knot nematodes | IARI, New Delhi |
| Pusa Hybrid-2* | 55 | Root-knot nematodes | IARI, New Delhi |
| Pusa Hybrid-4* | 42.5-45.0 | Root-knot nematodes (field tolerance) | IARI, New Delhi |
| Kalyanpur Angoorlata | 48-50 | Root-knot nematodes, Leaf curl (MR) | CSAUAT, Kanpur |
| Azad T-2 | 33-35 | Leaf curl (MR) | CSAUAT, Kanpur |
| Punjab NR-7 | 38-40 | Root-knot nematodes | PAU, Ludhiana |
| TH-2312 | 63 | Root-knot nematodes | PAU, Ludhiana |
| Utkal Urbasi | 38-40 | Bacterial wilt, Early blight (T) | OUAT, Bhubaneshwar |
| Utkal Pallavi | 37.5-38.0 | Bacterial wilt, Root-knot nematodes | OUAT, Bhubaneshwar |
| Utkal Pragyan | 41.0-42.5 | Bacterial wilt (T) | OUAT, Bhubaneshwar |
| Utkal Raja | 45-47 | Bacterial wilt (T) | OUAT, Bhubaneshwar |
| Hisar Lalit | 30.0-32.5 | Root-knot nematodes | HAU, Hisar |
| Solan Shagun* | 49-50 | Early blight (T), Buckeye rot (T) | YSPUHF, Solan |
| Swarna Baibhav | 70.0-72.5 | Bacterial wilt | HARP, Ranchi |
| Swarna Sampada* | 60-65 | Bacterial wilt, Early blight | HARP, Ranchi |
| Swarna Lalima | 50-55 | Bacterial wilt | HARP, Ranchi |
| Swarna Naveen | 50-55 | Bacterial wilt | HARP, Ranchi |
| Kashi Amrit | 62.0-63.5 | Leaf curl | IIVR, Varanasi |
| Kashi Vishesh | 43-45 | Leaf curl | IIVR, Varanasi |

| | | | |
|---|---|---|---|
| VRTH-1 | 110 | TLCV (T), Early blight (T) | IIVR, Varanasi |
| Mani Leima | 63-65 | Leaf curl (T) | ICARRC, Barapani |
| Gujarat Tomato-2 | 33-35 | Leaf curl (T), Early blight (T) | GAU, Anand |
| Narendra Tomato 3 | 35-38 | Root-knot nematodes (T), Early blight (T) | NDUAT, Faizabad |
| NDTS-2001-3 | | Leaf curl (T), Early blight (T) | NDUAT, Faizabad |
| NTDR-1 | 35-38 | Root-knot nematodes (T) | NDUAT, Faizabad |
| LE-415 | 40.0-42.5 | Bacterial wilt | |
| BT-136 | 38.5-40.0 | Bacterial wilt (T) | OUAT, Bhubaneshwar |
| Sel-1-6-1-4 | 56-58 | Root-knot nematodes | |

* Hybrids

### 17.2.2. Micro irrigation

A considerable savings in water, besides increased yields of tomato under drip irrigation as compared to conventional methods has been reported by Bafna *et al.* (1993) (Tables 17.22 and 17.23).

**Table 17.22.** Irrigation requirement of tomato crop under surface and drip methods

| Year | Amount of irrigation water applied (cm) | | Water saving by drip (%) | Increase in fruit yield by drip (%) |
|---|---|---|---|---|
| | Surface | Drip | | |
| 1987 | 77 | 40 | 48 | 32 |
| 1988 | 56 | 24 | 57 | 49 |
| Average | 67 | 32 | 53 | 40 |

**Table 17.23.** Yield performance of hybrid tomato under micro-irrigation

| Irrigation system | Fruit yield (MT/ha) |
|---|---|
| Micro tube | 81.6 |
| Drip | 78.3 |
| Furrow | 70.4 |
| Broad ridge | 62.9 |
| Drip + plastic mulch | 93.9 |
| Shade net house + misters | 102.5 |

The highest crop yield of about 51 MT/ha and 44% saving in irrigation water were obtained using the combination of drip irrigation at 0.4 PE and mulch of sugarcane trash. This treatment also gave maximum yield of 163 kg/ha/mm of water applied. The treatment combining drip irrigation with black plastic mulch reduced weed infestation by 95%, increased yield by 53% and resulted in a 44% saving in irrigation water compared to surface flooding without mulch (Shrivastava *et al.*, 1994).

Bhalla (1988) observed that black polythene mulch, trickle irrigation, and black polythene mulch + trickle irrigation increased the tomato fruit yield by 66, 70 and 123%, respectively than the control plants.

Raina *et al.* (1999) observed that drip irrigation gave significantly higher yield (16.63 MT/ha) compared with surface irrigation (11.95 MT/ha). Polythene mulch + drip irrigation further increased the yield to 23.25 MT/ha. Water use efficiency under drip irrigation alone, drip irrigation + polythene mulch and surface irrigation was 3.4, 4.8 and 1.6, respectively.

Singh *et al.* (2000) reported that the drip irrigation saved up to 78% irrigation water and increased tomato fruit yield up to 43% compared with the control.

On red sandy loam soils of Karnataka, drip irrigation resulted in higher yields of tomato as compared to furrow method. Increasing drip irrigation from 0.3 Epan level to 0.7 Epan increased yield from 54 to 71 MT/ha.

The cost-benefit ratio was 5.15 under drip irrigation in tomato as compared to 2.96 for flood method. The yield under drip (48 MT/ha) was 50% more in comparison to flood irrigated crop (32 MT/ha). Irrigation water saving was to the tune of 31.5% with drip.

Under Delhi conditions, an increase in yield of tomato cv. Pusa Ruby by 47.4% in drip irrigated plots over the furrow irrigated plots.

A 2 day drip irrigation frequency with 100% ET water application gave highest tomato fruit yields (35.66 MT/ha and 35.10 MT/ha during 2003 and 2004, respectively) compared to 8 day drip irrigation frequency which gave minimum fruit yields (24.10 MT/ha and 23.20 MT/ha during 2003 and 2004, respectively) (Thiyagarajan *et al.*, 2006).

Application of drip irrigation once in 3 days for 2 hours of operation at 1.2 kg/cm$^2$ pressure gave significantly higher yield (25% more) over check basin method of irrigation. The drip irrigation also resulted in 53.4% water saving and 62.7% more water use efficiency. TSS of tomato fruits was also higher (18.6%) in drip irrigation than the check basin irrigation (Yadav and Yadav, 2006).

### 17.2.3. Fertigation

A field experiment conducted to compare the effect of soil application of fertilizers and fertigation on tomato revealed that the treatment with 50% NK fertigation and drip with black polyethylene mulch resulted in the highest yield of 134.1 MT/ha under protected cultivation (Table 17.24) (Prabhakar *et al.*, 2001).

**Table 17.24.** Fertigation in tomato

| Treatment | Fruit yield (MT/ha) |
|---|---|
| Soil NPK + drip | 74.7 |
| Soil NPK + drip + black polythene mulch | 87.9 |
| ½ NK through fertigation + drip | 81.4 |
| ½ NK through fertigation + drip + black polythene mulch | 134.1 |
| Full NPK through fertigation + drip + black polythene mulch | 121.3 |

Application of 350: 350: 350 kg NPK/ha in the form of water soluble fertilizers through fertigation at fortnightly intervals is recommended for getting maximum productivity and best quality of tomato fruits (cv. Hybrid SUN 7611) under cost effective greenhouse (Khan *et al.*, 2001) (Table 17.25).

**Table 17.25.** Productivity of tomato cv. Hybrid SUN 7611 as influenced by sources and level of fertilizers

| Treatment | Fruit yield (MT/ha) |
|---|---|
| Normal fertilizers 80% Fertigation | 73 |
| Normal fertilizers 100% Fertigation | 101 |
| Normal fertilizers 120% Fertigation | 128 |
| Normal fertilizers 140% Fertigation | 144 |
| Water soluble fertilizers 80% Fertigation | 101 |
| Water soluble fertilizers 100% Fertigation | 133 |
| Water soluble fertilizers 120% Fertigation | 154 |
| Water soluble fertilizers 140% Fertigation | 168 |

Fertigation trial in tomato has demonstrated higher yields and better fertilizer use efficiency under fertigation as compared to conventional methods (Bafna *et al.*, 1993) (Tables 17.26 and 17.27).

**Table 17.26.** Influence of method of N-fertilizer application on yield and total N uptake by tomato

| Treatment | Fruit yield (MT/ha) | N-uptake (kg/ha) |
|---|---|---|
| Placement | 75.39 | 144.04 |
| Fertigation | 107.30 | 171.87 |

**Table 17.27.** Yield of tomato as influenced by different fertigation treatments

| Treatment | Fruit yield (MT/ha) |
|---|---|
| Control (furrow irrigation) | 34.9 |
| Control (drip irrigation) | 42.9 |
| 50% NK fertigation using urea and MOP | 40.0 |
| 50% NK fertigation using urea and Multi-K | 40.8 |

| | |
|---|---|
| 50% NPK fertigation using 19-9-19 urea and MAP | 43.8 |
| 100% NK fertigation using urea and Multi-K | 43.6 |
| 100% NPK fertigation using 19-9-19 urea and MAP | 45.7 |
| CD (P = 0.05) | 5.7 |

Fertigation experiments carried out at TNAU, Coimbatore, revealed that 50 kg each of NPK/ha through straight fertilizers as basal, remaining 250 kg NPK/ha as water soluble fertilizer through fertigation along with black polyethylene mulch registered the highest number of fruits/plant (71), yield per hectare (186 MT) and BCR (1.76) as compared to control in tomato hybrid SH 7711 (Table 17.28) grown under polyhouse (Natarajan *et al* . 2005).

**Table 17.28.** Effect of growing media, irrigation regime, fertigation and mulching on growth and yield of tomato hybrid SH 7711

| Treatment | Yield (MT/ha) | B/C ratio |
|---|---|---|
| Soil: FYM: Saw dust + 20 kPa + 50 kg/ha SF (basal) + 250 kg/ha WSF (fertigation) + Mulch | 113 | 1.58 |
| Soil: FYM: coir pith + 20 kPa + 50 kg/ha SF (basal) + 250 kg/ha WSF (fertigation) + Mulch | 186 | 1.76 |

Balasubramanian and Vadivelu (2009) reported that drip-fertigated tomato plants yielded 15.9 to 27.0% higher than drip and conventional systems, respectively. Fertigated plants produced tomato fruits having higher amounts of ascorbic acid, TSS, lycopene and carotene concentrations.

Tomato fertigated using liquid fertilizer (8: 8: 8) gave 44% higher yield as compared to soil applied furrow irrigated crop and saving of 25% fertilizers.

Highest tomato fruit yield of 177 MT/ha was obtained with 180 mg N/litre through fertigation.

### 17.2.4. Integrated nutrient management

Maximum fruit yield of tomato (35.4 MT/ha) and net income (Rs. 49770/ha) have been obtained with FYM at 20 MT/ha + full recommended dose of NPK (120: 60: 60 kg/ha) followed by green manuring with dhaincha + ½ dose of NPK, which could produce 33.2 MT/ha fruit yield and given the net income of Rs. 47200/ha. Highest benefit-cost ratio (3.45) has been found with the treatment green manuring with dhaincha + ½ dose of NPK. The FYM applied alone at 20 and 40 MT/ha could produce 14.5 and 16.6 MT/ha fruit yields, respectively. Dhaincha green manuring has been found superior to both the doses of FYM by producing a fruit yield of 18.3 MT/ha (Singh *et al.*, 2002).

Application of 100% RDF (200:100:100 kg NPK/ha) + FYM gave significantly high marketable fruit yield (37.90 MT/ha) (Chavan *et al.*, 2009).

The treatment combination of FYM at 10 MT/ha + vermicompost at 2.5 MT/ha recorded the highest fruit yield of 42.74 MT/ha compared to 40.09 MT/ha in treatment 100% RDF (Sathiyamurthy and Pugalendhi, 2009).

The availability of nutrients and their uptake were higher with the application of vermicompost. The vermins, enzymes and antibiotics can be effectively utilized for increasing the yield and quality of fruits in tomato. An integration of inorganics with vermicompost also gave higher yields (Table 17.29).

**Table 17.29.** Influence of vermicompost on yield of tomato.

| Treatment | Fruit yield (MT/ha) |
|---|---|
| FYM 25 MT/ha | 44 |
| Vermicompost 25 MT/ha | 54 |
| Vermicompost 37.5 MT/ha | 69 |
| Vermicompost 50 MT/ha | 71 |
| Vermicompost 100 MT/ha | 85 |
| Vermiculture *in situ* | 60 |
| FYM + Full NPK inorganics | 85 |
| Vermicompost 25 MT/ha + Full NPK inorganics | 108 |
| CD at (P = 0.05) | 17 |

Rizvi *et al.* (1990) reported highest number of fruits/plant (28.6) and fresh weight of fruit (70.95g) with the application of 2.25g *Aulosira fertilissma* ($N_2$ fixing *Cyanobacterium*) + 2.25g *Spirulina subsulsa* (non-$N_2$ fixing *Cyanobacterium*) + 0.5g DAP/pot. This represents a 522% increase in number of fruits and a 977% increase in yield over control.

Palaniappan *et al.* (1999) reported that application of 100% NK + 2 sprays of Polyfeed at 30 and 45 DAP and 3 sprays of Multi K at 60, 75 and 90 DAP gave the highest tomato yield of 75.15 MT/ha. Application of 75% NK + 2 sprays of Polyfeed + 3 sprays of Multi K gave more or less the similar yields thus saving 25% NK. The yield increase due to 2 sprays of Polyfeed was 6.25% and with 3 sprays of Multi K was 10.8% over no spray (Table 17.30).

**Table 17.30.** Effect of foliar application of Polyfeed and Multi K on tomato fruit yield

| Treatment | Fruit yield (MT/ha) | Marketable yield (MT/ha) |
|---|---|---|
| 100% NK alone | 59.93 | 44.95 |
| 100% NK + 2 sprays of Multi K | 66.73 | 50.05 |
| 100% NK + 3 sprays of Multi K | 63.41 | 47.56 |
| 100% NK + 2 sprays of Polyfeed | 61.37 | 46.03 |
| 100% NK + 2 sprays of Polyfeed + 2 sprays of Multi K | 64.60 | 48.45 |
| 100% NK + 2 sprays of Polyfeed + 3 sprays of Multi K | 71.15 | 53.36 |

| | | |
|---|---|---|
| 75% Multi K alone | 56.29 | 42.22 |
| 75% NK + 2 sprays of Multi K | 58.42 | 43.82 |
| 75% NK + 3 sprays of Multi K | 59.16 | 44.37 |
| 75% NK + 2 sprays of Polyfeed | 59.76 | 44.82 |
| 75% NK + 2 sprays of Polyfeed + 2 sprays of Multi K | 62.05 | 46.54 |
| 75% NK + 2 sprays of Polyfeed + 3 sprays of Multi K | 68.77 | 51.58 |
| CD (P = 0.05%) | 3.86 | --- |

Boron application at 25 DAP significantly increased tomato fruit yield over control. The highest yields were obtained in plots treated with foliar application of 2.5 kg Borax/ha (22.767 MT/ha) (Prasad *et al.*, 1997).

At the lowest level of N and P application (75 kg N/ha and 60 kg $P_2O_5$/ha), tomato fruit yield increment due to K addition at a rate of 60 kg $K_2O$/ha reached 29%.

Rajesh Kanna (1999) conducted an experiment in tomato hybrid COTH 1 at HC & RI, TNAU, Coimbatore, imposing different N levels (200, 250 and 300 kg/ha), P levels (200, 250 and 300 kg/ha) and K levels (200, 250 and 300 kg/ha) in all possible combinations and reported that economic yield per hectare with the highest BCR was observed with the application of 200: 300: 200 kg NPK ha-1 (Table 17.31). The entire quantity of P and K are given as basal while N is applied in three equal split doses at planting, 30 and 60 days after planting.

**Table 17.31.** Effect of NPK on yield characters of tomato hybrid COTH 1

| Treatment | No. of fruits/cluster | Yield (MT/ha) |
|---|---|---|
| NPK at 200: 300: 200 kg/ha | 6.21 | 117.30 |
| NPK at 300: 250: 200 kg/ha | 4.13 | 73.49 |

Another experiment on hybrid tomato COTH 2 conducted at Hort. College & Res. Inst., Tamil Nadu Agricultural University, Coimbatore, during 2006 (Anon, 2007) revealed that application of SOP at 100 kg/ha basal + SOP at 50 kg/ha top dressing on 30 DAP + SOP at 50 kg/ha top dressing on 60 DAP registered the highest number of fruits per plant, single fruit weight, yield per hectare, TSS, lycopene and firmness (Table 17.32).

**Table 17.32.** Effect of SOP on yield attributes of tomato hybrid COTH 2

| Treatment | Fruit yield (MT/ha) | B: C Ratio |
|---|---|---|
| Control (No $K_2Oo$) | 61.20 | 3.09 |
| 200 kg MOP (basal) | 70.83 | 3.54 |
| 200 kg SOP (basal) | 74.52 | 3.73 |
| 100 kg (basal) + SOP 100 kg (top dressing) on 30 DAP | 78.10 | 3.90 |
| SOP 100 kg (basal) + SOP 50 kg (top dressing) on 30 DAP +SOP 50 kg/ha (top dressing) on 60 DAP | 87.29 | 4.36 |

| SOP 100 kg (basal) + SOP 50 kg (top dressing) on 30 DAP + SOP 2% (foliar spray) on 60 DAP | 81.20 | 4.06 |
|---|---|---|
| SOP 100 kg (basal) +SOP 50 kg top dressing on 30 DAP + SOP 2% (foliar spray) on 60 DAP and 75 DAP | 83.85 | 4.19 |
| CD (P = 0.05) | 3.05 | --- |

The application of biogas slurry + FYM, neem cake + FYM and vermicompost were found to be far superior in tomato fruit yield over other treatments (Table 17.33). The good nutritive value of digested slurry, the nitrification regulative property and high sulfur content of neem cake and easy transfer of nutrient to the plant through the formation of drillospheres, were the reason for the higher yields of tomato (Renuka and Ravishankar, 2001).

**Table 17.33.** Effect of different organic amendments and biofertilizers on fruit number and yield of tomato

| Treatment | No. of fruits/plant | Fruit yield (MT/ha) |
|---|---|---|
| FYM | 12.32 | 29.03 |
| NPK | 12.47 | 18.44 |
| Vermicompost | 14.63 | 25.59 |
| Biogas slurry | 17.17 | 33.59 |
| Neem cake | 15.62 | 27.92 |
| *Azospirillum* | 13.63 | 28.62 |
| Phosphobacteria | 10.72 | 13.07 |
| FYM + NPK | 16.82 | 32.37 |
| FYM + Vermicompost | 15.00 | 35.81 |
| FYM + Biogas slurry | 17.81 | 47.66 |
| FYM + Neem cake | 18.64 | 38.92 |
| FYM + *Azospirillum* | 14.00 | 33.74 |
| FYM + Phosphobacteria | 14.81 | 23.62 |
| CD (P = 0.01) | 3.19 | 10.95 |

Prilled urea coated with neem oil was superior to prilled urea in hybrid tomato cv. Arka Shreshta and produced 21% higher yield on an overall basis (Kotur *et al.*, 2006).

### 17.2.5. Bioregulators

Spraying tomatotone (parachlorophenoxy acetic acid) improves fruit set to 70% and fruit size in summer grown tomato.

### 17.2.6. Mulching

Hooda *et al.* (1999) found that plots mulched with black polythene recorded significantly higher soil temperature and moisture per cent as compared with other mulch materials and the control. Black polythene mulch resulted in higher, early and total yield than rest of the treatments. Planting of tomato during 15th December and mulched with black polythene gave highest yield (Table 17.34).

**Table 17.34.** Effect of planting dates and mulching on yield of tomato cv. Sel-18

| Planting date | Black polythene | White polythene | Sugarcane trash | Control |
|---|---|---|---|---|
| Early marketable yield MT/ha) | | | | |
| 15th December | 10.293 | 8.467 | 6.234 | 4.827 |
| 30th December | 8.120 | 4.523 | 5.567 | 4.523 |
| Late marketable yield MT/ha) | | | | |
| 15th December | 29.033 | 26.200 | 15.667 | 13.000 |
| 30th December | 24.267 | 17.783 | 13.033 | 11.007 |
| Total marketable yield MT/ha) | | | | |
| 15th December | 40.026 | 34.667 | 21.901 | 17.827 |
| 30th December | 32.387 | 22.306 | 18.600 | 15.530 |

Elkner and Kaniszewski (1995) reported 24% higher tomato fruit yield under black polythene mulch as compared to unmulched control.

### 17.2.7. Protected cultivation

Crop yields are continuously increasing under protected condition (Fig. 17.2). For example, ten years ago, good tomato yields reached 8-12 kg/m$^2$ (between November and June) in unheated greenhouses, whereas in 1995, yields of 25-30 kg/m$^2$ in heated greenhouses are common, which means an increase of 100%. It is astonishing to learn that tomato yields up to 60 to 65 kg/m$^2$ are being obtained with a single crop in the Netherlands spread over a harvesting period of 9 months, corresponding to an 11 month growing cycle in the greenhouse. With tomatoes, the potential yield level can even reach 80 to 85 kg/m$^2$ in one long cycle. A regular crop of tomatoes almost throughout the year with a productivity of 300 to 400 MT/ha can be obtained under greenhouses in India. This is a clear indication of the tremendous yield potential of vegetable crop species under protected conditions.

Productivity is manifold in greenhouses in comparison to growing the vegetables in open field (Sanwal *et al.*, 2004) (Table 17.35).

**Table 17.35.** Performance of tomato varieties under polyhouse and open field conditions in NEH region (Barapani)

| Varieties | Polyhouse yield (MT/ha) | Open field yield (MT/ha) | % increase in yield |
|---|---|---|---|
| BT-117-5-3-1 | 342.00 | 115.00 | 197 |
| KT-10 | 283.60 | 117.40 | 138 |
| BT-10 | 294.00 | 111.65 | 163 |

| | | | |
|---|---|---|---|
| Arka Alok | 260.00 | 57.90 | 349 |
| BT-12 | 302.40 | 101.00 | 200 |
| Selection-2 | 233.00 | 73.83 | 215 |
| Selection-1 | 200.98 | 84.03 | 139 |
| KT-15 | 211.60 | 51.65 | 309 |
| H-24 | 243.17 | 58.75 | 314 |
| Arka Abha | 193.50 | 70.33 | 175 |

**Fig 17.2.** Protected cultivation of tomato.

The yield and number of tomatoes/unit area for the crop grown in the polyhouse was 78.75 and 53.4% higher, respectively compared with the control grown in the open. The average increase in irrigation water use efficiency in the crop grown inside the polyhouse was 33.3%. The quality of fruits from inside the polyhouse was considerably better as indicated by the values for TSS and ascorbic acid contents (Ahluwalia *et al.*, 1996).

Under polyhouse conditions, tomato cv. Naveen yielded 110 MT/ha and Pusa Hybrid 96.8 MT/ha. The control was eliminated due to frost (Chandra *et al.*, 2000) (Table 17.36).

**Table 17.36.** Yield of tomato grown under polyhouse conditions

| Tomato cv. | No. of fruits/plant | Fruit yield (kg/plant) | Fruit yield (MT/ha) |
|---|---|---|---|
| Naveen | 36.5 | 3.97 | 110.5 |
| Pusa Hybrid-2 | 28.5 | 3.55 | 98.6 |

The growing medium, irrigation regime, fertilizer application and mulching for tomato (hybrid SH 7711) under naturally ventilated polyhouse conditions were standardized (Table 17.37).

**Table 17.37.** Protected cultivation of tomato under naturally ventilated polyhouse

| Particulars | Tomato |
|---|---|
| Growing medium | Soil : FYM : composted coir pith (2: 1: 1). |
| Irrigation regime | 20 kPa |
| Fertilizer application | Basal-NPK each at 50 kg/ha through straight fertilizers |
| | Fertigation-NPK each at 250 kg/ha through water soluble fertilizer |
| Mulching | Black polyethylene sheet (50 micron) |
| Yield (MT/ha) | 176 |
| B/C ratio | 1.76 |

The influence of shade net and fertigation on growth and productivity of tomato were standardized (Table 17.38).

**Table 17.38.** Protected cultivation of tomato under shade net

| Particulars | **Tomato** |
|---|---|
| Hybrid | Ruchi |
| Shade | 35 per cent |
| Spacing | Paired row planting 80 x 40 x 60 cm (between pairs, rows & plants) |
| Fertilizer application | Basal: NK each at 50 kg/ha and P at 250 kg/ha through straight fertilizers<br>Fertigation: NK each at 200 kg/ha through straight fertilizers |
| Yield (MT/ha) | 96.0 |
| B/C ratio | 3.07 |

The module soil: compost: sand as substrate, 20 kPa irrigation regime, basal dose of 50 kg/ha NPK with straight fertilizers + fertigation at 250: 250: 250 kg/ha NPK with straight fertilizers, black polythene mulch and tomato cv. SH-7711 recorded significantly maximum fruit yield per plant (6.70 kg). This module was found to be the best package for higher production of tomato under naturally ventilated greenhouse (Nagre *et al.*, 2006b).

### 17.2.8. Integrated weed management

The treatment combining drip irrigation with black plastic mulch reduced weed infestation by 95%, increased tomato fruit yield by 53% and resulted in a 44% saving in irrigation water compared to surface flooding without mulch (Shrivastava *et al.*, 1994).

Gutal *et al.* (1992) reported that the 25$\mu$ black LDPE film reduced weed growth by 90%, saved soil moisture by conserving to the extent of 28% and increased tomato fruit yield by 55% compared with the control.

### 17.2.9. Integrated pest management

#### (i) Fruit borer, *Helicoverpa armigera*

Use of African marigold (*Tagetes erecta*) as a trap crop for the management of fruit borer on tomato involves planting one row of 45 day-old marigold seedlings after every 16 rows of 25 day-old tomato seedlings (Fig. 17.3) and spraying of *Ha* NPV at 250 LE/ha or 4% NSKE or 4% pulverized NSPE, 28 and 45 DAP coinciding with peak flowering. The IPM gave 26% more yield and 28% more returns than farmers' practice (Table 17.39) (Khaderkhan *et al.*, 1998a).

**Table 17.39.** Tomato yields and net returns realized in Farmers' practice and IPM practice

| Practice | Yield (MT/ha) | % increase | Net returns (Rs.) | % increase |
|---|---|---|---|---|
| Farmers' practice | 49.40 | --- | 47,359 | --- |
| IPM practice | 62.28 | 26 | 60,168 | 28 |

**Fig. 17.3.** Integrated management of tomato fruit borer using African marigold as a trap crop.

The economics of IPM in tomato is presented in Table 17.40.

**Table 17.40.** Economics of tomato IPM

| Treatment | Yield (MT/ha) | Net returns (Rs.) | Benefit: Cost ratio |
|---|---|---|---|
| IPM plots | 62.28 | 60168 | 1.53 |
| Non-IPM plots | 49.40 | 47359 | 1.08 |

### Validated IPM technology in tomato at different locations Bangalore, Karnataka

#### Nursery

- Prepare raised seed bed of 15 cm height.

- Solarize soil for 3 weeks using transparent polythene sheet of 45 $\mu m$ thickness.
- Grow leaf curl resistant hybrid/varieties like Avinash-2.
- Treat seed with *Trichoderma viride* at 4 g/kg of seed.
- Use nylon nets to prevent entry of white fly, etc.

**Main field**

- Apply neem cake at 250 kg/ha while planting or 20 DAP.
- Give wide spacing of 90 x 60 cm.
- Dip seedling roots in imidacloprid at 0.5 ml/l for 15 min before transplanting.
- Transplant 1 row of African marigold as trap crop for *Helicoverpa armigera* after every 14 rows of tomato.
- Spray imidacloprid/thiomethoxam at 0.5 ml/l at 15 DAP for white fly.
- Spray 5% NSKE at 15 DAP against leaf miner.
- Instal pheromone traps at 5/ha 20 DAP for monitoring of *H. armigera*.
- Monitor top 3 leaves for *H. armigera* eggs at flowering.
- Release of *Trichogramma pretiosum* at 1 lakh/ha 6 times after appearance of adults.
- Spray *Ha*NPV at 250 LE/ha or neem soap 3 times at 28, 35 and 42 DAP.
- Regularly collect and destroy damaged fruits.
- Spray endosulfan at 650 g a.i./ha against *H. armigera*.
- Destroy leaf curl and wilt affected plants.
- Spray mancozeb/captan at 0.2% for the control of early and late blight.

**Varanasi, Uttaranchal**

**Nursery**

- Prepare raised seed bed of 15 cm height.
- Solarize soil for 3 weeks using transparent polythene sheet of 45 $\mu m$ thickness.
- Treat seed with *Trichoderma viride* at 4 g/kg of seed.
- Spray 0.2% copper oxychloride.

**Main field**

- Spray imidacloprid at 0.5 ml/l at 15 DAT.

- Instal pheromone traps at 5/ha, for monitoring of *H. armigera.*
- Spray *Ha*NPV at 250 LE/ha.
- Release of *Trichogramma bactrae* at 1 lakh/ha 6 times.
- Regularly collect and destroy damaged fruits.
- Apply pesticides like thiomethoxam and mancozeb at 0.2% based on needs.

**Ranchi, Jarkhand**

**Nursery**

- Prepare raised seed bed of 15 cm height.
- Solarize soil for 3 weeks using transparent polythene sheet of 45 $\mu$m thickness.
- Grow leaf curl resistant hybrid/varieties like Avinash-2.
- Treat soil with FYM enriched with *Trichoderma viride* at 1 kg/MT.

**Main field**

- Spray imidacloprid at 0.5 ml/l at 15 DAT.
- Instal pheromone traps at 5/ha, for monitoring of *H. armigera.*
- Monitor top 3 leaves for *H. armigera* eggs at flowering.
- Release of *Trichogramma chilonis* at 1 lakh/ha 6 times after appearance of adults.
- Spray *Sl*NPV at 250 LE/ha twice at 15 days interval or *Ha*NPV for *H. armigera.*
- Regularly collect and destruction damaged fruits from time to time.
- Apply pesticides like endosulfan at 0.07% based on needs.

During the period 2001-04 IPM technology in tomato was validated and promoted in more than 40 ha area in 42 villages covering 88 families located 40 km from Bangalore. Similarly, near Varanasi also IPM technology has been validated in 8 villages in about 40 ha area covering 100 families in tomato. Near Ranchi, IPM technology has been validated and promoted in 20 villages with the support of 100 farming families covering an area of 40 ha together. In IPM validation studies conducted at three locations (Bangalore, Varanasi and Ranchi), IPM fields recorded higher tomato fruit yields of 74.038, 14.250 and 22.293 MT/ha as compared to 45.056, 18.772 and 18.772 MT/ha in non-IPM fields, respectively (Amerika Singh *et al.*, 2004) (Table 17.41). The cost of production of tomato by IPM was Rs.0.95/kg as against Rs. 2.30 in non-IPM plots.

**Table 17.41.** Yield and economics of IPM in tomato at different locations

| Centre | Yield (MT/ha) | Net returns (Rs.) | Benefit : Cost ratio |
|---|---|---|---|
| Bangalore - IPM | 74.03 | 249721 | 4.82 |
| - Non IPM | 45.05 | 69704 | 0.61 |
| Varanasi - IPM | 14.25 | 39917 | 3.30 |
| - Non IPM | 13.00 | 38167 | 2.02 |
| Ranchi - IPM | 22.29 | 56705 | 1.87 |
| - Non IPM | 18.77 | 41776 | 1.32 |

### 17.2.10. Integrated disease management

#### (i) Bacterial wilt, *Ralstonia solanacearum*

Even though soil application of *Trichoderma viride* at 5 kg/ha mixed in FYM and *Glomus mosseae* at 500 g/m$^2$ gave maximum fruit yield (142.1% and 141.1% increase over control), karanj cake along with bleaching powder was more effective in reducing the apparent bacterial growth rate at flowering stage, which is the most critical stage of bacterial wilt, in which 132.2% more yield was obtained which was at par with that of *T. viride* and *G. mosseae* (Sharma and Kumar, 2009) (Table 17.42).

**Table 17.42.** Effect of microbes, plant extracts, oil cakes and chemicals on bacterial wilt and yield of tomato cv. Pusa Ruby

| Treatment | % reduction of initial inoculum | % decrease over control | % increase in yield over control |
|---|---|---|---|
| *Glomus mosseae* | 41.4 | 39.41 | 141.1 |
| *Trichoderma viride* | 29.0 | 27.52 | 142.1 |
| *Azotobacter* + Phosphobacteria | 37.3 | 27.62 | 89.9 |
| Asafoetida + Turmeric extract | 39.3 | 44.01 | 71.2 |
| Onion extract | 32.7 | 38.09 | 71.2 |
| Garlic extract | 24.4 | 9.80 | 64.4 |
| Karanj cake + Bleaching powder | 32.5 | 34.87 | 123.4 |
| Bleaching powder + lime | 36.2 | 37.37 | 107.5 |
| Urea + lime | 30.8 | 22.72 | 52.6 |
| Control | 33.6 | --- | --- |

The combined treatment of *Pseudomonas fluorescens* at 1 x 10$^{10}$ cfu/ml (seed treatment) + green manuring with sunhemp was highly effective in reducing the bacterial wilt incidence to 6.75% and significantly increasing the fruit yield to 25.5 MT/ha as compared to high wilt incidence of 55.6% and low yield of 7.8 MT/ha recorded in control (Gopalakrishnan and Ajit Kumar, 2006).

### 17.2.11. Integrated nematode management

#### (i) Root-knot nematode, *Meloidogyne* spp.

Among sixteen isolates of *Aspergillus niger* tested for the management of *M. incognita* on tomato, soil application (2 x $10^9$ cfu/kg soil) and bare root dip treatment (10 x $10^9$ cfu/100 ml water) of SkNAn5 isolate was found most effective in suppressing the root-knot nematode population by 62% and in increasing the fruit yield by 42% (Anwar and Khan, 2010).

Seed treatment with *Pseudomonas fluorescens* at 10 g/kg seed 24 hours before sowing and seed treatment with *Azospirillum* at 100 g/kg just before sowing in nursery and main field treatment with *P. fluorescens* at 2 kg/ha, *Azospirillum* at 2 kg/ha and *Phosphobacterium* at 2 kg/ha was found effective in reducing nematode population in soil and root by 48.9% and 57.8%, respectively over control. The above treatment also gave 58.9% increase in fruit yield and benefit-cost ratio of 4.2 (Shanthi, 2010).

Tomato seedlings raised in nursery beds treated with *P. chlamydosopria* and *P. penetrans* when transplanted in the main field had significantly lower root-knot index, number of eggs per egg mass, nematode population in roots and soil. The above treatment also increased root colonization and egg parasitization with *P. chlamydosporia,* infection of *M. incognita* females with *P. penetrans* and increased tomato fruit yield. This method of integrated nematode management resulted in significant reduction in the amount of inocula of the bioagents required for the treatment of soil only in the nursery beds (Rao *et al.*, 1998) (Table 17.43).

**Table 17.43.** Effect of integration of *Pochonia chlamydosopria* with *Pateuria penetrans* on root galling, root colonization and egg parasitization by *P. chlamydosopria* and *Meloidogyne incognita* female parasitization by *P. penetrans* on tomato

| Treatment | Root-knot index | Yield (kg) / 6 $m^2$ | % eggs parasitized by *P. chlamydosopria* |
|---|---|---|---|
| *P. chlamydosopria* | 5.8 | 7.4 | 43 |
| *P. penetrans* | 6.6 | 6.3 | --- |
| *P. chlamydosopria* + *P. penetrans* | 4.7 | 8.2 | 57 |
| Control | 8.0 | 5.1 | --- |
| CD (P = 0.05) | 0.78 | 0.88 | 4.72 |

Integration of 3 summer ploughings at 10 days interval during May-June with spot application of aldicarb at 1.0 kg a.i./ha while transplanting healthy tomato seedlings of 8-weeks-old gave effective control of *M. javanica* (85.5 to 93.7% reduction in nematode population) and gave maximum fruit yield (45% increase over control) of tomato (Jain and Bhatti, 1989).

Treatment combination of deep ploughing (20 cm deep) during June (atmospheric temperature between 42-47°C) (92.3% reduction in *M. javanica* population) + aldicarb treated nursery (0.4 g a.i./m$^2$) + aldicarb application at 1 kg a.i./ha at transplanting of tomato seedlings gave maximum yield (7.8 kg/plot) with minimum root-knot index (1.3) and final population in comparison to 3.8 kg/plot fruit yield in check.

Application of neem cake in the nursery at 100 g/m$^2$ followed by carbofuran at 1 kg a.i./ha in the main field significantly reduced the soil population of *R. reniformis* and enhanced the fruit yield of tomato by 67% (Anitha and Subramanian, 1998).

*P. penetrans* at 1 x 10$^9$ spores/m$^2$ was found to be effective in reducing the nematode population (*M. incognita*) in the main field by 72.3% over control and improved plant growth, fruit yield (7.800 MT/ha compared to 3.325 MT/ha in control) and nutritional status of tomato fruits (Ravichandra and Reddy, 2008) (Table 17.44).

**Table 17.44.** Effect of bioagent, botanical and nematicide for the management of *Meloidogyne incognita* infecting tomato

| Treatment | No. of galls/plant | No. of egg masses/plant | Root-knot index | Yield MT/ha |
|---|---|---|---|---|
| Neemark – 2% at 1.25 ml/m$^2$ | 43.18 | 20.40 | 4.0 | 5.725 |
| Carbofuran – 3 kg a.i./ha | 52.80 | 19.78 | 4.0 | 5.125 |
| *P. penetrans* -1x10$^9$ spores/m$^2$ | 27.92 | 12.22 | 2.8 | 7.800 |
| Control | 122.54 | 72.04 | 5.0 | 3.325 |
| CD (P = 0.05) | 1.0715 | 1.29 | 0.64 | --- |

The control of *M. incognita* on tomato by *P. lilacinus* was tested on a large-scale basis in the province of Herrera, Panama. The fungus-treated plots had higher yield than did the control plots and those treated with carbofuran (Table 17.45) (Jatala, 1985).

**Table 17.45.** Biological control of Meloidogyne incognita by Paecilomyces lilacinus

| Treatment | Yield (kg/plant) |
|---|---|
| *P. lilacinus* | 2.13 |
| Carbofuran at 2.5 kg a.i./ha | 1.85 |
| Control | 1.52 |
| P = 0.05 | 0.38 |

In field, planting of tomato seedlings (raised in nursery beds amended with neem cake at 0.5 kg/m$^2$ + *P. penetrans* at 28 x 10$^4$ spores/m$^2$) in pits incorporated with *P. lilacinus* (at 0.5 g/plant) gave least root galling and nematode multiplication rate and increased fruit weight and yield of tomato (Parvatha Reddy *et al.*, 1997) (Table 17.46).

**Table 17.46.** Effect of neem cake, *Pasteuria penetrans* and *Paecilomyces lilacinus* on root galling, nematode multiplication rate and yield of tomato infected with *Meloidogyne incognita*

| Treatment | | Root-knot index | Yield (kg/6 m²) |
|---|---|---|---|
| Nursery (m²) | Main field (per plant) | | |
| Neem cake -1 kg | *P. lilacinus* – 0.5 g | 3.4 | 9.168 |
| Neem cake -1 kg | *P. penetrans* (28 x $10^4$ spores) | 3.2 | 9.312 |
| *P. lilacinus* – 20 g | *P. penetrans* (28 x $10^4$ spores) | 3.0 | 9.504 |
| *P. penetrans* (28 x $10^7$ spores) | *P. lilacinus* – 0.5 g | 2.9 | 9.624 |
| Neem cake – 0.5 kg + *P. lilacinus* – 10 g | *P. penetrans* (28 x $10^4$ spores) | 2.5 | 9.672 |
| Neem cake – 0.5 kg + *P. penetrans* (28 x $10^4$ spores) | *P. lilacinus* – 0.5 g | 2.0 | 9.984 |
| Neem cake – 0.5 kg + *P. lilacinus* – 10 g + *P. penetrans* (28 x $10^4$ spores) | --- | 2.6 | 9.600 |
| Control | --- | 4.6 | 8.352 |
| CD (P = 0.05) | | 0.14 | 0.100 |

The seedlings raised in nursery treated with neem cake (at 500 g/m²) and *Glomus mosseae* (at 250 g/m² containing 16 chlamydospores/g) when planted in the main field significantly reduced root galling and increased fruit yield, root colonization with *G. mosseae* and chlamydospore population in soil (Parvatha Reddy *et al.*, 1998b; Rao *et al.*, 1995) (Table 17.47).

**Table 17.47.** Effect of integration of *Glomus mosseae* with oil cakes on root galling, nematode reproduction, root colonization, spore density and yield of tomato infected with *Meloidogyne incognita*

| Treatment | Dose/plant | Root gall index | Yield (kg/plant) |
|---|---|---|---|
| Control | --- | 4.4 | 1.44 |
| Castor cake | 50 g | 3.9 | 2.00 |
| Karanj cake | 50 g | 4.1 | 2.06 |
| Neem cake | 50 g | 3.6 | 2.23 |
| *G. mosseae* | 500 cm³ soil | 3.4 | 2.19 |
| Castor cake + *G. mosseae* | 50 g + 500 cm³ soil/m³ | 3.4 | 2.53 |
| Karanj cake + *G. mosseae* | 50 g + 500 cm³ soil/m³ | 3.0 | 2.69 |
| Neem cake + *G. mosseae* | 50 g + 500 cm³ soil/m³ | 2.8 | 2.98 |
| Carbofuran | 0.1 g | 3.0 | 2.33 |
| CD (P = 0.05) | --- | 0.27 | 0.43 |

In field, planting of tomato seedlings raised in nursery beds treated with neem cake at 500 g/m² + *G. fasciculatum* 200 g/m² containing 15 spores/g in pits incorporated with *T. harzianum* (at 0.5 g/ plant with 4 x $10^8$ spores/g) was effective in increasing tomato fruit yield, root colonization with *G. fasciculatum* and egg parasitization with *T. harzianum* and significant reduction in root galling and multiplication rate of *M. incognita* (Parvatha Reddy *et al.*, 1998a) (Table 17.48).

**Table 17.48.** Effect of integration of neem cake with *Trichoderma harzianum*/*Glomus fasciculatum* on root galling, root colonization by *G. fasciculatum*, egg mass parasitization by *T. harzianum* and yield of tomato infected with *Meloidogyne incognita*

| Treatment | Dose/m² | Root-knot index | Yield (kg/plant) |
|---|---|---|---|
| Neem cake | 1 kg | 3.1 | 1.57 |
| *T. harzianum* | 100 g | 3.2 | 1.43 |
| *G. fasciculatum* | 6000 spores | 2.6 | 1.72 |
| Neem cake + *T. harzianum* | 500 g + 50 g | 2.4 | 2.22 |
| Neem cake + *G. fasciculatum* | 500 g + 3000 spores | 2.0 | 2.64 |
| Control | --- | 4.2 | 1.10 |
| CD (P = 0.05) | --- | 0.21 | 0.38 |

Application of carbofuran at 1 kg a.i./ha in the nursery beds followed by neem cake at 400 kg/ha in the main field increased yield and reduced the gall index (Singh and Gill, 1998) (Table 17.49).

**Table 17.49.** Integrated management of root-knot nematodes in tomato

| Treatment/Dose | % Reduction in root galling | % increase in yield |
|---|---|---|
| Carbofuran (1 kg a.i./ha) + Neem cake at 400 kg/ha | 77.00 | 61.50 |
| Carbofuran (1 kg a.i./ha) + Urea (23.8 kg/ha) + Neem cake at 200 kg/ha | 67.90 | 49.20 |
| Phenamiphos (1 kg a.i./ha) + Neem cake at 400 kg/ha | 74.90 | 39.70 |
| Phenamiphos (1 kg a.i./ha) + Urea (23.8 kg/ha) + Neem cake at 200 kg/ha | 67.90 | 31.43 |

### (ii) Reniform nematode

Soil amendment with fresh glyricidia leaves, sawdust and chicken manure controlled the root-knot and reniform nematode population on tomato (Castillo, 1985). The yield increases and the nematode suppressive effect of fresh chicken manure surpassed even phenamiphos (Table 17.50).

**Table 17.50.** Effect of soil amendment with organic materials on *M. incognita* and *R.reniformis* population and tomato fruit yield.

| Treatment | No. of nematodes/300 ml soil & 1 g roots No. of weeks after treatment | | | Fruit yield (MT/ha) |
|---|---|---|---|---|
| | 2 | 6 | 11 | |
| Phenamiphos at 10 kg a.i./ha | 96 | 366 | 465 | 31.1 |
| Rice straw at 10 MT/ha | 215 | 500 | 1921 | 29.1 |
| Glyricidia leaves at 10 MT/ha | 201 | 533 | 1885 | 27.3 |
| Saw dust at 10 MT/ha | 342 | 859 | 958 | 25.3 |
| Chicken manure at 10 MT/ha | 17 | 169 | 1078 | 32.6 |
| Control | 552 | 816 | 2515 | 22.2 |

## 17.3. BRINJAL, *SOLANUM MELONGENA*

### 17.3.1. High yielding varieties/hybrids

**Table 17.51.** Increased productivity in high yielding varieties/F1 hybrids of brinjal

| HYV/F1 hybrid | Yield MT/ha | Developed at |
|---|---|---|
| Pusa Purple Long | 28-31 | IARI, New Delhi |
| Pusa Kranti | 35 | IARI, New Delhi |
| Pusa Anupam* | 37 | IARI, New Delhi |
| Pusa Bindu | 30 | IARI, New Delhi |
| Pusa Shyamla | 39-41 | IARI, New Delhi |
| Pusa Uttam | 42-45 | IARI, New Delhi |
| Pusa Upkar | 65-67 | IARI, New Delhi |
| Pusa Anmol | 50-40 | IARI, New Delhi |
| Pusa Ankur | 35 | IARI, New Delhi |
| Pusa Hybrid-5* | 52-54 | IARI, New Delhi |
| Pusa Hybrid-6* | 45-47 | IARI, New Delhi |
| Pusa Hybrid-9* | 67.5-70.0 | IARI, New Delhi |
| Arka Navneet* | 70-71 | IIHR, Bangalore |
| Arka Sheel | 38-40 | IIHR, Bangalore |
| Arka Anand | 60-65 | IIHR, Bangalore |
| Arka Kusumakar | 40 | IIHR, Bangalore |
| Arka Shirish | 39-41 | IIHR, Bangalore |
| Arka Nidhi | 48.5 | IIHR, Bangalore |
| Arka Keshav | 45 | IIHR, Bangalore |
| Arka Neelkanth | 43 | IIHR, Bangalore |
| Swarna Prathibha | 47-50 | HARP, Ranchi |
| Swarna Shyamali | 50.0-52.5 | HARP, Ranchi |
| Swarna Shree | 55-60 | HARP, Ranchi |
| Swarna Ajay* | 70-72 | HARP, Ranchi |
| Swarna Mani | 75-77 | HARP, Ranchi |
| Swarna Mohit | 70-80 | HARP, Ranchi |
| Kalpataru* | 48-50 | |
| Rajendra Baingan | 30 | RAU, Pusa |
| Punjab Chamkila | 43-45 | PAU, Ludhiana |
| BH-1 | 51-53 | OUAT, Bhubaneshwar |
| BH-2 | 51.5-53.0 | OUAT, Bhubaneshwar |
| Gujarat Oblong Brinjal-1 | 47-49 | GAU, Anand |
| GBH-1 | 56-58 | GAU, Anand |
| Pant Samrat | 39.0-41.5 | GBPUAT, Pantnagar |
| Pant Brinjal Hybrid-1* | 60-62 | GBPUAT, Pantnagar |

| | | |
|---|---|---|
| Aruna | 41.5-43.0 | PDKV, Akola |
| Anand Brinjal Small Round-2 | 41.0-42.5 | GAU, Anand |
| Anand Brinjal Hybrid-1* | 56-58 | GAU, Anand |
| Anand Brinjal Hybrid-2* | 60.0-62.5 | GAU, Anand |
| Narendra Hybrid Brinjal -1* | 45-50 | NDUAT, Faizabad |
| Narendra Hybrid Brinjal -3* | 50-55 | NDUAT, Faizabad |
| DBSR-91 | 56-58 | |
| Kashi Prakash | 65.0-67.5 | IIVR, Varanasi |
| Kashi Taru | 70-72 | IIVR, Varanasi |
| Kashi Sandesh* | 80-82 | IIVR, Varanasi |
| Kashi Ganesh* | 80-82 | IIVR, Varanasi |
| Shyamal* | 70-75 | |
| KS-331 | 40.0-42.5 | |
| Green Long | 41.0-42.5 | |
| Hybrid Arbh-786* | 80 | |
| CO-2 | 35 | TNAU, Coimbatore |
| MDU-1 | 34-35 | TNAU, Coimbatore |
| PKM-1 | 34.75 | HC&RI, TNAU, Periyakulam |
| KKM-136-38 | 36-38 | HRS, TNAU, Pechiparai |
| Hisar Shyamal | 40-42.5 | HAU, Hisar |
| JC-1 | 40-45 | IGKV, Jabalpur |
| JC-2 | 48-50 | IGKV, Jabalpur |
| Krishna | 48 | MPKV, Rahuri |
| RHRBH-1 | 57.6 | MPKV, Rahuri |
| RHRBH-2 | 40.7 | MPKV, Rahuri |
| RHRBH-3 | 44 | MPKV, Rahuri |
| PLR 1 | 25.10 | TNAU, Coimbatore |
| KKM 1 | 37 | TNAU, Coimbatore |
| PPI(B) 1 | 50 | TNAU, Coimbatore |
| COBH 1 | 56.4 | TNAU, Coimbatore |
| PLR(B) 1 | 42 | TNAU, Coimbatore |

* Hybrids

**Table 17.52.** Increased productivity in high yielding varieties/F1 hybrids of brinjal with pest/disease resistance (Swamy and Sadashiva, 2004)

| Improved cultivar | Yield (MT/ha) | Resistance to pest/ disease | Developed at |
|---|---|---|---|
| Arka Anand | 60-65 | Bacterial wilt | IIHR, Bangalore |
| Arka Nidhi | 48.5-50.0 | Bacterial wilt | IIHR, Bangalore |
| Arka Keshav | 45-47 | Bacterial wilt | IIHR, Bangalore |
| Arka Neelkanth | 43 | Bacterial wilt | IIHR, Bangalore |
| Pusa Purple cluster | 30.0-32.5 | Bacterial wilt (MR) | IARI, New Delhi |

| | | | |
|---|---|---|---|
| Pusa Bhairav | 30 | Phomopsis fruit rot | IARI, New Delhi |
| Pusa Purple Round | 27.5-30.0 | Little leaf, fruit and shoot borer | IARI, New Delhi |
| Pusa Anupam | 37-40 | Bacterial wilt | IARI, New Delhi |
| Azad Kranti | 31-33 | Little leaf (T) | |
| Hisar Pragati | 34-36 | Little leaf (T) | HAU, Hisar |
| Pant Brinjal Hybrid-1* | 60-62 | Bacterial wilt, Phomopsis blight | GBPUAT, Pantnagar |
| BH-2* | 51-53 | Fruit borer (T) | OUAT, Bhubaneshwar |
| Utkal Madhuri | 32-34 | Bacterial wilt, Fusarium wilt | OUAT, Bhubaneshwar |
| Utkal Tarini | 34.5-36.5 | Bacterial wilt | OUAT, Bhubaneshwar |
| Utkal Keshari | 32.5-34.0 | Bacterial wilt (T), Fusarium wilt (T) | OUAT, Bhubaneshwar |
| Utkal Jyoti | 35.0-37.5 | Bacterial wilt (T) | OUAT, Bhubaneshwar |
| JC-1 | 40-45 | Little leaf, Phomopsis blight, Fruit borer | IGKV, Jabalpur |
| JC-2 | 48-50 | Little leaf, Phomopsis blight, Fruit borer, Fusarium wilt | IGKV, Jabalpur |
| Azad B-1 | 30-35 | Little leaf (MR), Phomopsis fruit rot & blight (MR) | CSAUAT, Kanpur |
| Azad Kranti | 31-33 | Little leaf (T), Phomopsis fruit rot (T) | CSAUAT, Kanpur |
| Phule Harit | 35-37 | Fruit borer (T), Little leaf (T) | MPKV, Rahuri |
| Krishna (Phule Hybrid-2)* | 45-47 | Fruit borer (T), Little leaf (T) | MPKV, Rahuri |
| Swarna Pratibha | 47-50 | Bacterial wilt | HARP, Ranchi |
| Swarna Shyamali | 50.0-52.5 | Bacterial wilt | HARP, Ranchi |
| Swarna Abhilamb | 50-60 | Bacterial wilt | HARP, Ranchi |
| Swarna Shree | 55-60 | Bacterial wilt (MR) | HARP, Ranchi |
| Swarna Ajay* | 70-72 | Bacterial wilt, Phomopsis blight | HARP, Ranchi |
| Narendra Hybrid Brinjal-1* | 45-50 | Little leaf | NDUAT, Faizabad |
| Narendra Hybrid Brinjal-2* | 60.0-62.5 | Little leaf (MR), Phomopsis blight (MR), Fruit borer (MR) | NDUAT, Faizabad |
| BB-64 | 35.0-37.5 | Bacterial wilt | OUAT, Bhubaneshwar |
| Annamalai | 31.0-32.5 | Aphids | AU, Annamalai Nagar |
| GBH-1 | 56-58 | Fruit borer (T), Little leaf (T) | GAU, Anand |

* Hybrids; T – Tolerant; MR – Moderately resistant

### 17.3.2. Fertigation

Microjet irrigation + 100% recommended dose of solid soluble fertilizers recorded higher fruit yield (Table 17.53) (Shinde *et al.*, 2002).

**Table 17.53.** Fertigation in brinjal

| Treatment | Fruit yield (MT/ha) |
|---|---|
| Check basin with normal fertilizers as per recommendation | 28.18 |
| Microjet + Recommended dose of conventional fertilizers | 38.87 |
| Microjet + 75% Recommended dose of solid soluble fertilizers | 40.68 |
| Microjet + 100% Recommended dose of solid soluble fertilizers | 41.51 |
| Drip + Recommended dose of conventional fertilizers | 33.32 |
| Drip + 75% Recommended dose of solid soluble fertilizers | 32.36 |
| Drip + 100% Recommended dose of solid soluble fertilizers | 36.11 |
| Drip + 125% Recommended dose of solid soluble fertilizers | 40.65 |

### 17.3.3. Integrated nutrient management

Prabhu (2001) reported that the highest yield in brinjal hybrid COBH 1 was obtained at 200: 100 kg NP ha-1 (Table 17.54).

**Table 17.54.** Effect of N and P on yield characters of brinjal hybrid COBH 1

| Treatment | Yield (MT/ha)* | | |
|---|---|---|---|
| | P 50 | P 75 | P 100 |
| N 100 | 15.98 | 18.17 | 22.11 |
| N 125 | 24.05 | 26.24 | 27.68 |
| N 150 | 29.71 | 33.51 | 37.02 |
| N 175 | 41.24 | 46.87 | 58.42 |
| N 200 | 46.06 | 57.56 | 58.51 |

* Common dose of 100 kg $K_2O$/ha

Five foliar sprays of NPK (19: 19: 19) recorded the highest fruit yield of 41.01 MT/ha followed by five foliar sprays of NPK (17: 10: 27) which recorded 38.98 MT/ha (Yassin and Kanthaswamy, 2009). The increased yield due to foliar application of water soluble fertilizer NPK five times might have supplied adequate nutrients easily in correct proportion during the growth period of crop compared to soil application.

The highest yield of brinjal fruits of 37.6 MT/ha was obtained when using 75 kg $K_2O$/ha as compared with the yield of 23.3 MT/ha harvested at the zero K treatment.

The treatment consisting of *Azospirillum brasiliense* strain B-14 in combination with N at 50 kg/ha gave 49.84% increase in brinjal (cv.

Muktokesi) fruit yield during first year and 50.70% increase during second year (Bhattacharya, 2006).

Fruit yield varied from 22.3 MT/ha in control in low fertility gradient plots (50: 60: 60 kg NPK/ha) to 81.3 MT/ha in high fertility plot with highest level of N, P and K (200: 120: 120 kg NPK/ha) in hybrid brinjal cv. BWBR-3 (Ganeshmurthy and Hebbar, 2006).

### 17.3.4. Bioregulators

Hooda *et al.* (1985) reported higher brinjal fruit yield with cytozyme seed treatment at 10% and foliar sprays at 1.25%.

Foliar spray of mixatoll at 4 mg/litre four weeks after transplanting has been found to give the highest brinjal fruit yield (Shukla and Prabhakar, 1988).

### 17.3.5. Integrated weed management

Studies conducted in N.G. Ranga Agricultural University, Hyderabad (Andhra Pradesh) revealed that mulching with black polythene (25 $\mu$) gave 10% increase in yield and 90% weed control.

### 17.3.6. Integrated pest management

#### (i) Shoot and fruit borer, *Leucinodes orbonalis*

For the management of brinjal shoot and fruit borer, combination of clipping of shoots affected by the borer at weekly interval followed by neem cake application (250 kg/ha) at 30 days after planting and 4% pulverized NSPE/1% neem soap/1% pongamia soap sprays at 60, 75, 90 and 105 days after planting was most effective with 22% borer damage as compared to 46% in control (Krishnamoorthy and Krishna Kumar, 2002) (Table 17.55).

**Table 17.55.** Incidence of shoot and fruit borer of brinjal under different treatments

| Treatments | Cumulative % of borer infestation | Yield (kg/plot) |
|---|---|---|
| Soil appln. of neem cake at 250 kg/ha at 30, 60 and 90 DAT. | 30.05 | 5.96 |
| Shoot clipping at weekly interval + neem cake at 30 DAT + 4% NSKE spray at 60, 75, 90 and 105 DAT. | 22.04 | 6.72 |
| Barrier crop of maize sown 10 days before brinjal planting (2 rows at 30 cm from brinjal and maize at 5 cm form plant to plant) + shoot clipping at weekly interval + 4% NSKE spray at 45, 60, 75 and 90 DAT. | 23.51 | 6.53 |
| Barrier crop as above + shoot clipping + endosulfan 700 g a.i./ha alternated with cypermethrin at 50 g a.i./ha at 45, 60, 75 and 90 DAT. | 29.44 | 9.01 |

| | | |
|---|---|---|
| Cypermethrin at 50 g a.i./ha at 45, 60, 75 and 90 DAT. | 36.62 | 9.59 |
| 4% NSKE spray at 45, 60, 75 and 90 DAT. | 28.54 | 14.33 |
| Control. | 45.90 | 5.33 |
| CD (P = 0.05). | 2.71 | 3.40 |

An IPM for brinjal fruit and shoot borer involving soil application of neem cake has been developed. The economics of IPM is presented in Table 17.56 (Krishnamoorthy and Krishna Kumar, 2002).

**Table 17.56.** Economics of IPM of brinjal fruit and shoot borer

| Treatment | Borer incidence (%) | Yield (MT/ha) | Net returns (Rs/ha) |
|---|---|---|---|
| Neem cake IPM | 8 | 18.33 | 47033 |
| Non-IPM (Farmers' practice) | 40 | 9.37 | (-) 8343 |

Minimum brinjal fruit infestation by *Leucinodes orbonalis* (3.74%) and maximum fruit yield (16.410 MT/ha) was recorded in treatment *Bacillus thuringiensis* var. *kurstaki* (*Btk*) + methomyl (1.0 ml + 0.8 g/litre of water) as against 12.34% fruit damage and 9.977 MT/ha fruit yield in untreated control. Treatment of *Btk* + endosulfan (1.0 ml + 0.75 ml/litre of water) was on par with the above treatment (Qureshi *et al.*, 1998) (Table 17.57).

**Table 17.57.** Bioefficacy of *Bacillus thuringiensis* var. *kurstaki* (*Btk*) and insecticides for the management of *Leucinodes orbonalis* on brinjal

| Treatment | Dose/litre of water | Fruit damage (%) | Fruit yield (MT/ha) |
|---|---|---|---|
| *Btk* (Dipel 8 L) | 1.5 ml | 10.33 | 10.549 |
| *Btk* (Dipel 8 L) | 2.0 ml | 8.78 | 12.073 |
| *Btk* + Endosulfan 35 EC | 1.0 ml + 0.75 ml | 4.74 | 15.666 |
| *Btk* + Methomyl 40 SP | 1.0 ml + 0.8 g | 3.74 | 16.410 |
| Endosulfan 35 EC | 1.5 ml | 6.10 | 12.939 |
| Methomyl 40 SP | 1.6 g | 5.60 | 14.983 |
| Control (Untreated check) | --- | 12.34 | 9.977 |
| CD (P = 0.05) | --- | 3.07 | 2.000 |

Neem oil at 4% recorded the minimum infestation of brinjal shoot and fruit borer (9.07%) which was on par with endosulfan 0.07% + neem oil 2% (9.49%), endosulfan 0.07%, endosulfan 0.07% (9.56%) + NSKE 5% (9.98%) and NSKE 5% (10.89%). The maximum fruit yield was obtained in neem oil 4% (24.48 MT/ha) which was on par with endosulfan 0.07% + neem oil 2% (23.53 MT/ha), carbaryl 0.05% (23.15 MT/ha) and endosulfan 0.07% (23.13 MT/ha) (Raja *et al*, 1998a) (Table 17.58).

**Table 17.58.** Effect of neem products and insecticides on the shoot and fruit borer infestation and yield of brinjal

| Treatment | Fruit borer damage (%) | % decrease over control | Marketable yield (MT/ha) | % increase over control |
|---|---|---|---|---|
| Endosulfan 0.035% | 16.6 | 61.38 | 20.15 | 54.28 |
| Endosulfan 0.07% | 9.56 | 77.76 | 23.13 | 77.11 |
| Carbaryl 0.05% | 10.89 | 74.66 | 23.15 | 77.26 |
| Neem oil 2% | 14.18 | 67.01 | 21.60 | 65.39 |
| Neem oil 4% | 9.07 | 78.90 | 24.48 | 87.44 |
| NSKE 5% | 10.89 | 74.66 | 21.85 | 67.30 |
| Endosulfan 0.07%+ Neem oil 2% | 9.49 | 77.92 | 23.53 | 80.17 |
| Endosulfan 0.07% + NSKE 5% | 9.98 | 76.78 | 21.86 | 67.38 |
| *Trichogramma chilonis* release at fortnightly interval | 16.18 | 62.35 | 20.30 | 55.44 |
| Control | 42.98 | --- | 13.06 | --- |
| CD (P = 0.05) | 2.83 | --- | 1.61 | --- |

The fruit weight per micro-plot (4 $m^2$) treated with *Steinernema thermophilum* and *S. glaseri* was 3.482 kg and 2.280 kg, respectively, as compared to 1.212 kg in control. The percent increase in fruit weight in micro-plots treated with *S. thermophilum* was found to be 60.18% and in those treated with *S. glaseri* was 57.84% over control. The weight of fruits per treated micro-plots with *S. thermophilum* and *S. glaseri* was found to be 640 g and 567 g, respectively as compared to 202 g in control (Rathour *et al.*, 2010).

### (ii) Epilachna beetle, *Henosepilachna vignitioctopunctata*

Neem oil 2% + endosulfan 0.035% reduced the epilachna grub population by 57.3% and was comparable to endosulfan 0.07% alone (53.9%) (Rajendran, 1998). The maximum fruit yield was obtained in endosulfan 0.07% alone (5.99 MT/ha) which was on par with endosulfan 0.07% + neem oil 2% (5.64 MT/ha) (Rajendran, 1998) (Table 17.59).

**Table 17.59.** Effect of neem oil and endosulfan on epilachna beetle on brinjal

| Treatment | Epilachna grubs/ 5 plants | % reduction over control | Fruit yield (MT/ha) |
|---|---|---|---|
| Neem oil 1% | 8.75 | 20.2 | 4.12 |
| Neem oil 2% | 7.90 | 27.3 | 4.56 |
| Neem oil 3% | 6.29 | 41.1 | 4.98 |
| Neem oil 4% | 5.30 | 49.2 | 5.49 |
| Neem oil 2% + Endosulfan 0.035% | 4.60 | 57.3 | 5.64 |
| Endosulfan 0.035% | 4.70 | 53.9 | 5.99 |

| | | | |
|---|---|---|---|
| Teepol | 10.30 | 13.8 | 4.41 |
| Control | 11.30 | 11.3 | 4.24 |
| CD (P = 0.05) | 0.48 | --- | 0.62 |

## Validated IPM technology in brinjal at Ghaziabad

### Nursery

- Raised seed bed of 10-15 cm height to avoid flooding of bed.
- Soil solarization for three weeks during June using polythene sheets of 45 $\mu$m thickness.
- Soil treatment with *Trichoderma viride* at 100 g/kg FYM. Enrichment of *T. viride* for 3 weeks before mixing in soil.

### Main field

- Erection of bird perches at 25/ha for facilitation of predation of insects.
- Soil application of neem cake at 250 kg/ha.
- Installation of delta traps and yellow sticky traps at 5/ha for hopper and white fly.
- Pheromone traps installed at 12/ha for mass trapping as well as monitoring of *Leucinodes orbonalis*.
- Soil application of neem cake at 250 kg/ha along the plant rows 30 DAP.
- Three sprays of 5% NSKE against leaf hoppers, aphids, mites depending upon the appearance of the pests.
- Six releases of egg parasitoid, *Trichgramma chilonis* at 1 lakh/ha at weekly interval for shoot and fruit borer.
- Collection and destruction of egg masses, larvae and adults of hadda beetle.
- Clipping of borer damaged shoots and collection and destruction of damaged fruits.
- Rouging out of little leaf affected plants at monthly interval.
- One spray each of imidacloprid at 0.5 ml/litre and cypermethrin at 1 ml/litre in the season.

In brinjal crop, IPM technology has been validated in about 3 ha area in Raispur village near Ghaziabad during 2003-04. IPM fields gave higher yields of 61.3 MT/ha as compared to 50.9 MT/ha in non-IPM fields (Amerika Singh *et al.*, 2004) (Table 17.60).

**Table 17.60.** Yield and economics of IPM in brinjal

| Parameter | IPM | Non IPM | % increase |
|---|---|---|---|
| Yield (MT/ha) | 61.30 | 50.90 | 20.43 |
| Net returns (Rs/ha) | 62700 | 28043 | 20.01 |
| Benefit: cost ratio | 2.27 | 1.02 | 122.55 |

### 17.3.7. Integrated nematode management

#### (i) Root-knot nematode

When brinjal seedlings raised in nursery treated with *P. chlamydosporia* at 50 g/m$^2$ were transplanted in field, there was reduction in root galling index (5.2 compared to 7.8 in control), number of egg masses in 5 g root (20 compared to 49 in control), number of juveniles in 5 g root (62 compared to 162 in control) and number of juveniles in 100 ml soil (67 compared to 152 in control) and increased plant growth and yield in eggplant at harvest (Rao *et al.*, 2003) (Table 17.61).

**Table 17.61.** Effect of *Pochonia chlamydosporia* for the management of *Meloidogyne incognita* infecting brinjal

| Treatment/Dose (g/m$^2$) | Root-knot index | Yield (kg/6 m$^2$) | % egg parasitization by bioagent | Root colonization (cfu/g) | Spore density (cfu/g soil) |
|---|---|---|---|---|---|
| *P. chlamydosporia* – 25 | 6.5 | 7.4 | 42.59 | 38480 | 32580 |
| *P. chlamydosporia* – 50 | 5.2 | 8.2 | 59.35 | 52690 | 47870 |
| *P. chlamydosporia* – 100 | 5.1 | 8.5 | 65.37 | 54680 | 45690 |
| Control | 7.8 | 5.2 | --- | --- | --- |
| CD (P = O.05) | 0.37 | 0.86 | 9.23 | 4676.34 | 4216.87 |

Integrating summer ploughing with soil solarization with polythene mulching effectively reduced *M. javanica* population in brinjal (Jain and Gupta, 1991) (Table 17.62).

**Table 17.62.** Integrated management of root-knot nematodes by summer ploughing/ fallowing and nursery bed treatment in brinjal.

| Treatment | Reduction in nematode population (%) | | Yield in kg/m$^2$ | |
|---|---|---|---|---|
| | Hisar | Vellayani | Hisar | Vellayani |
| Ploughing + Fallowing for 15 days | 87.00 | 66.20 | 28.98 | --- |
| Ploughing + Covering with polythene sheet for 15 days | 94.90 | 77.00 | 18.84 | 78.00 |
| No ploughing + No covering | 77.60 | --- | --- | --- |

Application of aldicarb at 1.0 kg a.i./ha in the nursery beds along with neem cake in main field at 400 kg/ha increased yield and reduced root galling (Singh and Gill, 1998) (Table 17.63).

**Table 17.63.** Integrated management of root-knot nematodes in brinjal with neem cake and nematicides.

| Treatment | % Reduction in gall index | % increase in yield |
|---|---|---|
| Aldicarb (1.0 kg a.i./ha) + Neem cake (400 kg/ha) | 92.90 | 58.70 |
| Carbofuran (1.0 kg a.i./ha) + Neem cake (400 kg/ha) | 68.10 | 68.10 |

Application of *Pseudomonas fluorescens* at 10g/m$^2$ + carbofuran at 0.15g a.i./m$^2$ effectively controlled initial root-knot nematode (*M. incognita*) population in the nursery which in turn protected the brinjal crop in the main field also and increased the yield by 17 to 22% (Sheela *et al.*, 2002).

Nursery treatment with carbofuran at 0.3g a.i./m$^2$ along with main field treatment with ploughing + exposing the field and ploughing + covering with polythene sheets for 15 days improved the plant growth and yield of brinjal by 20-32% (Sheela *et al.*, 2002).

Soil application of poultry manure at 10 MT/ha a week prior to transplanting + *P. lilacinus* ($10^9$/conidia/g) at 25 kg/ha spore dust with carrier at the time of transplanting; or neem cake at 2 MT/ha a week prior to transplanting + *P. lilacinus* ($10^9$/conidia/g) at 25 kg/ha spore dust with carrier at the time of transplanting; or carbofuran at 2 kg/ha in 2 equal splits (one at the time of transplanting and the other after 75 days) + *P. lilacinus* ($10^9$/conidia/g) at 25 kg/ha spore dust with carrier at the time of transplanting considerably reduced root galling and also gave higher brinjal fruit yield over control (Vyas *et al.*, 2009) (Table 17.64).

**Table 17.64.** Effect of different organic amendments, bioagent and nematicide for the management of *Meloidogyne incognita* infecting brinjal

| Treatment | Gall index (0-5 scale) | Yield (MT/ha) | ICBR |
|---|---|---|---|
| *P. lilacinus* at 25 kg/ha + Poultry manure at 10 MT/ha | 2.0 | 42.088 | 1: 18.5 |
| *P. lilacinus* at 25 kg/ha + Mustard cake at 2 MT/ha | 2.0 | 44.801 | 1: 7.7 |
| *P. lilacinus* at 25 kg/ha + Neem cake at 2 MT/ha | 1.8 | 41.745 | 1: 5.8 |
| *P. lilacinus* at 25 kg/ha + Carbofuran at 2 kg a.i./ha | 1.9 | 41.498 | 1: 13.6 |
| *P. lilacinus* at 25 kg/ha | 2.5 | 39.941 | 1: 77.6 |
| Poultry manure at 10 MT/ha | 2.3 | 40.000 | 1: 18.5 |
| Mustard cake at 2 MT/ha | 1.9 | 41.810 | 1: 6.7 |
| Neem cake at 2 MT/ha | 1.9 | 39.954 | 1: 5.2 |
| Carbofuran at 2 kg a.i./ha | 2.0 | 40.530 | 1: 14.9 |
| Control | 3.9 | 28.107 | --- |

In field, transplanting of brinjal seedlings raised in nursery treated with neem cake at 400 g/m$^2$ along with *G. mosseae* (at 500 g/m$^2$ containing 26-32 chlamydospores/g) and *P. lilacinus* (at 2 liters/m$^2$ containing 6 x $10^5$ spores/ml) gave maximum reduction in root galling, egg mass production, fecundity and final nematode population in soil and roots. The above treatment also gave highest fruit yield, root colonization with *G. mosseae* and *P. lilacinus* and egg parasitization with *P. lilacinus* (Rao and Parvatha Reddy, 2001) (Table 17.65).

**Table 17.65.** Effect of integration of *Glomus mosseae, Paecilomyces lilacinus* and neem cake on root galling, root colonization, egg parasitization and yield of egg plant infected with *Meloidogyne inc ognita*

| Treatment | Rot-knot index | Yield (kg)/6 m$^2$ | Parasitization of eggs by *P. lilacinus* (%) |
|---|---|---|---|
| *G. mosseae* | 7.2 | 6.2 | --- |
| Neem cake | 6.6 | 5.6 | --- |
| *P. lilacinus* | 7.0 | 5.3 | 31 |
| Neem cake + *G. mosseae* | 5.9 | 7.3 | --- |
| Neem cake + *P. lilacinus* | 5.7 | 6.9 | 36 |
| *G. mosseae* + *P. lilacinus* | 5.4 | 6.5 | 40 |
| Neem cake + *G. mosseae* + *P. lilacinus* | 4.8 | 7.8 | 51 |
| Control | 8.7 | 5.3 | --- |
| CD (P = 0.05) | 0.49 | 0.89 | 4.52 |

Integration of AM fungus *G. fasciculatum* at 25 g/m$^2$ (600 spores) with carbofuran at 0.5 kg a.i./ha gave significant increase in plant growth characters and yield with corresponding decrease in root galling and egg mass production, followed by *T. harzianum* at 2 g/kg soil + carbofuran at 0.5 kg a.i./ha (Saikia and Borah, 2008) (Table 17.66).

**Table 17.66.** Effect of integration of bioagents and carbofuran for the management of *Meloidogyne incognita* infecting brinjal

| Treatment | % increase in yield | No. of galls/plant | Egg masses/plant |
|---|---|---|---|
| *G. fasciculatum*- 25 g/m$^2$ | 70.12 | 51.73 | 29.30 |
| *P. penetrans* – 1 g/kg soil | 56.49 | 77.86 | 61.93 |
| *T. harzianum* -2 g/kg soil | 55.84 | 138.93 | 116.86 |
| Carbofuran – 2 kg a.i./ha | 43.50 | 40.76 | 33.83 |
| *G. fasciculatum*- 25 g/m$^2$ + Carbofuran – 0.5 kg a.i./ha | 108.44 | 20.63 | 12.30 |
| *P. penetrans* – 1 g/kg soil + Carbofuran – 0.5 kg a.i./ha | 68.83 | 33.63 | 22.43 |
| *T. harzianum* -2 g/kg soil + Carbofuran – 0.5 kg a.i./ha | 80.51 | 24.06 | 16.43 |

| | | | |
|---|---|---|---|
| Control | --- | 366.10 | 211.30 |
| CD (P = 0.05) | --- | 4.944 | 17.33 |

### 17.3.8. Integrated management of disease complex

#### i) Root-knot nematode, *Meloidogyne incognita* and bacterial wilt, *Ralstonia solanacearum* disease complex

Eggplant is prone to many soil borne diseases among which the bacterial wilt (*R. solanacearum*) in combination with root-knot nematode (*M. incognita*) takes heavy toll every year all over the world (Naik, 2004) (Fig. 17.4).

Results related to root-knot and wilt disease complex management in eggplant under field conditions revealed that the combination formulation of *T. harzianum* and *P. chlamydosporia* performed very well. The yield was 2.024 kg/3m$^2$ followed by combination of *P. fluorescens* + *P. chlamydosporia* (1.782 kg) and *T. harzianum* + *P. fluorescens* (1.680 kg). There was significant difference between these two treatments. Root gall index (RGI) and wilt disease incidence due to *Ralstonia solanacearum* (WDI) was minimum in combination treatment of *T. harzianum* + *P. chlamydosporia* followed by *P. fluorescens* + *P. chlamydosporia* and *T. harzianum* + *P. fluorescens* where the RGI was 1.8, 1.81 and 1.93, respectively and that of WDI was 22.18%, 25.13% and 27.28%, respectively (Naik, 2004).

**Fig. 17.4.** Root-knot nematode and bacterial wilt disease complex in brinjal.

Integration of two summer ploughing during May-June, half recommended dose each of carbofuran (0.75 kg a.i./ha), neem cake (7.5 g/spot), streptocycline (250 ppm at 30 ml/spot) and full dose of *T. harzianum* (150 g/spot) improved plant growth parameters and fruit yield with corresponding decrease in the nematode reproduction rate (*M. incognita*), bacterial population in the soil and bacterial wilt incidence (*R. solanacearum*) in brinjal (Hussain and Bora, 2008) (Table 17.67).

**Table 17.67.** Effect of integration of different practices for the ma..agement of disease complex in brinjal

| Treatment | No. of galls/root system | No. of egg masses/ root system | % wilt incidence | Yield (kg/4m²) |
|---|---|---|---|---|
| 2 Summer ploughings + full dose of carbofuran (1.5 kg a.i./ha) | 65 | 59 | 27.77 | 3.21 |
| 2 Summer ploughings + full dose of neem cake (15 g/spot) | 71 | 69 | 27.77 | 3.10 |
| 2 Summer ploughings + full dose of streptocycline (500 ppm at 30 ml/spot) | 128 | 112 | 38.88 | 2.43 |
| 2 Summer ploughings + full dose of *T. harzianum* (150 g/spot) | 67 | 63 | 27.77 | 3.10 |
| 2 Summer ploughings + ½ dose carbofuran + full dose *T. harzianum* | 35 | 26 | 5.53 | 3.90 |
| 2 Summer ploughings + ½ dose neem cake + full dose *T. harzianum* | 57 | 47 | 11.10 | 3.30 |
| 2 Summer ploughings + ½ dose streptocycline + full dose *T. harzianum* | 59 | 55 | 24.99 | 3.20 |
| 2 Summer ploughings + ½ dose carbofuran + ½ dose neem cake + ½ dose streptocycline | 28 | 20 | 5.53 | 4.20 |
| 2 Summer ploughings + ½ dose carbofuran + ½ dose neem cake + ½ dose streptocycline + full dose *T. harzianum* | 24 | 17 | 5.53 | 4.90 |
| Control | 362 | 207 | 66.66 | 1.6 |
| CD (P = 0.05) | 6.69 | 6.75 | 11.15 | 0.83 |

## 17.4. CHILLI, *CAPSICUM ANNUUM*

### 17.4.1. High yielding varieties/hybrids

**Table 17.68.** Increased productivity in high yielding varieties/F1 hybrids of chilli

| High yielding variety/ hybrid | Green yield (MT/ha) | Dry yield (MT/ha) | Developed at |
|---|---|---|---|
| Pusa Jwala | 7.5-9.0 | 1.8 | IARI, New Delhi |
| Pusa Sadabahar | 9.5 | 2.0 | IARI, New Delhi |
| DH-76-6 | --- | 2 | IARI, New Delhi |
| Arka Suphal | 23-25 | 3-4 | IIHR, Bangalore |
| Arka Meghna | --- | 5-6 | IIHR, Bangalore |
| Phule Jyoti | 9.25 | --- | MPKV, Rahuri |
| Musalwadi | 11-12.5 | 1.93 | MPKV, Rahuri |

| | | | |
|---|---|---|---|
| Agnirekha | 25 | --- | MPKV, Rahuri |
| Andhra Jyoti | --- | 2-3 | RARS, Lam |
| Bhagyalakshmi | --- | 1.6-4.5 | RARS, Lam |
| X-235 | --- | 3.7 | RARS, Lam |
| LCA-305 (LAM-305) | --- | 2.0-2.5 | RARS, Lam |
| LCA-353 | --- | 7.5 | RARS, Lam |
| Prashanth | 15-17 | --- | RARS, Lam |
| CO-1 | --- | 2.11 | TNAU, Coimbatore |
| CO-2 | 11-12.5 | --- | TNAU, Coimbatore |
| CO-3 | 15-18 | --- | TNAU, Coimbatore |
| PKM-1 | --- | 2.36 | TNAU, Coimbatore |
| MDU 1 | --- | 1.809 | TNAU, Coimbatore |
| K 1 | --- | 1.8 | TNAU, Coimbatore |
| K 2 | --- | 1.9 | TNAU, Coimbatore |
| Chanchal | --- | 1.5-2.0 | CSAUAT, Kanpur |
| Punjab Lal | 10-12 | --- | PAU, Ludhiana |
| Punjab Surkh | --- | 2.0-2.5 | PAU, Ludhiana |
| Jawahar Mirch 218 | 9-10 | 2-3 | JNKVV, Jabalpur |
| Jayanthi | --- | 2-3 | PDKV, Akola |
| Gujarat Vegetable Chilli-111 | 11-12.5 | --- | AAU, Anand |
| Gujarat Vegetable Chilli-101 | 14-15 | --- | AAU, Anand |
| Kashi Anmol | 20.0-22.5 | --- | IIVR, Varanasi |
| Kashi Surkh | 22-24 | 1.4-1.5 | IIVR, Varanasi |
| Anugraha | 27-30 | --- | KAU, Vellanikkara |
| RCH-1 | --- | 2-3 | ARS, Durgapura |
| Kashmir Long-1 | --- | 1.6-1.8 | SKUA&T, Srinagar |
| Utkal Ava | 12-14 | 3.5-5.0 | OUAT, Bhubaneshwar |
| Hybrid Chilli* | 42.5-45.0 | --- | |
| Jwalan | 12-16 | --- | |
| Ulka 681 $F_1$ | 11-21 | --- | |
| Tejaswini | 8.95 | --- | |

**Table 17.69.** Increased productivity in high yielding varieties/F1 hybrids of chilli with pest/disease resistance (Swamy and Sadashiva, 2004)

| Improved cultivar | Yield (MT/ha) | Resistance to pest/disease | Developed at |
|---|---|---|---|
| Arka Suphala | 25 (green)<br>3 (dry) | Powdery mildew (T) | IIHR, Bangalore |
| Arka Harita (F1) | 38.2 (green)<br>5.86 (dry) | Powdery mildew, Viruses | IIHR, Bangalore |

| | | | |
|---|---|---|---|
| Arka Meghana (F1) | 33.5 (green) 5.0 (dry) | Powdery mildew, Viruses | IIHR, Bangalore |
| Pusa Jwala | 1.8 (dry) | CMV, TMV, Leaf curl complex, Thrips (T), Mites (T), Aphids (T) | IARI, New Delhi |
| Pusa Sadabahar | 2.0 (dry) | CMV, TMV, Leaf curl complex | IARI, New Delhi |
| Chanchal | 1.5-2.0 (dry) | Leaf curl (T) | CSAUAT, Kanpur |
| Bhaskar | 1.5-1.7 (dry) | Thrips (T), Mites (T), Aphids (T) | RARS, Lam |
| LCA 305 | 2.0-3.5 (dry) | Leaf curl (T) | RARS, Lam |
| Punjab Lal | 10-12 (dry) | CMV, Leaf curl, Fruit rot (MR), Dieback (MR) | PAU, Ludhiana |
| Punjab Surkh | 2.0-2.5 (dry) | Fruit rot (T) | PAU, Ludhiana |
| Musalwadi | 11.0-12.5 (green) | Powdery mildew (T), Dieback (T) | MPKV, Rahuri |
| Phule Suryamukhi | 2.8-3.0 (dry) | Leaf curl (T) | MPKV, Rahuri |
| Kashmir Long-1 | 16-18 (green) | Fusarium wilt (MR) | SKUA&T, Srinagar |
| Utkal Ava | 12-14 (green) 3.5-5.0 (dry) | Bacterial wilt | OUAT, Bhubaneshwar |
| Kashi Surkh* | 22-24 (green) 1.4-1.5 (dry) | Thrips (T), Mites (T), Viruses (T) | IIVR, Varanasi |
| Pant C 1 | 1.2-1.5 (dry) | Mosaic (T), Leaf curl (T) | GBPUAT, Pantnagar |
| Jawahar Mirch 218 | 9-10 (green) 2-3 (dry) | Leaf curl (T), Fruit rot (T) | JNKVV, Jabalpur |

* Hybrids, T – Tolerant, MR- Moderately Resistant

### 17.4.2. Micro irrigation

According to Prabhakar and Hebbar (1998), micro-irrigation systems produced 17-29% higher marketable yield in chilli and among them micro-tube irrigation system resulted in significant reduction in investment costs. Scheduling irrigation at 40 mm depth at 0.75 IW: CPE ratio produced the highest red chilli yield of 4.704 MT/ha. This treatment also had the highest water use efficiency of 96.6 kg/ha/cm.

Singh *et al.* (2000) reported that the quantity of irrigation water used in chilli through drip system was only 41.7 cm compared with 108.0 cm through surface method. Yield obtained through drip and surface methods was 6.09 and 4.23 MT/ha, respectively. It shows that drip irrigation method resulted in water saving of 61.7% and yield increase of 20.75%, respectively.

Studies conducted at Tamil Nadu Agricultural University, Coimbatore on chilli cv. K-1 revealed that saving of 62% of water by drip irrigation was possible. The yield of chilli was increased by 25% and reduced weed infestation by 50%.

The yields of chilli with drip were substantially higher during summer (6.08 MT/ha) than in winter (3.98 MT/ha) and were superior to yields under furrow irrigation (4.23 and 3.73 MT/ha in summer and winter, respectively).

An higher yield of green fruit (7.385 MT/ha) in chilli crop with better quality fruits was obtained under drip irrigation scheduled at 0.6 Epan as compared to drip at 0.3 Epan as well as furrow irrigation at 0.3 and 0.6 Epan under Bangalore conditions. With paired row planting at 30-60-30 x 45 cm spacing, the yields were further increased to 8.990 MT/ha.

### 17.4.3. Fertigation

The highest yield was obtained with 100% recommended NPK through fertigation followed by 70% recommended N + 80% recommended P and K through fertigation (Table 17.70) (Tumbare and Bhoite, 2002).

**Table 17.70.** Fertigation in chilli

| Treatment | Fruit yield (MT/ha) |
|---|---|
| 100% RDF + Surface irrigation | 4.77 |
| 100% RDF + Drip irrigation | 5.01 |
| 100% RDF as fertigation | 6.01 |
| 70% N & 80% PK as fertigation | 5.58 |
| 70% RDF as fertigation | 4.98 |
| 50% N & 80% PK as fertigation | 4.76 |

Drip irrigation at 100% PE + 100% N and K through fertigation, recorded 2.162, 2.364 and 2.125 MT/ha dry pod yield in I crop, II crop and III crop, respectively. The treatment recorded 67% increased yield over surface irrigation at 0.90 IW/CPE ratio + entire NPK as soil application; and was on par with drip irrigation at 75% PE + 100% N and K fertigation (67.5%) and drip irrigation at 50% PE + 100% N and K fertigation (65.8%). Fertigation alone recorded 50.6%, 66.8% and 58.6% increased dry pod yield under drip irrigation at 75%, 100% and 125% PE over surface irrigation at 0.90 IW/CPE ratio + entire NPK as soil application (Murali Krisnasamy *et al.*, 2006a).

### 17.4.4. Integrated nutrient management

Application of vermicompost along with recommended dose of fertilizers gave better yield (8.36 MT/ha) as compared to FYM + recommended dose of fertilizers (7.41 MT/ha) (Table 17.71).

**Table 17.71.** Effect of combined application of organic amendments and fertilizers on yield of chilli

| Treatment | Fruit yield (MT/ha) |
|---|---|
| FYM 25 at MT/ha + Full NPK | 7.41 |
| Vermicompost at 25 MT/ha + Full NPK | 8.36 |

Application of FYM at 5 MT/ha + vermicompost at 1 MT/ha with 50% RDF had significantly higher yield (1037 kg/ha) (Shivaprasad *et al.*, 2009) (Table 17.72).

**Table 17.72.** Effect of organic and inorganic sources of nutrients on yield of chilli

| Treatment | Yield (kg/ha) |
|---|---|
| FYM at 5 MT/ha + Vermicompost at 1 MT/ha | 967 |
| FYM at 10 MT/ha + Vermicompost at 1 MT/ha | 919 |
| FYM at 5 MT/ha + Vermicompost at 1 MT/ha + 50% RDF | 1037 |
| FYM at 5 MT/ha + Vermicompost at 1 MT/ha + 100% RDF | 1034 |
| FYM at 10 MT/ha + 100% RDF | 1029 |

Arbuscular mycorrhizal fungi such as *Glomus fasciculatum* and *G. macrocarpum* improved the ionic P uptake in chilli and increased yield (Table 17.73).

**Table 17.73.** Effect of arbuscular mycorrhizal fungi on P uptake and yield of chilli

| Arbuscular mycorrhizal fungi | % increase in P uptake | % increase in yield | Reference |
|---|---|---|---|
| *Glomus fasciculatum* | 31.25 | 19.4 | Sreenivasa & Krishnaraj, 1992 |
| *G. macrocarpum* | 34.00 | 23.0 | Sreenivasa, 1994 |

Studies on the use of SOP vs MOP revealed the efficacy of SOP even at 60 kg/ha in increasing the number of fruits, yield, quality and benefit cost ratio (Ananthi, 2002) (Table 17.74).

**Table 17.74.** Effect of different potassium treatments on yield of chilli

| Treatment | Yield (MT/ha) | BCR |
|---|---|---|
| NPK at 60: 30: 30 as Urea, SSP & MOP + FYM at 25 MT/ha | 1.99 | 2.08 |
| 60 kg $K_2O$ ha-1 as SOP | 5.09 | 5.11 |
| CD (5%) | 0.47 | --- |

Dry chilli yield and net returns (1.058 MT/ha and Rs. 18,475, respectively) were highest with 100 kg N/ha (half applied after establishment of seedlings and the remaining three weeks later) and irrigation level of 1.0 IW:

CPE. However, yield was not significantly lower with 80 kg N/ha (Gulati *et al.*, 1995).

Application of recommended dose of N along with *Azospirillum* application as seed and root dip method in chilli cv. Pant C-1 recorded the highest dry fruit yield (140.25 g/plant) (Kanthaswamy and Selvaraj, 2006).

### 17.4.5. Bioregulators

The highest chilli fruit yield (26.825 MT/ha) was obtained with 90 ppm NAA + 1% urea compared with 8.430 MT/ha yield in control (Katwal and Saraf, 1990).

Foliar application of ethrel increased the yield of red ripe fruits by 32.8 to 46.8%. Picking at the green stage increased the fruit yield by 68.8% compared with picking at red ripe stage (Singh *et al.*, 1992).

### 17.4.6. Intercropping

The yield of chilli was not affected by the intercrops. The chilli yields were in the order of chilli monocrop-46.76 MT/ha, chilli + onion-46.53 MT/ha (Kadalli *et al.*, 1988). Three rows of onion reduced the yields of chilli, but early maturing variety of onion yield increased from 3.406 MT/ha to 5.465 MT/ha from three rows plot (Elangovan *et al.*, 1982).

### 17.4.7. Mulching

Mulching in chilli with dry grass or plastic film resulted in 13 and 4.6% increase in fruit yield as compared to non-mulched crop (Prabhakar and Hebbar, 1998).

### 17.4.8. Integrated weed management

In chilli, best control of broad leaf weeds was achieved by oxyfluorfen at 0.3 kg/ha (92%). The above treatment also showed improvement in crop growth and fruits per plant giving the maximum dry chilli yield (2.21 MT/ha) compared with other weed control treatments (Singh *et al.*, 1995).

### 17.4.9. Integrated pest management

#### i) Fruit borer, *Helicoverpa armigera*

Planting of one row of African marigold after every 18 rows of chilli was very efficient in controlling the larval population of *H. armigera* on chilli. There was a significant reduction in mean larval population of *H. armigera* (1.10 larvae/plant). The percentage fruit damage of 8.88% was lowest and the fruit yield was highest (2.822 MT/ha of dry chillies) in this treatment (Shivaramu and Kulkarni, 2001).

### (ii) Thrips, *Scirtothrips dorsalis*

Pongamia oil 2 ml when mixed with 0.5 g acephate + 2 ml sticker gave excellent control of thrips in chillies, even during summer (Krishnamoorthy and Krishna Kumar, 2002). The thrips damage rating was reduced to 20.67 in the above treatment as compared to 45.67 in control. Similarly, the yield was also increased in treated plots over control (Table 17.75).

**Table 17.75.** Thrips damage and yield under different management programmes in chilli (Summer 2002)

| Treatments | Doses | Rating of thrips damage/5 leaves | Yield (kg/plot) |
|---|---|---|---|
| Acepahte + Pongamia oil | 0.5 g+2 ml | 20.67 | 1.69 |
| Dimethoate + Neem oil | 2 ml+2ml | 30.00 | 1.26 |
| Pongamia oil | 20 ml | 45.67 | 1.05 |
| Neem oil | 20 ml | 40.67 | 3.50 |
| Dimetoate | 2 ml | 25.00 | 0.90 |
| Neem seed powder | 4% | 40.00 | 0.05 |
| Acephate | 0.75 g | 10.67 | 1.35 |
| Control | --- | 45.67 | 0.02 |
| CD (P = 0.05) | --- | 7.33 | 0.10 |

## 17.4.10. Integrated disease management

### i) Leaf curl virus (Integrated management of insect-borne viruses)

At Coimbatore, Soil application of carbofuran at 1 kg a.i./ha + covering the nursery bed with nylon net of 400 mesh followed by 2 sprays of 2% neem oil at 15 and 35 DAP recorded lesser virus incidence and highest yield with cost: benefit ratio of 1: 3.3.

At Sabour, Use of nylon net and soil application of carbofuran at 1 kg a.i./ha in seed bed and application of carbofuran at 1.5 kg a.i./ha in the main field a week after transplanting + 3 sprays of nuvacron at 1 ml/litre was most effective in managing the disease with cost: benefit ratio of 1: 5.64.

## 17.5. SWEET PEPPER, *CAPSICUM ANNUUM*

## 17.5.1. High yielding varieties/hybrids

**Table 17.76.** Increased productivity in high yielding varieties/F1 hybrids of sweet pepper

| High yielding variety | Yield (MT/ha) | Developed at |
|---|---|---|
| Pusa Kesar | 30 | IARI, New Delhi |
| Pusa Meghali | 25 | IARI, New Delhi |

| | | |
|---|---|---|
| Pusa Deepti | 25-30 | IARI, New Delhi |
| California Wonder | 12.5-15.0 | IARI, New Delhi |
| Arka Mohini | 25.0-27.5 | IIHR, Bangalore |
| Arka Basant | 15-20 | IIHR, Bangalore |
| Arka Gaurav | 18-25 | IIHR, Bangalore |
| Solan Hybrid | 30-35 | YSPUHF, Solan |
| Nishat-1 | 20.0-22.5 | SKUA&T, Srinagar |
| DARL-202* | 34-37 | DARL, Pithoragarh, Uttarakhand |
| KTCPH-3 | 22.5-25.0 | |

* Hybrids

**Table 17.77.** Increased productivity in high yielding varieties/F1 hybrids of sweet pepper with disease resistance

| Improved cultivar | Yield (MT/ha) | Resistance to disease | Developed at |
|---|---|---|---|
| Arka Gurav | 18-25 | Bacterial wilt (T) | IIHR, Bangalore |
| DARL-202* | 34-37 | *Fusarium* wilt | DARL, Pithoragarh, Uttarakhand |

* T - Tolerant

### 17.5.2. Micro irrigation

The performance of paprika was good with all micro-irrigation systems tested as compared to that under furrow method of irrigation. The yields were higher with 0.75 Epan loss replenishment as compared to those with 0.5 Epan replenishment level in all irrigation methods tested (Table 17.78). The margin of reduction in yields with decreased irrigation regime was quite substantial in furrow irrigation than with drip irrigation (Prabhakar and Hebbar, 1998).

**Table 17.78.** Yield and water use of paprika under different irrigation systems

| Irrigation system | Dry chilli yield (MT/ha) | | Water use efficiency (kg/ha/mm) | |
|---|---|---|---|---|
| | 0.75 Epan | 0.50 Epan | 0.75 Epan | 0.50 Epan |
| Micro tube | 1.099 | 0.879 | 28.2 | 34.1 |
| Chapin | 0.925 | 0.740 | 23.7 | 28.7 |
| Drip | 1.502 | 1.101 | 38.5 | 42.7 |
| Micro-sprinkler | 1.229 | 1.059 | 33.3 | 41.0 |
| Furrow | 0.837 | 0.637 | 21.4 | 24.7 |

Capsicum yields with trickle irrigation were higher (7.4MT/ha) than those under sprinkler irrigation (5.9 MT/ha) although similar soil moisture tensions were maintained under both the systems.

Drip irrigation scheduled at 0.6 Epan gave higher yield (7.36 MT/ha) than furrow irrigation scheduled at 0.6 Epan (6.08 MT/ha).

Paprika performed well under drip irrigation producing 0.4 MT/1000 $m^2$ with 180 $m^3$ of water.

### 17.5.3. Fertigation

Application of 300: 300: 300 kg NPK/ha in the form of water soluble fertilizers through fertigation at fortnightly intervals is recommended for getting maximum productivity and best quality of sweet pepper fruits under cost effective greenhouse (Khan *et al.*, 2001) (Table 17.79).

**Table 17.79.** Productivity of sweet pepper cv. Hybrid Green Gold as influenced by sources and level of fertilizers

| Treatment | Fruit yield (MT/ha) |
|---|---|
| Normal fertilizers 80% fertigation | 75 |
| Normal fertilizers 100% fertigation | 82 |
| Normal fertilizers 120% fertigation | 88 |
| Water soluble fertilizers 80% fertigation | 86 |
| Water soluble fertilizers 100% fertigation | 94 |
| Water soluble fertilizers 120% fertigation | 96 |

### 17.5.4. Protected cultivation

Sasikala *et al.* (2007) reported that package consisting of soil: FYM: coir pith (2: 1: 1) as growing medium, irrigation at 20 kPa, INM with 50 kg each of NPK ha-1 as basal with straight fertilizer and 150 kg each of NPK ha-1 through fertigation with water soluble fertilizer and mulching in capsicum hybrid 'Indra' recorded the highest number of fruits per plant (48.3), average fruit weight (148 g), yield per hectare (144 MT) and B/C ratio (3.72) (Nagre *et al.*, 2006a) (Table 17.80).

**Table 17.80.** Effect of growing media, irrigation regime, fertigation and mulching on growth and yield of capsicum hybrid 'Indra'

| Treatment | Yield (MT/ha) | B/C ratio |
|---|---|---|
| Soil: FYM: Coir pith (2: 1: 1) + 20 kPa + 50kg NPK/ha SF (basal) + 150 kg NPK/ha WSF (fertigation) + Mulch (black polythene) | 144 | 3.72 |
| Soil: FYM: Saw dust (2: 1: 1) + 20 kPa + 50kg NPK/ha SF (basal) + 150 kg NPK/ha WSF (fertigation) + Mulch (black polythene) | 116 | 3.21 |

Fertigation with plastic mulch gave maximum fruit yield (Table 17.81).

**Table 17.81.** Performance of greenhouse capsicum in relation to mulching and method of fertilization

| Treatment | Fruit yield (MT/ha) |
|---|---|
| Soil application without mulch | 22.37 |
| Soil application with plastic mulch | 29.41 |
| Fertigation without plastic mulch | 33.37 |
| Fertigation with plastic mulch | 49.13 |

**Shade net technology for sweet pepper and paprika**

Standardized influence of shade on growth and productivity in sweet pepper and paprika is presented in Table 17.82.

Variety : Sweet pepper – Indra
Paprika – Kt – PL – 19

Shade : 35% as against earlier 50%

Spacing : Paired row planting 80 x 40 x 60 cm (between pairs, rows and plants)

**Table 17.82.** Economics of shade net technology for sweet pepper and paprika

| Particulars | Shade net (35%) | | Open | |
|---|---|---|---|---|
| | Sweet pepper | Paprika | Sweet pepper | Paprika |
| Yield (MT/ha) | 95.32 | 9.32 | 15.3 | 1.35 |
| B: C ratio | 3.85 | 2.66 | 1.58 | 1.26 |

Out of seven *Capsicum* hybrids, the hybrid 'Bharat' was significantly higher in having marketable yield (1.118 kg/plant) and total yield (1.170 kg/plant) as compared to other hybrids (0.468 – 0.910 kg/plant marketable and 0.511-930 kg/plant total yield). Fruits were more uniform, larger in size, and mature one month earlier to conventional cultivation.

Application of 50% inorganic + 50% organic nutrients recorded significantly higher capsicum fruit yield/m$^2$ (3.78 kg) compared to application of 100% inorganic and 100% organic nutrients. Fruit yield/m$^2$ was significantly higher (3.93 kg) in 1m$^2$ bed with spacing of 45 x 30 cm (4.4 plants/m$^2$) since more number of plants were accommodated as compared to 0.6m$^2$ bed with a spacing of 45 x 45 cm (3.7 plants/m$^2$) (3.44 kg) (Mantur *et al.*, 2006).

### 17.5.5. Integrated pest management

IPM comprising of three releases of *Trichogramma japonicum* followed by sequential application of *Ha*NPV at 250 LE/ha, 3% neem oil and need based application of 0.07% endosulfan gave effective control of pests and increased the fruit yield (22.4 MT/ha) as compared to non-IPM fields (18.2 MT/ha) (Sachan *et al.*, 2006).

### 17.5.6. Integrated nematode management

#### i) Root-knot nematodes, *Meloidogyne* spp.

When capsicum seedlings raised in nursery treated with *P. lilacinus* + *P. chlamydosporia* were transplanted in the main field, there was reduction in root gall index by 51% and increase in fruit yield by 44% compared to control (Naik, 2004) (Table 17.83).

**Table 17.83.** Effect of integration of bioagents for the management of *Meloidogyne incognita* and fruit yield of capsicum.

| Treatment | Dose/m$^2$ | Root-knot index (1-10 scale) | Yield in (kg/2m$^2$) |
|---|---|---|---|
| *P. chlamydosporia* | 50 g ($10^6$ cfu/g) | 6.20 | 2.58 |
| *T. harzianum* | 50 g ($10^6$ cfu/g) | 7.20 | 2.36 |
| *P. lilacinus* | 50 g ($10^6$ cfu/g) | 6.70 | 2.61 |
| *P. chlamydosporia* + *P. lilacinus* | 25 g + 25 g | 4.35 | 3.28 |
| *T. harzianum* + *P. lilacinus* | 25 g + 25 g | 4.75 | 2.86 |
| *P. chlamydosporia* + *T. harzianum* | 25 g + 25 g | 5.35 | 2.07 |
| *P. chlamydosporia* + *T. harzianum* + *P. lilacinus* | 17 g + 17 g + 17 g | 5.10 | 2.05 |
| Carbofuran | 25 g | 7.80 | 2.30 |
| Control | --- | 8.75 | 2.04 |
| CD (P = 0.05) | --- | 0.44 | 0.77 |

Capsicum seedlings raised in nursery treated with *P. fluorescens* at 50g/kg seed (seed treatment) combined with nursery bed treatment with *P. chlamydosporia* at 50g/m$^2$ when transplanted in field, significantly reduced nematode population both in soil and roots, root galling and increased root colonization by bioagents, propagule density in soil, parasitization of eggs and fruit yield. The main purpose of these studies was to raise capsicum seedlings that are colonized by *P. chlamydosporia* before transplanting so that they could carry the bio-agent to the field. During this process the field soil would be enriched with the propagules of eco-friendly component of management (here it is a bio-agent) in two or three seasons (Rao *et al.*, 2004) (Table 17.84).

**Table 17.84.** Effect of integration of bioagents on plant growth and management of *Meloidogyne incognita* infecting capsicum under field conditions

| Treatment | Root-knot index (1-10 scale) | Yield (kg/ 6 m$^2$) | % eggs parasitized by *P. chlamydosporia* |
|---|---|---|---|
| *P. chlamydosporia*-Nursery treat (25 g/m$^2$) | 6.3 | 4.4 | 41.00 |
| *P. chlamydosporia*-Nursery treat (50 g/m$^2$) | 5.6 | 4.7 | 45.00 |

| | | | |
|---|---|---|---|
| *P. fluorescens*-Seed treat (50g/kg) | 6.0 | 5.1 | --- |
| *P. fluorescens*-Seed treat (50g/kg) | 4.8 | 5.3 | 40.00 |
| *P. fluorescens*-Seed treat (50g/kg) + *P. chlamydosporia*-Nursery treat (50 g/m$^2$) | 4.4 | 5.5 | 42.67 |
| Control | 8.1 | 4.0 | --- |
| CD (P = 0.05) | 0.45 | 0.28 | 3.78 |

### 17.5.7. Integrated management of disease complex

#### (i) Root-knot nematode, *Meloidogyne incognita* and bacterial wilt, *Ralstonia solanacearum* disease complex

Capsicum is prone to many soil borne diseases among which the bacterial wilt (*R. solanacearum*) in combination with root-knot nematode (*M. incognita*) takes heavy toll every year all over the world (Naik, 2004).

In capsicum, based on the field performance of bio-agents, Naik (2004) found that combination of *T. harzianum* along with *P. fluorescens* increased the yield (3320g/3m$^2$ plot) followed by combination of *T. harzianum* and *P. chlamydosporia* (2960 g/3m$^2$). The least gall index was present where the combination of *T. harzianum* + *P. fluorescens* was used (1.5) followed by *T. harzianum* + *P. chlamydosporia* (1.7) and *P. fluorescens* + *P. chlamydosporia* (1.7). But among combination treatment of bio-agents, all the treatments were on par. Results related to the percent bacterial wilt disease (*Ralstonia solanacearum)* incidence also showed similar trend in which *T. harzianum* + *P. fluorescens* performed very well and the percent incidence was 16.90 followed by *T. harzianum* + *P. chlamydosporia* (17.20).

Combined application of neem based formulations of *P. fluorescens* and *P. chlamydosporia* / *T. harzianum* at 40 g/m$^2$ in nursery beds and transplanting these seedlings in the main field resulted in significant reduction in disease index and root-knot index in capsicum to the tune of 70% and increased the crop yield by 37% (Rao *et al.*, 2002) (Table 17.85). Combinations of the bioagents did not affect the colonization of the individual bioagents on the roots and hence the transplants carried the bioagents to the main field. This has resulted in the effective management of the pathogens involved in the disease complex.

**Table 17.85.** Effect of integration of neem-based bioagents on plant growth and management of root-knot and bacterial wilt disease complex and yield of capsicum.

| Treatment | Seedling weight (g) | Root-knot index (1-10) | Disease index (1-9) | Yield (kg/4m$^2$) |
|---|---|---|---|---|
| Seed treat. with *P. fluorescens* | 421 | 5.6 | 6.4 | 4.3 |
| Seed treat. with neem-based *P. fluorescens* | 428 | 5.2 | 6.7 | 4.7 |
| Nursery treat. with *P. fluorescens* | 435 | 4.6 | 5.4 | 4.8 |

| | | | | |
|---|---|---|---|---|
| Nursery treat. with *P. chlamydosporia* | 364 | 4.4 | 7.3 | 3.2 |
| Nursery treat. with *T. harzianum* | 374 | 4.8 | 7.0 | 3.4 |
| Nursery treat. with *P. fluorescens* + *P. chlamydosporia* | 463 | 4.1 | 5.2 | 4.0 |
| Nursery treat. with *P. fluorescens* + *T. harzianum* | 493 | 3.8 | 3.5 | 5.1 |
| Control | 340 | 8.7 | 8.2 | 2.6 |
| CD (P = 0.05) | 27.20 | 0.49 | 0.38 | 0.25 |

Capsicum seedlings raised in nursery treated with soil application of organically developed *P. lilacinus* (50 g/m$^2$ of formulated product containing $10^6$ spores/g) along with of *B. pumilis* ($10^8$ cfu/ml) when transplanted in field significantly reduced root galling and wilt incidence, and increased root colonization, propagule density in soil, egg parasitization and fruit yield in the main field (Naik *et al.*, 2003) (Table 17.86).

**Table 17.86.** Effect of integration of bioagents on root galling, wilt incidence, root colonization, propagule density and yield of sweet pepper under field conditions

| Treatment | Root-knot index | Mortality (%) | Yield (kg/6m$^2$) |
|---|---|---|---|
| Nursery bed treatment with *P. lilacinus* (25 g/m$^2$) | 6.5 | 68.90 | 4.8 |
| Nursery bed treatment with *P. lilacinus* (50 g/m$^2$) | 5.3 | 52.45 | 5.4 |
| Nursery bed treatment with *B. pumilis* | 7.0 | 34.62 | 7.5 |
| Nursery bed treatment with *P. lilacinus* (25 g/m$^2$) + Nursery bed treatment with *B. pumilis* | 4.3 | 25.86 | 8.2 |
| Nursery bed treatment with *P. lilacinus* (50 g/m$^2$) + Nursery bed treatment with *B. pumilis* | 4.1 | 17.69 | 8.5 |
| Control | 8.4 | 91.45 | 0.5 |
| CD (P = 0.05) | 0.58 | 8.94 | 0.35 |

### ii) Root-knot nematode, *Meloidogyne incognita* and foot and root rot, *Phytophthora capsici* disease complex

Capsicum seed treatment with consortia of *P. fluorescens* (3 strains) at the rate of 10 g/kg and subsequent application of consortia formulation of *P. fluorescens* at the rate of 15 g/m$^2$, significantly reduced *M. incognita* in capsicum roots by 74%; reduced the incidence of *Phytophthora* foot and root rot and increased the yield by 26%. Cost benefit ratio (calculated for the additional cost of the bio-pesticides and additional returns accrued by the application of the bio-pesticide) was 1: 3.8 (Rao *et al.*, 2009).

## 17.6. OKRA, *ABELMOSCHUS ESCULENTUS*

### 17.6.1. High yielding varieties/hybrids

**Table 17.87.** High yielding varieties/hybrids of okra

| HYV/F1 hybrid | Yield MT/ha | Developed at |
|---|---|---|
| Pusa Sawani | 15.0-17.5 | IARI, New Delhi |
| Pusa A-4 | 14 | IARI, New Delhi |
| Arka Abhay | 18-20 | IIHR, Bangalore |
| Arka Anamika | 20 | IIHR, Bangalore |
| Phule Kirti | 37-39 | MPKV, Rahuri |
| Phule Utkarsha | 30.0-32.5 | MPKV, Rahuri |
| Sheetala Uphar* | 26.5-28.0 | IIVR, Varanasi |
| Sheetala Jyoti* | 22-24 | IIVR, Varanasi |
| Kashi Mahima* | 23-25 | IIVR, Varanasi |
| Kashi Vibhuti | 17-18 | IIVR, Varanasi |
| Kashi Pragati | 18-19 | IIVR, Varanasi |
| Kashi Bhairav | 16-18 | IIVR, Varanasi |
| AOH 262 | 78 | GAU, Ahmedabad |
| AOH 263 | 89.4 | GAU, Ahmedabad |
| Anjitha | 14.6 | KAU, Thrissur |
| Manjima | 16 | KAU, Thrissur |
| CO 2 | 16.51 | TNAU, Coimbatore |
| CO 3 | 16-18 | TNAU, Coimbatore |
| COBhH 1 | 22.10 | TNAU, Coimbatore |
| Azad Bhindi-1 | 27.5-30.0 | CSAUAT, Kanpur |
| Saloni | 18.4 | |
| Evergreen | 17.9 | |
| Makhamali | 13.8 | |
| P-13 | 25-30 | |

* Hybrids

**Table 17.88.** Increased productivity in high yielding varieties/F1 hybrids of okra with pest/disease resistance

| Improved cultivar | Yield (MT/ha) | Resistance to pest/disease* | Developed at |
|---|---|---|---|
| Arka Abhay | 18-20 | YVMV | IIHR, Bangalore |
| Arka Anamika | 20 | YVMV | IIHR, Bangalore |
| Pusa Sawani | 12 | YVMV | IARI, New Delhi |
| Pusa A-4 | | YVMV, Fruit borer | IARI, New Delhi |

| | | | |
|---|---|---|---|
| Punjab Padmini | 10.0-12.5 | YVMV, Jassids (T), Fruit borer (T) | PAU, Ludhiana |
| Punjab-7 | 9.5-11.0 | YVMV | PAU, Ludhiana |
| Varsha Uphar | 10-12 | YVMV | HAU, Hisar |
| CO-3 | 16-18 | YVMV (MR) | TNAU, Coimbatore |
| Sheetala Uphar | 26.5-28.0 | YVMV, ELCV | IIVR, Varanasi |
| Sheetala Jyoti | 22-24 | YVMV, ELCV | IIVR, Varanasi |
| Kashi Pragati | 13-15 | YVMV, ELCV | IIVR, Varanasi |
| Kashi Vibhuti | 18-20 | YVMV, ELCV | IIVR, Varanasi |
| Kashi Bhairav | 16-18 | YVMV, ELCV | IIVR, Varanasi |
| Kashi Lila | 15-17 | YVMV | IIVR, Varanasi |
| Kashi Mahima | 23-25 | YVMV, ELCV | IIVR, Varanasi |
| Kashi Mohini | 13-15 | ELCV | IIVR, Varanasi |
| Kashi Mangali | 16.0-17.5 | ELCV | IIVR, Varanasi |
| Phule Utkarsha | 30.0-32.5 | YVMV (T) | MPKV, Rahuri |
| HBH-142 | 12-14 | YVMV | CCSHAU, Hisar |
| AOH 263 | | YVMV | GAU, Ahmedabad |
| Anjitha | 14.6 | YVMV | KAU, Thrissur |
| Manjima | 16 | YVMV | KAU, Thrissur |

* YVMV – Yellow Vein Mosaic Virus, ELCV – Enation Leaf Curl Virus, T – Tolerant, MR – Moderately Resistant

### 17.6.2. Crop rotation

In crop rotation, significant increase in yield of kharif okra by 28.4% was obtained due to previous crops of cauliflower in *rabi* and cluster bean in summer.

### 17.6.3. Micro irrigation

The highest okra fruit yield (11.73 MT/ha) was obtained with drip irrigation at 0.8 PEF, which was about 33.9% and 45.5% higher than the surface irrigation and drip irrigation at 0.4 PEF. The above treatment also gave highest seasonal net income (Rs.74, 459) and highest benefit-cost ratio (4.84) (Patel *et al.*, 2006).

### 17.6.4. Fertigation

The highest yield of 9.784 MT/ha was obtained with fertigation at 100% recommended dose of fertilizers applied through fertigation (Tumbare *et al.*, 1999) (Table 17.89). There was an increase of 8.04% in the yield of okra in 100% N fertigated treatment.

**Table 17.89.** Nitrogen fertigation in okra

| Nitrogen | Fruit yield (MT/ha) | % increase over conventional |
|---|---|---|
| 100% fertigation | 9.784 | 8.04 |
| 50% fertigation | 9.309 | 2.79 |
| 33% fertigation | 9.561 | 5.58 |
| 100% conventional | 9.056 | --- |

Five sprays of water soluble fertilizers (NPK 19: 19: 19) recorded the maximum okra fruit yield (298.83 g/plant) followed by 5 sprays of NPK 17: 10: 27 (265.04 g/plant). The increase in yield in the above treatments could be attributed to the increase in number of fruits produced per plant (Sundaram *et al.*, 2006).

### 17.6.5. Integrated nutrient management

Combined application of FYM with other organic manures such as neem cake, poultry manure, green leaf and enriched compost produced better okra yield as compared to FYM alone (Table 17.90).

**Table 17.90.** Effect of organic manures on yield and quality of okra.

| Treatment | Fruit yield (MT/ha) | Crude protein (%) | Ascorbic acid (mg /100g fresh fruit) | Shelf life (days) |
|---|---|---|---|---|
| FYM | 10.590 | 17.04 | 21.49 | 5.17 |
| FYM + Poultry manure | 13.538 | 17.92 | 21.78 | 5.37 |
| FYM + Neem cake | 15.848 | 17.88 | 21.82 | 5.68 |
| FYM + Green leaf | 15.278 | 17.86 | 21.80 | 5.54 |
| FYM + Enriched compost | 12.671 | 17.78 | 21.85 | 5.65 |
| CD (P = 0.05) | 1.802 | 0.014 | 0.025 | 0.05 |

Significant increase in yield was observed with respect to treatments involving vermicompost prepared from banana leaves alone, banana leaves and neem cake (6: 1 w/w), banana leaves and neem leaves (6: 1 w/w) and banana leaves and clerodendrone leaves (6: 1 w/w) with 23, 17, 24 and 17%, respectively (Jiji and Dale, 1999) (Table 17.91).

**Table 17.91.** Effect of vermicompost on the yield of okra

| Treatment | Fruit yield (g/plant) |
|---|---|
| POP recommendations | 196.67 |
| Vermicompost from banana waste alone | 243.33 |
| Vermicompost containing neem cake | 231.66 |
| Vermicompost containing neem leaves | 245.00 |
| Vermicompost containing clerodendrone leaves | 231.66 |
| CD (P = 0.05) | 24.19 |

*In situ* vermiculture at 250 worms/pot was as effective as the package of practices recommendation (Jiji and Dale, 1999) (Table 17.92).

**Table 17.92.** Effect of *in situ* vermiculture on fruit yield of okra

| Treatment | Fruit yield (kg/4 $m^2$) |
|---|---|
| POP recommendations | 5.393 |
| Vermicompost substituting FYM + Full inorganic fertilizers as per POP | 6.193 |
| *In situ* vermiculture (150 worms/plot) | 4.220 |
| *In situ* vermiculture (250 worms/plot) | 4.723 |
| Vermicompost substituting FYM + N requirement as per POP through vermicompost | 4.200 |
| CD (P = 0.05) | 0.892 |

Five foliar sprays of NPK (19: 19: 19) recorded the highest fruit yield of 21.63 MT/ha followed by five foliar sprays of NPK (17: 10: 27) which recorded 20.67 MT/ha (Kanthaswamy and Yassin, 2009). The increased yield due to foliar application of water soluble fertilizer NPK five times might have supplied adequate nutrients easily in correct proportion as and when needed (split spray) by the crop compared to soil application. Because of balanced nutrition throughout the growth period might have helped in increasing the photosynthetic efficiency of the plant thereby increasing the yield.

Combined application of 75% RDF + vermicompost at 4 MT/ha + neem cake at 1 MT/ha + biofertilizers (microbial consortium with *Azotobacter*, phosphobacteria, *Trichoderma* + *Glomus leptotrichum* in equal proportion) at 10 kg/ha recorded maximum yield of 14.07 MT/ha (Kempe Gowda *et al.*, 2009).

The organic treatment combination of FYM at 10 MT/ha + vermicompost at 2.5 MT/ha recorded the highest fruit yield (12.995 MT/ha) compared to 10.575 MT/ha in 100% RDF (Sathiyamurthy and Pugalendhi, 2009).

The marketable yield of okra increased by 13% and 23% over the control NP treatment due to potash fertilizer application at 30 and 60 kg $K_2O$/ha, respectively.

Application of 52 MT/ha of fly ash or sewage sludge individually or co-application of fly ash and sewage sludge in 50: 50 proportion increased the fruit yield of okra by 18.5, 61.9 and 64.0% over no solid waste 'control'. The increase in fruit yield of okra was attributed to elevated concentrations of both macro and micro elements in different parts. There was efficient use of nutrients during the early stage of crop up to 45 days after sowing, which contributed to higher fruit yield and biomass (Yeledahalli *et al.*, 2006).

The okra hybrid HBH-142 gave maximum marketable fruit yield (18.05 MT/ha) when spaced at 60 x 30 cm and fertilized with 50% more than recommended RDF (150 kg N and 90 kg $P_2O_5$/ha) followed by the hybrid HRB 108-2 at 60 x 20 cm spacing when fertilized with 25% more than recommended RDF (12.07 MT/ha) compared to okra cv. Varsha Uphar (7.97 MT/ha) (Yadav *et al.*, 2006).

Combined inoculation of *Azotobacter* and *Azospirillum* along with full dose of inorganic fertilizers in okra crop resulted in significantly higher fruit yield (6.03 MT/ha) (Pattanayak and Rao, 2006).

### 17.6.6. Sequential cropping

Studies conducted by IIHR, Bangalore, in peri-urban areas have shown that the adoption of sequential cropping with vegetables can increase crop land use efficiency from 300 to 500% and benefit : cost ratio from 2.0 to 2.5 (Table 17.93).

**Table 17.93.** Crop land use efficiency and benefit : cost ratio in sequential inter-cropping with vegetables in peri-urban areas of Bangalore, Karnataka.

| Cropping system | Crop land use efficiency | Benefit: cost ratio |
|---|---|---|
| Okra – Capsicum – Muskmelon | 300 | 2.0 |
| (Okra + Beans) – Capsicum – Muskmelon | 380 | 2.0 |
| (Okra + Beans) – (Capsicum + Onion) – Muskmelon | 410 | 2.3 |
| (Okra + Beans) – (Capsicum + Onion) – (Muskmelon + Radish) | 500 | 2.5 |

### 17.6.7. Mulching

Use of paddy straw mulch in okra gave good yield performance (Table 17.94).

**Table 17.94.** Increasing productivity of okra by use of mulching in summer season under rain fed conditions

| Treatment | Plant height (cm) | No of nodes | No. of irrigations | Yield (MT/ha) | B: C Ratio |
|---|---|---|---|---|---|
| Irrigation problem in non-mulching field (Farmer's practice) | 93.98 | 13.9 | 12 | 4.564 | 2.22 |
| Irrigation + paddy straw mulching | 102.87 | 16.1 | 8 | 6.366 | 3.39 |
| Irrigation + dry leaf mulching | 99.31 | 15.2 | 10 | 5.929 | 3.14 |

### 17.6.8. Integrated pest management

#### (i) Fruit borer, *Earias vitella*

The fruit damage was reduced to a minimum of 7.22% in endosulfan 0.07% + NSKE 5% which was on par with neem oil 4% (7.53%) and endosulfan 0.07% + neem oil 2% (9.29%). The marketable fruit yield was maximum in neem oil 4% (13.77MT/ha) which was on par with endosulfan 0.07% (13.40 MT/ha), endosulfan 0.07% + NSKE 5% (13.17 MT/ha), NSKE 5%

(13.08 MT/ha) and neem oil 2% (12.98 MT/ha) (Raja *et al.*, 1998b) (Table 17.95).

**Table 17.95.** Effect of neem products and endosulfan on fruit borer and yield of okra

| Treatment | Fruit borer infestation (%) | % reduction over control | Marketable fruit yield (MT/ha) |
|---|---|---|---|
| Endosulfan 0.07% | 10.55 | 61.48 | 13.40 |
| Endosulfan 0.07% + neem oil 2% | 9.29 | 66.08 | 13.28 |
| Neem oil 2% | 10.95 | 60.02 | 12.98 |
| Neem oil 4% | 7.53 | 72.51 | 13.77 |
| NSKE 5% | 10.64 | 61.15 | 13.08 |
| Endosulfan 0.07% + NSKE 5% | 7.22 | 73.64 | 13.17 |
| *Trichogramma chilonis* release at fortnightly interval | 13.91 | 49.22 | 12.20 |
| Control | 28.39 | --- | 10.67 |
| CD (P = 0.05) | 2.68 | --- | 1.12 |

Module II (First spray of monocrotophos at flowering followed by two subsequent sprays of a combination of *Bacillus thuringiensis* var. *kurstaki* and methomyl at fortnightly intervals) recorded minimum fruit borer infestation (4.21%) and maximum fruit yield (4.067 MT/ha) as against 41.59% fruit borer damage and 1.658 MT/ha fruit yield in the untreated control. However, statistically comparable results (fruit borer damage – 5.30% and yield – 4.009 MT/ha) were obtained when the second spray of *Btk* – methomyl (module II) was replaced by nimbecidine (module III) (Mathur *et al.*, 1998) (Table 17.96).

**Table 17.96.** Bioefficacy of different IPM modules for management of *Earias vitella* and yield of okra

| Module | Treatment | Dose/ha | % fruit damage | Fruit yield (MT/ha) |
|---|---|---|---|---|
| I | i) Monocrotophos 36 SL | 1.0 Litre | 8.8 | 3.41 |
| | ii) *Btk* (Dipel-8L) + Endosulfan 35 EC | 1.0 Litre + 0.625 Litre | | |
| | iii) *Btk* (Dipel-8L) + Endosulfan 35 EC | 1.0 Litre + 0.625 Litre | | |
| II | i) Monocrotophos 36 SL | 1.0 Litre | 4.21 | 4.067 |
| | ii) *Btk* (Dipel-8L) + Methomyl 40 SP | 1.0 Litre + 0.625 Litre | | |
| | iii) *Btk* (Dipel-8L) + Methomyl 40 SP | 1.0 Litre + 0.625 Litre | | |
| III | i) Monocrotophos 36 SL | 1.0 Litre | 5.30 | 4.009 |
| | ii) Nimbecidine (Azadirachtin-300 pm) | 1.5 Litre | | |
| | iii) *Btk* (Dipel-8L) + Methomyl 40 SP | 1.0 Litre + 0.625 Litre | | |
| IV | i) Monocrotophos 36 SL | 1.0 Litre | 12.42 | 2.583 |
| | ii) Endosulfan 35 EC | 1.25 Litre | | |

| | | | | |
|---|---|---|---|---|
| | iii) Endosulfan 35 EC | 1.25 Litre | | |
| V | i) Nimbecidine (Azadirachtin-300 pm) | 1.5 Litre | 16.73 | 2.100 |
| | ii) *Beauveria bassiana* (Dispel) | 1.5 Litre | | |
| | iii) *Btk* (Dipel-8L) | 1.5 Litre | | |
| VI | Control (Untreated check) | | | |
| | CD (P = 0.05) | --- | 2.24 | 0.424 |

Soil application of oiled neem cake (with 8% oil) at 250 kg/ha when combined with sprays of 1% neem soap (NS), 4% neem seed powder extract (NSPE), 1% *Bacillus thuringiensis* (*Bt*) reduced the okra fruit borer (*Earias vitella*) and leaf hopper (*Amrasca biguttula biguttula*) significantly and increased marketable okra fruit yield in all the three seasons as compared to spray treatments of NS, NSPE, *Bt* and indoxacarb alone (Table 17.97).

**Table 17.97.** Effect of neem products on fruit borer, leaf hopper and yield of okra

| Treatment | Fruit borer incidence | Plant hopper incidence | Marketable yield (MT/ha) |
|---|---|---|---|
| Neem cake – 250 kg/ha | 15.09 | 26.92 | 9.37 |
| Neem cake – 250 kg/ha + Neem seed powder | 9.19 | 19.83 | 9.81 |
| Neem cake – 250 kg/ha + Neem soap | 9.77 | 19.50 | 9.89 |
| Neem cake – 250 kg/ha + *Bacillus thuringiensis* | 8.74 | 17.50 | 11.06 |
| Neem cake – 250 kg/ha + Indoxacarb | 9.23 | 26.33 | 10.54 |
| Neem seed powder | 16.97 | 35.75 | 9.58 |
| Neem soap | 16.49 | 38.92 | 9.84 |
| *Bacillus thuringiensis* | 22.85 | 42.58 | 10.97 |
| Indoxacarb | 15.00 | 39.17 | 10.73 |
| Control | 26.61 | 58.92 | 7.20 |
| CD (P = 0.05) | 4.28 | 1.36 | 2.66 |

## Validated IPM technology in okra at Ghaziabad

- Planting of yellow vein mosaic virus (YVMV) resistant hybrids *viz.* Sun-40 and Makhmali.
- Sowing of sorghum/maize as border crop.
- Installation of yellow sticky polythene traps smeared with castor oil, and delta traps set up for white fly and other small sucking pests.
- Erection of bird perches at 25/ha for facilitating predation of borer larvae.
- Installation of pheromone traps at 5/ha for monitoring *Earias vitella*.
- Three sprays of 5% NSKE for hopper, white fly and mites starting at 28 DAS.

- Five releases of *Trichogramma chilonis* at 1 lakh/ha starting from 42 DAS at weekly interval.
- Rouging out YVMV affected plants from time to time.
- Three to four sprays of imidacloprid, cypermethrin, chlorpyriphos and fenvalerate insecticides.

In okra crop, IPM technology has been validated in about 3 ha area in Raispur village near Ghaziabad during 2003-04. IPM fields gave higher yields of 10.305 MT/ha as compared to 7.246 MT/ha in non-IPM fields (Amerika Singh *et al.*, 2004) (Table 17.98).

**Table 17.98.** Yield and economics of IPM in okra

| Parameter | IPM plots | Non IPM plots | % increase |
|---|---|---|---|
| Yield (MT/ha) | 10.30 | 7.24 | 42 |
| Net returns (Rs/ha) | 64797 | 34678 | 86 |
| Benefit: cost ratio | 1.28 | 0.72 | 77 |

### 17.6.9. Integrated nematode management

#### i) Root-knot nematodes, *Meloidogyne* spp.

*M. incognita* was responsible for 28.08 to 90.90% loss in fruit yield of okra, while *M. javanica* caused 20.20 to 41.20% loss in yield.

Soaking of okra seeds in 10% castor cake suspension mixed with spores of *P. lilacinus* (1.5 x $10^6$ spores/ml) for 30 min. and sowing in soil drenched with 10% castor cake suspension at 20 litres/6 $m^2$ was effective in reducing root galling, final nematode population of *M. incognita* and increasing the fruit yield, root colonization, propagule density in soil and parasitization of eggs by *P. lilacinus* (Rao *et al.*, 1997) (Table 17.99).

**Table 17.99.** Effect of integration of bioagents and botanicals for the management of *Meloidogyne incognita* infecting okra

| Treatment | Root-knot index | Yield (kg/6$m^2$) | % egg parasitization |
|---|---|---|---|
| Seed treat. with *P. lilacinus* | 7.6 | 5.2 | 43 |
| Seed treat. with castor cake suspension | 7.2 | 5.0 | --- |
| Soil drenching with castor cake suspension | 7.5 | 5.4 | --- |
| Seed treat. with *P. lilacinus* + castor cake suspension | 5.5 | 5.7 | 46 |
| Seed treat. with *P. lilacinus* & castor cake suspension + soil drenching with castor cake suspension | 4.7 | 6.8 | 53 |
| Seed treat. with castor cake suspension + soil drenching with castor cake suspension | 5.9 | 6.3 | --- |
| Castor cake – 1 MT/ha | 5.6 | 6.5 | --- |
| Castor cake – 2 MT/ha | 4.6 | 7.0 | --- |
| Carbofuran – 1 kg a.i./ha | 5.2 | 6.4 | --- |

| | | | |
|---|---|---|---|
| Carbofuran – 2 kg a.i./ha | 4.5 | 6.7 | --- |
| Control | 9.2 | 4.7 | --- |
| CD (P = 0.05) | 0.87 | 0.72 | 2.24 |

*Bacillus circulans* and *B. macerans* and their combinations were effective for the management of root-knot nematodes and increased the fruit yield ranging from 27 to 31% (Sheela and Jiji, 1999) (Table 17.100).

**Table 17.100.** Effect of *Bacillus circulans* and *B. macerans* on the yield and nematode population in okra

| Treatment | Fruit yield (kg/plant) | % increase over control | Root-knot nema popn. |
|---|---|---|---|
| *B. macerans* (1.3 x $10^6$ cells/ml) | 0.456 | 27.02 | 34 |
| *B. circulans* (1.2 x $10^6$ cells/ml) | 0.472 | 31.48 | 35 |
| *B. macerans* + *B. circulans* | 0.467 | 30.10 | 32 |
| Carbofuran (1.0 kg a.i./ha) | 0.589 | 64.07 | 37 |
| Carbofuran (0.5 kg a.i./ha) | 0.446 | 24.23 | 32 |
| Neem leaf (15.0 MT/ha) | 0.389 | 8.36 | 42 |
| Neem leaf (7.5 MT/ha) | 0.382 | 6.41 | 57 |
| Untreated control | 0.359 | --- | 79 |
| CD (P = 0.05) | 0.109 | --- | 5.1 |

### 17.6.10. Integrated management of disease complex

#### i) Root-knot nematode, *Meloidogyne incognita* and wilt, *Fusarium oxysporum* f. sp. *vasinfectum* disease complex

*Meloidogyne-Fusarium* disease complex has been considered important on many crops including okra leading to reduction in its productivity.

Soil application of 25 g/m$^2$ of *Pseudomonas fluorescens* (2 x $10^6$ cfu/g) or *Pochonia chlamydosporia* (2 x $10^6$ cfu/g) enriched deoiled neem cake has proved to be an effective treatment in combating the damage caused by *M. incognita* and *F. oxysporum* f. sp. *vasinfectum* to the tune of 68 and 57%, respectively. This treatment also increased the yield of okra fruits by 24% under field conditions (Chaaya *et al.*, 2010).

## 17.7. ONION, *ALLIUM CEPA*

### 17.7.1. High yielding varieties/hybrids

**Table 17.101.** High yielding varieties/hybrids of onion

| High yielding variety/hybrid | Yield (MT/ha) | Developed at |
|---|---|---|
| Pusa White Flat | 32.5-35.0 | IARI, New Delhi |
| Pusa White Round | 30.0-32.5 | IARI, New Delhi |
| Pusa Madhavi | 30 | IARI, New Delhi |
| Pusa Spanish | 28 | IARI, New Delhi |
| Pusa Red | 27.4 | IARI, New Delhi |

| | | |
|---|---|---|
| Pusa Ratnar | 30 | IARI, New Delhi |
| Early Grano | 50-60 | IARI, New Delhi |
| Arka Kalyan | 47-50 | IIHR, Bangalore |
| Arka Niketan | 42-45 | IIHR, Bangalore |
| Arka Pragathi | 40 | IIHR, Bangalore |
| Arka Pitambar | 35.0-37.5 | IIHR, Bangalore |
| Arka Kirthiman * | 45 | IIHR, Bangalore |
| Arka Lalima* | 50 | IIHR, Bangalore |
| Arka Bindu | 32.5-35.0 | IIHR, Bangalore |
| Hisar-2 | 31-33 | HAU, Hisar |
| Hisar-3 | 31.6-32.0 | HAU, Hisar |
| Punjab Selection | 30-.0-32.5 | PAU, Ludhiana |
| Punjab White | 31-33 | PAU, Ludhiana |
| Punjab Red Round | 28-30 | PAU, Ludhiana |
| Punjab-48 | 30 | PAU, Ludhiana |
| Kalyanpur Red Round | 25-30 | CSAUAT, Kanpur |
| Agrifound Dark Red | 37.5-40.0 | NHRDF, Nashik |
| Agrifound Light Red | 30.0-32.5 | NHRDF, Nashik |
| Akola Safed | 27.5-30.0 | PKV, Akola |
| Phule Safed | 23.1 | MPKV, Rahuri |
| Phule Samarth | 50-52 | MPKV, Rahuri |
| N-2-4-1 | 30.0-32.5 | Dept. of Agri., Maharashtra |
| N 257-9-1 | 25.0 | Dept. of Agri., Maharashtra |
| N 53 | 34.7 | Dept. of Agri., Maharashtra |
| Bhima Raj | 25-30 | NRCOG, Rajgurunagar |
| Bhima Super | 40-45 | NRCOG, Rajgurunagar |
| Bhima Red | 48-52 | NRCOG, Rajgurunagar |
| Udaipur 102 | 30-35 | RAU, Rajasthan |
| ALR 1987-93 | 29.7 | |
| Line No. 102 | 35.5 | |
| Basant 780 | 29.8 | |
| ADR | 33.7 | |
| Rajasthan Onion-1 | 32.5-36.0 | |
| Arpita | 35-40 | |

* Hybrids

**Table 17.102.** Onion varieties resistant to pests/diseases

| High yielding variety/hybrid | Yield (MT/ha) | Resistance to pests/diseases | Developed at |
|---|---|---|---|
| CO-4 | 20.0-22.5 | Thrips (MR) | TNAU, Coimbatore |
| CO-2 | 12 | Thrips (MR), *Alternaria* blight (MR) | TNAU, Coimbatore |

| | | | |
|---|---|---|---|
| CO-3 | 16 | Thrips (MR) | TNAU, Coimbatore |
| Arka Kalyan | 47-50 | Purple blotch (MR) | IIHR, Bangalore |
| Punjab Naroya | 37.5-40.0 | Purple blotch (T), Thrips (T), *Heliothis* (T) | PAU, Ludhiana |
| Phule Suvarna | 30-31 | Purple blotch (MR) | MPKV, Rahuri |
| Phule Samarth | 30-31 | Purple blotch (MR) | MPKV, Rahuri |

### 17.7.2. Micro irrigation

Drip system of irrigation in onion increases the yield and saves water up to 55% as conveyance, percolation and evaporation losses are eliminated. The pest and disease incidence is reduced under drip irrigation, because water is supplied just for plant requirement and thereby better control of humidity is achieved. The incidence of weeds was also found to be less under drip irrigation. In onion, early maturity, uniform growth and ripening are observed when irrigated by drip system (Soman, 2002).

The yield recorded under micro-irrigation was 13.5 MT/acre whereas furrow irrigated crop yielded only 7.0 MT/acre at Akola, Maharashtra (Table 17.103).

**Table 17.103.** Economics of onion cultivation with drip irrigation system at Akola

| Parameters | Drip | Conventional |
|---|---|---|
| Water consumption (mm) | 775 | 1410 |
| Yield (MT/acre) | 13.5 | 7.0 |
| Net income (Rs.) | 54,988 | 13,700 |
| Benefit: cost ratio | 2.20 | 1.27 |
| Water use efficiency (kg/ha mm) | 43.03 | 12.26 |

The onion bulb yields were increased by 16.0%, 18.3% and 18.8% respectively, in micro sprinkler irrigation with 60%, 80% and 100% PE compared to surface irrigation.

### 17.7.3. Fertigation

In onion, when fertigation with water soluble fertilizers is practiced, there is a saving of fertilizer up to 25% and the fertilizer use efficiency is improved. Both water and nutrient distribution is more uniform in the soil when applied through drip system.

Drip irrigation and fertigation with water soluble fertilizer resulted in a very high yield of 24 MT/acre as against the regular yield of 12 MT/acre under conventional irrigation in farmer's field at Nashik, Maharashtra (Table 17.104).

**Table 17.104.** Economics of onion cultivation with drip irrigation system at Nashik

| Parameters | Drip | Conventional |
|---|---|---|
| Water consumption (mm) | 775 | 1410 |
| Yield (MT/acre) | 24 | 12 |
| Net income (Rs.) | 165,852 | 74,700 |
| Benefit: cost ratio | 3.31 | 2.24 |
| Water use efficiency (kg/ha mm) | 76.49 | 21.02 |

Fertigation of N and K (P applied as basal fertilizer) resulted in 19.4% increased yield over soil application. Fertigation with 100% N and K recorded a mean increase of 20.5% over soil application, while 50% N and K through soil application and 50% as fertigation recorded 18.9% increased yield over soil application of entire recommended dose (Murali Krishnasamy *et al.*, 2006b).

The highest gross yield (25.224 MT/ha) and marketable yield (24.229 MT/ha) of onion bulbs were recorded in the treatment comprised of 100% NPK as water soluble fertilizer applied through drip irrigation which was on par with other fertigation treatments also. The reduction in fertilizer doses by 30-40% did not reduce the yield (Singh *et al.*, 2006).

### 17.7.4. Integrated nutrient management

Application of vermicompost along with 50% recommended dose of fertilizers in onion gave maximum yield (Table 17.105).

**Table 17.105.** Effect of manures on the yield of onion.

| Treatment | Yield (MT/ha) |
|---|---|
| Recommended dose of chemical fertilizers (RDF) | 37.67 |
| FYM – 8 MT/ha | 37.90 |
| Vermicompost – 8 MT/ha | 41.40 |
| 50% RDF + Vermicompost – 8 MT/ha | 48.70 |

Application of NPK at 100: 50: 50 + foliar spray of mancozeb at 0.25% (against *Stemphylium* blight and purple blotch) + deltamethrin at 0.095% (against thrips) gave significantly highest gross yield (31.885 MT/ha) of onion bulb and exportable yield (29.735 MT/ha) followed by treatment of FYM at 20 MT/ha + sulphur at 50 kg/ha as basal dose + spray of *Trichoderma viride* at 0.4% + *Bacillus thuringiensis* at 0.1% with gross yield (25.162 MT/ha) and exportable yield (22.016 MT/ha) (Gupta *et al.*, 2009).

The onion plants provided with *Azospirillum* + 100% NPK (125: 50: 125 kg NPK/ha) showed maximum uptake of N, P and K (157.88: 75.36: 70.46 kg NPK/ha) and gave highest bulb yield (25.029 MT/ha) under irrigated condition during *kharif* season in Central Dry Zone of Karnataka (Mahantesh *et al.*, 2009a). The same treatment also produced highest bulb yield (23.298 MT/ha) under rainfed condition during kharif season (Mahantesh *et al.*, 2009b).

Potash fertilizer use at a rate of 30-90 kg $K_2O$/ha considerably increased the yield of onion (up to 14-23%) over the control NP treatment. With balanced fertilization, onion bulbs noticeably increased in size. Of the treated plots, where K-fertilizer was applied at 60 kg $K_2O$/ha produced the highest average bulb diameter of 8.5 cm as compared with 6.9 cm at the zero K treatment. Moreover, the storage behaviour of onion was appreciably improved due to K application. After 60-days storage at room temperature, the percentage of rotted bulbs was 17% when using potash fertilizers at a rate of 90 kg $K_2O$/ha compared to 31% rotted bulbs in the zero K treatment.

Application of 50% RDF + 50% vermicompost at 5 MT/ha significantly gave higher marketable seed yield (1.467 MT/ha) which was on par with 50% RDF + 50% micoferti at 2 MT/ha (1.408 MT/ha) and 50% RDF + 50% FYM at 20 MT/ha (1.367 MT/ha) (Patil *et al.*, 2006).

Maximum onion bulb yield of 12.27 MT/ha was recorded in vermicompost at 4 MT/ha + 50% RDF treatment (Patil *et al.*, 2006).

Foliar application of organic sea weed extract at 30, 45 and 60 days after transplanting significantly improved the yield in onion (Table 17.106) (Sankar *et al.*, 2001).

**Table 17.106.** Effect of organic sea weed extract on yield of onion

| Treatment | Marketable yield (MT/ha) | % of A grade bulbs |
|---|---|---|
| Spray at 30 DAP | 43.03 | 47.2 |
| Spray at 45 DAP | 43.24 | 47.8 |
| Spray at 60 DAP | 43.52 | 44.2 |
| Spray at 75 DAP | 43.37 | 43.5 |
| Spray at 30 + 45 DAP | 44.12 | 58.2 |
| Spray at 30 + 45 + 60 DAP | 48.14 | 64.4 |
| Spray at 30 + 45 + 60 + 75 DAP | 47.57 | 58.3 |
| Control | 42.61 | 44.0 |

### 17.7.5. Intercropping

Onion gave significantly higher yield (2.88 MT/ha) in 1: 4 ratio with maize crop at Varanasi (Anon, 1990).

### 17.7.6. Integrated disease management

Katan *et al.* (1980) reported an increase of 60-125% in onion yield in solarized fields which were earlier infected with pink root rot (*Pyrenochaeta terrestris*).

## 17.8. GARLIC, *ALLIUM SATIVUM*

### 17.8.1. High yielding varieties/hybrids

**Table 17.107.** High yielding varieties/hybrids of garlic

| High yielding variety/hybrid | Yield (MT/ha) | Developed at |
|---|---|---|
| Agrifound White | 13-15 | NHRDF, Nashik |

| | | |
|---|---|---|
| Agrifound Parvati | 17.5-22.5 | NHRDF, Nashik |
| Yamuna Safed-2 | 15-20 | NHRDF, Nashik |
| Yamuna Safed-3 | 17.5-20.0 | NHRDF, Nashik |
| Yamuna Safed-4 | 17.5-20.0 | NHRDF, Nashik |
| G-1 | 15.0-17.5 | NHRDF, Nashik |
| G-41 | 13 | NHRDF, Nashik |
| G-323 | 17.5-20.0 | NHRDF, Nashik |
| Ooty 1 | 17.1 | TNAU, Coimbatore |
| Godhavari (Sel 2) | 11.750 | |
| Sweta (Sel 10) | 12.033 | |
| IC-49383 | 13.700 | |
| Jamnagar Local | 7.633 | |
| GHC-1 | 27.5-30.0 | |

**Table 17.108.** Garlic varieties resistant to pests/diseases

| High yielding variety/ hybrid | Yield (MT/ha) | Resistance to pests/diseases | Developed at |
|---|---|---|---|
| Yamuna Safed | 15.0-17.5 | Purple blotch (T), Thrips (T), *Stemphylium* blight (T) | NHRDF, Nashik |
| G-1 | 15.0-17.5 | Purple blotch (T), *Stemphylium* blight (T) | NHRDF, Nashik |
| Ooty 1 | 17.1 | Thrips, Stem & bulb nematode (*Ditylenchus dipsaci*) | TNAU, Coimbatore |

### 17.8.2. Micro irrigation

Among irrigation schedules, IW/CPE of 1.5 (13 + 3 irrigations) gave significantly highest garlic bulb yield (9.688 MT/ha) (Sinsinwar *et al.*, 2006).

### 17.8.3. Fertigation

Application of 100% recommended dose of NPK (150: 50: 80 kg NPK/ha) in the form of water soluble fertilizers through drip irrigation gave better gross garlic yield (12.665 MT/ha), marketable yield (11.487 MT/ha), grade A bulb yield (41.78%), grade B bulb yield (47.79%) and lowest C grade bulb yield (10.44%) (Verma and Gupta, 2009).

Application of 100% recommended dose of N + *Azospirillum* + phosphate solubilizing bacteria was found superior in relation to garlic yield (26.186 MT/ha), gross return (Rs.437,200), net return (Rs.352,352) and benefit-cost ratio (5.15) for commercial cultivation of garlic cv. G-282 under agro-climatic conditions of Uttar Pradesh (Bahadur *et al.*, 2009).

### 17.8.4. Integrated nutrient management

Application of vermicompost along with 50% recommended dose of fertilizers in garlic gave maximum yield (Table 17.109).

**Table 17.109.** Effect of manures on the yield of garlic.

| Treatment | Yield (MT/ha) |
|---|---|
| Recommended dose of chemical fertilizers (RDF) | 5.63 |
| FYM – 8 MT/ha | 6.80 |
| Vermicompost – 8 MT/ha | 10.60 |
| 50% RDF + Vermicompost – 8 MT/ha | 11.60 |

Application of 120: 50: 50 kg NPK + $ZnSO_4$ at 12.5 kg and biofertilizers (*Azospirillum* and *Phosphobacterium*) at 2 kg each per hectare gave increased yield (Table 17.110) (Mahendran and Kumar, 1996).

**Table 17.110.** Effect of biofertilizers on garlic yield

| Treatment | Yield (MT/ha) |
|---|---|
| 120: 50: 50 kg NPK and $ZnSO_4$ at 12.5 kg/ha | 5.52 |
| 120: 50: 50 kg NPK and $ZnSO_4$ at 12.5 kg/ha + biofertilizers (*Azospirillum* and *Phosphobacterium*) at 2 kg each/ha | 6.35 |

Application of sunnhemp at 5 MT/ha produced the highest garlic bulb yield (7.082 MT/ha) and was on par with poultry manure at 2.5 MT/ha (6.837 MT/ha) and vermicompost at 5 MT/ha (6.830 MT/ha) compared to control (2.665 MT/ha) (Shasidhar *et al.*, 2006).

Application of 80 kg N, 60 kg P with spacing of 15 x 10 cm produced highest garlic bulb yield (11.2 MT/ha) followed by application of 80 kg N, 50 kg P with spacing of 15 x 10 cm (10.4 MT/ha) (Dahiya *et al.*, 2009).

Among sulphur levels, 75 kg sulphur/ha gave significantly higher and maximum garlic bulb yield (9.438 MT/ha) (Sinsinwar *et al.*, 2006).

### 17.8.5. Integrated weed management

Application of pendimethalin at 0.75 kg a.i./ha three DAS combined with one hand weeding at 45 DAS recorded the least weed population with highest garlic bulb yield (5.856 MT/ha) during *rabi* season. During *rabi* season, Oxyfluorfen at 0.15 kg a.i./ha three DAS combined with one hand weeding at 45 DAS recorded the least weed population with highest garlic bulb yield (3.666 MT/ha) (Puttaraju *et al.*, 2009).

## 17.9. LEEK, *ALLIUM PORRUM*

### 17.9.1. Integrated nutrient management

Leek plants under treatment combination of 60% recommended dose of N + *Azotobacter* (basal) recorded maximum yield (43.441 MT/ha) compared to recommended dose of fertilizer (full inorganic) (20.194 MT/ha). The treatment was also economical and improved the fertility status of the soil. There is a chance of saving 40% inorganic N through application of biofertilizer (Bhattacharya *et al.*, 2009).

## 17.10. CARROT, *DACUS CAROTA*

### 17.10.1. High yielding varieties

**Table 17.111.** Increased productivity in high yielding varieties/F1 hybrids of carrot

| HYV/F1 hybrid | Yield MT/ha | Developed at |
|---|---|---|
| Pusa Kesar | 25 | IARI, New Delhi |
| Pusa Meghali | 25-26 | IARI, New Delhi |
| Nantes | 32.5-35.0 | IARI, RS, Katrain |
| Early Nantes | 30-35 | IARI, RS, Katrain |
| Ooty-1 | 49.10 | HRS, TNAU, Ooty |
| Hisar Gairic | 28.8 | HAU, Hisar |
| Hisar Selection-1 | 30-31 | HAU, Hisar |
| Chaman | 25-27 | SKUAT, Srinagar |
| Shalimar Carot-1 | 22.5-24.0 | |
| Hybrid Number-1* | 40-45 | |

**Table 17.112.** Carrot varieties resistant to pests/diseases

| HYV/F1 hybrid | Yield MT/ha | Resistance to | Developed at |
|---|---|---|---|
| Ooty-1 | 49.1 | Powdery mildew | HRS, TNAU, Ooty |
| Arka Suraj | | Powdery mildew (T),<br>Root-knot nematode (T) | IIHR, Bangalore |

### 17.10.2. Integrated nematode management

#### i) Root-knot nematode, *Meloidogyne incognita*

Soil application of *Paecilomyces lilacinus* (2 x $10^6$ cfu/g) at 2.5 kg + FYM at 2.5 MT/ha gave minimum number of galls and final nematode population (120 and 12.5, respectively) compared to control (440 and 235, respectively). The highest yield was also recorded in the above treatment (9.34 MT/ha) over control (6.54 MT/ha) (Somasekhara and Ravichandra, 2010).

#### ii) Reniform nematode, Rotylenchulus reniformis

Seed treatment of carrot with *P. fluorescens* at 20 g/kg significantly reduced the reniform nematode population both in soil and roots and increased root colonization by the bioagent and yield (Rao and Shylaja, 2004) (Table 17.113).

**Table 17.113.** Effect of seed treatment with *Pseudomonas fluorescens* for the management of *Rotylenchulus reniformis* infecting carrot

| Treatment/ Dose/ kg seed | No. of nemas/ 10 g roots | No. of nemas/ 100 ml soil | Yield (kg/4 m²) | Root colonization (cfu/g) |
|---|---|---|---|---|
| *P. fluorescens* – 5 g | 69 | 88 | 6.7 | 21745 |
| *P. fluorescens* – 10 g | 38 | 56 | 7.8 | 24574 |
| *P. fluorescens* – 20 g | 35 | 48 | 8.0 | 23789 |
| Control (Untreated) | 92 | 123 | 5.2 | --- |
| CD (P = 0.05) | 13.67 | 17.34 | 1.47 | 124374. |

### 17.10.3. Integrated management of disease complex

#### i) Root-knot nematode, *Meloidogyne incognita* and wilt, *Erwinia carotovora* s. sp. *carotovora* disease complex

The neem cake enriched with *Pseudomonas fluorescens* and *Paecilomyces lilacinus* applied at 10 g/m² increased the root colonization of both the bioagents and reduced the incidence of *M. incognita* and *E. carotovora* s. sp. *carotovora* by 68 and 56%, respectively. There was also a significant increase in the yield of carrot to the tune of 23% (Sowmya *et al.*, 2010).

## 17.11. RADISH, *RAPHANUS SATIVUS*

### 17.11.1. High yielding varieties

**Table 17.114.** Increased productivity in high yielding varieties/F1 hybrids of radish

| HYV/F1 hybrid | Yield MT/ha | Developed at |
|---|---|---|
| Pusa Desi | 30 | IARI, New Delhi |
| Pusa Reshmi | 35 | IARI, New Delhi |
| Pusa Himani | 32.5 | IARI, New Delhi |
| Pusa Chetki | 20-25 | IARI, New Delhi |
| Japanese White | 37.5 | IARI, RS, Katrain |
| Arka Nishant | 20-30 | IIHR, Bangalore |
| White Icicle | 22.5-25.0 | |
| Chinese Pink | 25.0-27.5 | |
| Punjab Pasand | 21.4 | PAU, Ludhiana |
| Punjab Safed | 23-26 | PAU, Ludhiana |
| Punjab Ageti | 30.0-32.5 | PAU, Ludhiana |
| CO-1 | 20-25 | HRS, TNAU, Ooty |
| Kashi Sweta | 48-50 | IIVR, Varanasi |
| Kashi Hans | 43-45 | IIVR, Varanasi |

### 17.11.2. Micro irrigation

The maximum root yield (24.58 MT/ha) was observed when irrigation was given at 7 days interval. The maximum root yield (30.89 MT/ha) was observed in 4th November sowing (Singh *et al.*, 2006).

## 17.12. BEET ROOT, *BETA VULGARIS*

### 17.12.1. High yielding varieties

**Table 17.115.** Increased productivity in high yielding varieties/F1 hybrids of beet root

| HYV/F1 hybrid | Yield MT/ha | Developed at |
|---|---|---|
| Detroit Dark Red | 25-30 | IARI, New Delhi |
| Crimson Globe | 20-25 | IARI, New Delhi |
| Ooty 1 | 31.45 | HRS, TNAU, Ooty |

## 17.13. TURNIP, *BRASSICA RAPA* VAR. *RAPIFERA*

### 17.13.1. High yielding varieties

**Table 17.116.** Increased productivity in high yielding varieties/F1 hybrids of turnip

| HYV/F1 hybrid | Yield MT/ha | Developed at |
|---|---|---|
| Purple Top White Globe | 27.5-32.5 | IARI, New Delhi |
| Pusa Chandrima | 35 | IARI, New Delhi |
| Pusa Swarnima | 35 | IARI, New Delhi |
| Snowball | 30-35 | IARI, New Delhi |

## 17.14. CABBAGE, *BRASSICA OLERACEA* VAR. *CAPITATA*; CAULIFLOWER, *B. OLERACEA* VAR. *BOTRYTIS* AND KNOL-KHOL, *B. OLERACEA* VAR. *GONGYLODES*

### 17.14.1. High yielding varieties/hybrids

**Table 17.117.** High yielding varieties/hybrids of cabbage, cauliflower and knol-khol

| Crop | High yielding variety/ hybrid | Yield (MT/ha) | Developed at |
|---|---|---|---|
| Cabbage | Pusa Drum Head | 21-22 | IARI, RS, Katrain |
| | Pusa Mukta | 30 | IARI, RS, Katrain |
| | Pusa Ageti | 33 | IARI, RS, Katrain |
| | Pusa Synthetic | 35 | IARI, RS, Katrain |
| | Golden Acre | 17.5-20.0 | IARI, RS, Katrain |
| | Pride of India | 21.5-24.0 | YSPUHF, Solan |
| | Palampur Green | 38-40 | HPKV, Palampur |
| | Kinner Red | 25.0-27.5 | HPKV, Palampur |

| | | | |
|---|---|---|---|
| | Sri Ganesh Gol* | 85.0-87.5 | |
| | Suvarna* | 27.5-30.0 | |
| | Quisto | 30.0-32.5 | |
| Cauliflower | Pusa Hybrid-2* | 25-27 | IARI, New Delhi |
| | Pusa Synthetic | 22.5 | IARI, RS, Katrain |
| | Pusa Shubra | 20.0-22.5 | IARI, New Delhi |
| | Pusa Snowball-16 | 15-20 | IARI, RS, Katrain |
| | Pusa Snowball-1 | 25 | IARI, RS, Katrain |
| | Pusa Snowball-2 | 27.5 | IARI, RS, Katrain |
| | Pusa Snowball K-1 | 30 | IARI, RS, Katrain |
| | Pusa Sharad | 26-28 | IARI, New Delhi |
| | Arka Kanthi | 22-25 | IIHR, Bangalore |
| | Hisar-1 | 25.0-27.5 | HAU, Hisar |
| | Ooty 1 | 46.4 | HRS, TNAU, Ooty |
| | Summer King* | 20.0-22.5 | |
| Knol-khol | White Vienna | 18-20 | IARI, RS, Katrain |
| | Large Green | 22.5-25.0 | IARI, RS, Katrain |
| | Palam Tender Knob | 25.0-27.5 | HPKV, Palampur |

* Hybrids

**Table 17.118.** Cabbage and cauliflower varieties resistant to diseases

| Crop | High yielding variety/hybrid | Yield (MT/ha) | Resistance to pests/diseases | Developed at |
|---|---|---|---|---|
| Cauliflower | Pusa Shubra | 25-30 | Black rot | IARI, New Delhi |
| | Pusa Snowball K-1 | 20.0-22.5 | Black rot | IARI, RS, Katrain |
| | Pusa Hybrid-2* | 25-27 | Downy mildew | IARI, New Delhi |
| | Pusa Kartik Sankar* | 14.5-16.0 | Downy mildew | IARI, RS, Katrain |
| Cabbage | Pusa Mukta | 20.0-22.5 | Black rot | IARI, RS, Katrain |
| | Pusa Drum Head | 21-22 | Black rot | IARI, RS, Katrain |

* Hybrids

### 17.14.2. Microirrigation

#### Cabbage

Drip irrigation with 0.8 PEF resulted in highest cabbage yield of 25.45 MT/ha, which was at par with drip irrigation at 0.6 PEF. As compared to the surface irrigation, the yield was 29.5% higher. The above treatment also gave highest net seasonal income (Rs. 43,802) and highest benefit-cost ratio (3.21) (Patel *et al.*, 2006).

#### Cauliflower

A considerable savings in water, besides increased yields of cauliflower under drip irrigation as compared to conventional methods has been reported by Ahluwalia *et al.* (1993) (Table 17.119).

**Table 17.119.** Yield and water use by cauliflower under drip and surface irrigation

| Irrigation method | Yield of heads (MT/ha) | Irrigation applied (cm) | Yield/unit water applied (kg/ha/cm) |
|---|---|---|---|
| Drip | 17.55 | 12.0 | 1462 |
| Surface | 10.88 | 15.4 | 706 |

The yield of cauliflower was 31.70 and 25.51 MT/ha in micro sprinkler (MI) and surface irrigation (SI), respectively with an increase of 24.31% in MI over SI. Water use efficiency obtained in MI and SI were 1.43 and 0.79 MT/ha-cm, respectively. Water saving of 31.2% was recorded in MI over SI (Bansod and Shukla, 2006a).

### 17.14.3. Integrated nutrient management

#### Cabbage

Application of FYM at 20 MT/ha was effective in increasing the cabbage head size (391.27 $cm^2$) and head yield (76.116 MT/ha) as compared to 339.40 $cm^2$ (head size) and 68.045 MT/ha (head yield) in recommended dose of NPK (120: 60: 60) (Table 17.120).

**Table 17.120.** Effect of different sources of nutrients on growth and yield of cabbage.

| Source of nutrients | Head size (sq. cm.) | Yield (MT/ha) |
|---|---|---|
| FYM at 5 MT/ha | 359.61 | 69.264 |
| FYM at 10 MT/ha | 344.17 | 69.155 |
| FYM at 15 MT/ha | 345.06 | 71.745 |
| FYM at 20 MT/ha | 391.27 | 76.116 |
| Press mud at 5 MT/ha | 358.54 | 57.451 |
| Press mud at 10 MT/ha | 335.69 | 61.635 |
| Press mud at 15 MT/ha | 350.76 | 66.519 |
| Press mud at 20 MT/ha | 350.32 | 70.931 |
| NPK (120: 60: 60 kg/ha) | 339.40 | 68.045 |
| CD at 5% | 22.5 | 4.555 |

Soil treated with 50% vermicompost + 50% RDF (60: 30: 30 NPK kg/ha) gave maximum cabbage head yield of 60.2 MT/ha (Ghosh *et al.*, 2009).

With increasing potash fertilizer application from 0 to 90 kg $K_2O$/ha, cabbage yield grew continuously from 28 to 196 MT/ha. K-fertilizer use increased the girth and especially the weight of cabbage heads. For instance, the average head weight was 1.49 kg at the treatment in which the highest rate of 90 kg $K_2O$/ha was applied. Without potash fertilizer application, the average head weight reached only 1.16 kg or 0.33 kg less compared to the K 90 treatment.

Foliar spray of 100 ppm zinc (as zinc sulphate) thrice at 10 days interval gave highest cabbage head yield (43.155 MT/ha) followed by foliar spray of 100 ppm manganese (as manganese sulphate) thrice at 10 days interval (41.825 MT/ha) (Nandi *et al.*, 2006).

Plots amended with 1.5 MT/ha of copper mine ore waste gave 37.29 MT/ha of cabbage heads (Hiremath *et al.*, 2006).

### Cauliflower

Mangal *et al.* (1983) reported that cauliflower yields rose from 37.5 kg/9 $m^2$ in plots receiving the lowest N rates (50 kg/ha) and the lowest irrigation level (100 mm CPE) to 56.1 kg/9 $m^2$ in plots receiving the highest N rates (130 kg/ha) and the highest irrigation level (440 mm CPE).

In cauliflower demonstrations conducted during two seasons, the additional yield due to K application was 5.3-7.8 MT/ha depending on the season of observation. The percentage of yield increase with K addition was 17-24%.

Plots amended with 1.5 MT/ha of copper mine ore waste gave 22.8 MT/ha of cauliflower heads (Hiremath *et al.*, 2006).

The maximum curd yield of cauliflower (26.074 MT/ha) along with maximum benefit-cost ratio (5.25) was recorded with the application of *Azospirillum* + 75% N and full recommended dose of $P_2O_5$ and $K_2O$. There was 20% increase in yield over recommended dose of NPK (Chaurasia *et al.*, 2006).

## 17.14.4. Intercropping

To improve the productivity of vegetable crops, several intercropping combinations have been developed at the Indian Institute of Horticultural Research, Bangalore. Intercropping of cabbage with coriander or fenugreek enhanced the productivity of soil significantly. The net returns were higher from cabbage-fenugreek followed by cabbage-coriander intercrops (Mathura Rai, 1999) (Table 17.121).

**Table 17.121.** Intercrop combinations and their productivity levels

| Intercrop combinations | Yield (MT/ha) | Net returns (Rs/ha) |
|---|---|---|
| Cabbage + Fenugreek | 49.91 | 71778 |
| Cabbage | 43.04 | 61333 |
| Fenugreek | 14.46 | 15469 |
| Cabbage + Coriander | 49.93 | 71580 |

## 17.14.5. Integrated pest management

### i) Diamondback moth (DBM), *Plutella xylostella*

**(a) Botanicals:** The performance of powder formulation of 'Soluneem' was on par with 4% NSKE in controlling DBM. Sprays of neem soap or pongamia soap both at 1% were highly effective in reducing DBM (Table

17.122), aphids (*Brevicoryne brassicae*) and leaf webber (*Crocidolomia binotalis*). Further, the soap sprays significantly increased the yield of cabbage and cauliflower and were relatively safe to the parasitoid, *Cotesia plutellae* (Srinivasan and Krishnamoorthy, 1993; Krishnamoorthy *et al.*, 1998; Krishnamoorthy and Krishna Kumar, 2002).

**Table 17.122.** Efficacy of neem and pongamia soaps on DBM and yield of cabbage

| Treatment | DBM incidence/plant | Yield (MT/ha) |
|---|---|---|
| NSKE | 3.00 | 109.60 |
| Bt | 3.00 | 92.67 |
| Soluneem | 3.33 | 99.60 |
| Neem oil | 9.00 | 66.53 |
| Pongamia oil | 7.67 | 73.33 |
| Neem soap | 2.33 | 111.33 |
| Pongamia soap | 4.67 | 116.67 |
| Control | 36.33 | 43.87 |

**(b) Use of neem and pongamia soaps:** Under field conditions, neem and pongamia soap sprays were highly effective in reducing DBM, aphids (*Brevicoryne brassicae*) and leaf webber (*Crocidolomia binotalis*) and were at par with NSKE. They were superior to many commercial synthetic insecticides like dichlorvos, cypermethrin, combination formulation of cypermethrin and profenfos in controlling DBM. Further, the soap sprays significantly increased the yield of cabbage. On cauliflower too, the neem and pongamia soap sprays were effective and increased the yield significantly. The data on DBM, its parasitoid *Cotesia plutellae* and yield of cabbage and cauliflower are given in Tables 17.123 and 17.124. The soap sprays were also found to be relatively safe to *C. plutellae* (Krishnamoorthy and Krishna Kumar, 2002).

**Table 17.123.** Effect of different botanicals on yield and the incidence of larval parasitoid *Cotesia plutellae* in cabbage (2001-2002)

| Treatments | Dose (%) | Yield (MT/ha) | | *Cotesia plutellae* cocoons/10 plants |
|---|---|---|---|---|
| | | Summer | Winter | |
| Neem seed powder | 4 | 88.00 | 51.00 | 3.66 |
| Neem soap | 0.5 | 60.92 | 37.73 | 1.66 |
| Neem Soap | 0.75 | 85.00 | 38.73 | 1.33 |
| Neem soap | 1.0 | 83.89 | 51.00 | 2.33 |
| Pongamia soap | 0.5 | 70.00 | 37.73 | 2.00 |
| Pongamia soap | 0.75 | 74.11 | 47.00 | 0.33 |
| Pongamia soap | 1.0 | 87.78 | 52.33 | 1.33 |
| Diclorvos | 0.1 | 48.89 | - | 0.00 |
| *Bacillus thuringiensis* | 0.1 | 79.33 | 41.46 | 0.00 |

| | | | | |
|---|---|---|---|---|
| Cypermethrin | 0.0125 | 51.89 | 27.20 | 0.00 |
| Control | - | 33.78 | 13.53 | 2.67 |
| CD (P = 0.05) | - | 16.34 | 6.13 | 2.14 |

**Table 17.124.** Effect of different botanicals on Diamondback moth population and yield of cauliflower (2001-2002)

| Treatments | Dose (%) | DBM/10 plants | Yield (MT/ha) |
|---|---|---|---|
| Neem soap | 0.5 | 4.33 | 6.18 |
| Neem soap | 0.75 | 5.00 | 5.27 |
| Neem soap | 1.0 | 4.33 | 7.83 |
| Pongamia soap | 0.5 | 6.67 | 3.23 |
| Pongamia soap | 0.75 | 3.00 | 5.85 |
| Pongamia soap | 1.0 | 1.33 | 7.90 |
| NSPE | 4 | 0.00 | 7.62 |
| *Bacillus thuringiensis* | 0.1 | 5.33 | 7.53 |
| Cypermethrin | 0.0125 | 13.33 | 4.82 |
| Control | --- | 33.00 | 3.24 |
| CD (P = 0.05) | | 3.69 | 2.73 |

**(c) Integrated methods:** IPM using Indian mustard as a trap crop involves planting of paired rows of mustard after every 25 rows of cabbage/ cauliflower and spraying of 4% NSKE at primordial formation. Two more sprays of 4% NSKE may be given at 10-15 day interval after the first spray. The IPM gave 60% more yield and 152% more returns than pure cabbage crop (Table 17.125) (Khaderkhan *et al.*, 1998b). IPM controls diamondback moth, leaf webber (*Crocidolomia binotalis*); stem borer (*Hellula undalis*); aphids (*Brevicorne brassicae, Hyadaphis erysimi*) and bug (*Bagrada cruciferarum*) (Srinivasan and Krishna Murthy, 1991) (Fig. 17.5).

**Table 17.125.** Cabbage yields and net returns realized in Farmers' practice and IPM practice

| Practice | Yield (MT/ha) | % increase | Net returns (Rs.) | % increase |
|---|---|---|---|---|
| Farmers' practice | 20 | --- | 19,817 | --- |
| IPM practice | 32 | 60 | 49,251 | 152 |

Economics of IPM in cabbage is presented in Table 17.126 (Krishnamoorthy *et al.*, 2003).

**Table 17.126.** Economics of IPM in cabbage

| Treatment | Yield (MT/ha) | Net returns (Rs/ha) | Benefit: cost ratio |
|---|---|---|---|
| IPM plots | 55 | 30085 | 2.42 |
| Non-IPM plots (Farmers practice) | 35 | 5090 | 0.83 |

**Fig. 17.5.** IPM in cabbage and cauliflower using Indian mustard as a trap crop.

## Validated IPM technology in cabbage at different locations Bangalore, Karnataka

### Nursery

- Preparation of raised seed beds up to 15 cm height.
- Soil drenching with 0.2% copper oxychloride.
- Seed treatment with *Trichoderma harzianum* at 4 g/kg.
- Use of nylon net to avoid entry of sucking pests.
- Spray *Bt* at 0.5 ml/l one day before planting.
- Spray Dithane M- 45/Ridomil at 0.2% against downy mildew.

### Main field

- Growing of 2 rows of mustard after every 25 rows of cabbage as trap crop at the time of planting.

- Wider spacing of 60 x 45 cm.
- Installation of light traps at 5/ha for trapping adult DBM.
- Spray *Bacillus thurungiensis* at 1 g/l if DBM is noticed early or spray 5% NSKE/1% neem soap/pongamia soap at primordial stage. Repeat 3-4 sprays.
- Removal of disease affected basal leaves from time to time.
- Need based sprays of chlorothalanil/mancozeb at 0.2% for *Alternaria* and blitox + streptocycline for black rot.

**Varanasi, Uttaranchal**

**Nursery**

- Preparation of raised seed beds up to 15 cm height.
- Soil solarization for 3 weeks with 45 $\mu$m polythene sheet.
- Seed treatment with *Trichoderma harzianum* at 4 g/kg.
- Use of nylon net to avoid entry of sucking pests.

**Main field**

- Spray *Bt* at 1 g/l if one DBM is noticed per plant.
- Spray 5% NSKE at primordial stage i.e. 15-20 DAP. Repeat 3-4 sprays.
- Release of *Trichogramma bactrae* at 0.75 lakh/ha at weekly interval.
- Removal of disease affected heads and bottom leaves from time to time.
- Need based application of pesticides like mancozeb at 0.2%.

**Ranchi, Jharkhand**

**Nursery**

- Preparation of raised seed beds up to 15 cm height.
- Soil solarization for 3 weeks with 45 $\mu$m polythene sheet.
- Soil treatment with FYM enriched with *T. harzianum* at 1 kg/MT.

**Main field**

- Spray Neemarin at 5% when DBM appeared early.
- Spray *Bt* at 1 g/l if one DBM is noticed per plant.
- Spray 1% pongamia soap at 15-20 DAP.
- Installation of pheromone traps at 5/ha for *Spodoptera litura* monitoring.
- Need based application of fungicides like chlorothalanil at 0.2% for *Alternaria*.

- Spray of insecticides like endosulfan at 0.07% or triazophos at 0.05%.

**Table 17.127.** Yield and economics of IPM in cabbage at different locations

| Centre | Yield (MT/ha) | Net returns (Rs.) | Benefit: Cost ratio |
|---|---|---|---|
| Bangalore - IPM | 62.01 | 80934 | 3.04 |
| - Non IPM | 58.01 | 40661 | 0.61 |
| Varanasi - IPM | 24.08 | 37500 | 3.70 |
| - Non IPM | 21.17 | 33333 | 1.89 |
| Ranchi - IPM | 34.75 | 231386 | 2.80 |
| - Non IPM | 28.76 | 156900 | 1.84 |

During the period 2001-04 IPM technology in cabbage was validated and promoted in more than 40 ha area in 42 villages covering 88 families located 40 km from Bangalore. Similarly, near Varanasi also IPM technology has been validated in 8 villages in about 40 ha area covering 100 families. Near Ranchi, IPM technology in cabbage has been validated and promoted in 20 villages with the support of 100 farming families covering an area of 40 ha. In IPM validation studies conducted at three locations (Bangalore, Varanasi and Ranchi), IPM fields recorded higher yields of 62.013, 24.080 and 34.750 MT/ha as compared to 58.013, 21.170 and 28.760 MT/ha, respectively in non-IPM fields (Amerika Singh *et al.*, 2004) (Table 17.127).

### 17.14.6. Integrated management of disease complex

#### (i) Club root and root-knot disease complex

Chemicals, PGPR strains (*Pseudomonas fluorescens, Bacillus subtilis*) and fungal biocontrol agents (*Trichoderma viride, T. harzianum*) were found to be effective in reducing the nematode-fungal disease complex in cabbage and increased the yield (Loganathan *et al.*, 2001) (Table 17.128).

**Table 17.128.** Efficacy of bioformulations and chemicals against club root disease–root-knot nematode complex in cabbage under field conditions.

| Treatment | Club root index | Root-knot index | Yield (MT/ha) |
|---|---|---|---|
| *T. viride* | 50.0 (45.00)* | 2.66 | 39.99 |
| *T. viride* | 40.0 (39.22) | 2.66 | 44.44 |
| *T. viride* | 20.0 (26.56) | 2.00 | 53.33 |
| *T. viride* | 38.0 (38.05) | 2.33 | 45.77 |
| *T. viride* | 36.5 (37.16) | 2.66 | 46.66 |
| *T. harzianum* | 36.2 (36.97) | 3.00 | 48.88 |
| *P. fluorescens* | 46.0 (42.7) | 2.00 | 44.44 |
| *B. subtilis* | 43.0 (40.97) | 3.33 | 41.33 |
| *P. fluorescens* | 23.0 (28.66) | 1.33 | 59.10 |
| *P. fluorescens* | 45.0 (42.13) | 2.00 | 47.40 |

| | | | |
|---|---|---|---|
| Carbendazim | 18.0 (25.10) | 3.66 | 51.11 |
| Carbofuran | 46.0(42.70) | 2.00 | 45.77 |
| Cabendazim + Carbofuran | 11.5 (19.81) | 2.00 | 54.66 |
| Control | 60.0 (50.79) | 4.00 | 33.77 |

* Values in parenthesis are transformed values.

## 17.15. PEA, *PISUM SATIVUM*

### 17.15.1. High yielding varieties

**Table 17.129.** High yielding varieties of pea

| High yielding varieties | Yield (MT/ha) | Developed at |
|---|---|---|
| Arka Ajit | 10 | IIHR, Bangalore |
| Arka Kartik | 11 | IIHR, Bangalore |
| Arkel | 10.0-12.5 | IARI, New Delhi |
| Bonneville | 10-12 | IARI, New Delhi |
| Lincoln | 11-13 | IARI, RS, Katrain |
| Ageti Matar-7 | 23-25 | |
| NDVP-8 | 13-16 | |
| NDVP-10 | 11.5-12.5 | |
| Kashi Nandini | 11-12 | IIVR, Varanasi |
| Kashi Udai | 10-11 | IIVR, Varanasi |
| Kashi Mukti | 12-14 | IIVR, Varanasi |
| Kashi Shakti | 14-16 | IIVR, Varanasi |
| Jawahar Matar-1 | 12.0-12.5 | JNKVV, Jabalpur |
| Punjab-88 | 15-16 | PAU, Ludhiana |
| JP-83 | 12-13 | |
| Azad P-2 | 11.5-12.0 | CSAUAT, Kanpur |
| VL Matar 3 | 10-11 | VIPKAS, Almora |
| Vivek Matar 8 | 11-12 | VIPKAS, Almora |
| Ooty 1 | 12.9 | HRS, TNAU, Ooty |

**Table 17.130.** Pea varieties resistant to pests/diseases

| High yielding varieties | Yield (MT/ha) | Resistance to pest/disease | Developed at |
|---|---|---|---|
| Kashi Mukti | 12-14 | Powdery mildew | IIVR, Varanasi |
| Swarna Mukti | 12-14 | Powdery mildew (T) | HARP, Ranchi |
| Pant Uphar | 10.0-10.5 | Stem fly | GBPUAT, Pantnagar |
| Jawahar Peas-4 | 11-12 | Powdery mildew | JNKVV, Jabalpur |
| Priya | 15.0-17.5 | Powdery mildew | |
| Arka Ajit | 10 | Powdery mildew, Rust | IIHR, Bangalore |

| | | | |
|---|---|---|---|
| Arka Kartik | 11 | Powdery mildew, Rust | IIHR, Bangalore |
| Arka Sampoorna | 8 | Powdery mildew, Rust | IIHR, Bangalore |
| Pusa Pragati | 7 | Powdery mildew | IARI, New Delhi |
| VL Matar 3 | 10-11 | Powdery mildew | VIPKAS, Almora |
| Vivek Matar 8 | 11-12 | Powdery mildew (T), White rot (T) | VIPKAS, Almora |
| Vivek Matar 9 | 9-11 | Powdery mildew (T), White rot (T) | VIPKAS, Almora |
| Palam Priya | | Powdery mildew, Leaf miner (T) | HPKV, Palampur |

T- Tolerant

### 17.15.2. Micro irrigation

The drip irrigation at 8 litres/m²/day and plant density of 20 plants/m² gave maximum pea pod yield of 8.33 MT/ha and 8.35 MT/ha, respectively. Thus it can be concluded that drip irrigation (8 litres/m²/day) and plant density (20 plants/m²) can increase the productivity of garden peas (Kempe Gowda *et al.*, 2009).

### 17.15.3. Fertigation

Fertigation trial in pea has demonstrated higher yields under fertigation as compared to conventional methods (Malik *et al.*, 1994) (Table 17.131).

**Table 17.131.** Effect of different methods of fertilizer application on green pod yield of pea

| Treatment | Green pod yield (MT/ha) | % yield response |
|---|---|---|
| Control | 5.38 | --- |
| No irrigation + Broadcast of fertilizer | 7.15 | 32.9 |
| Drip + Broadcast of fertilizer | 8.09 | 50.4 |
| Surface + Broadcast of fertilizer | 8.02 | 49.1 |
| Drip + Fertilizer through drip | 9.70 | 80.3 |

Application of recommended dose of K through fertigation increased the pea yield to the maximum extent (Table 17.132).

**Table 17.132.** Yield response to fertilizer under different methods of K application and irrigation in pea

| Irrigation/fertilizer | Yield response (%) | |
|---|---|---|
| | K at 50% RDF | K at 100% RDF |
| No irrigation | 8.51 | 18.55 |
| Floor irrigation + Fertilizer through broadcast | 12.31 | 22.94 |
| Drip irrigation + Fertilizer through broadcast | 15.92 | 31.75 |
| Drip irrigation + Fertilizer through drip | 23.12 | 45.47 |

### 17.15.4. Integrated nutrient management

Maximum pod yield (9.396 MT/ha), net return (Rs. 66122/ha) and benefit-cost ratio (2.4) was obtained when FYM at 1 MT/ha + poultry manure at 1 MT/ha along with mulch was applied followed by FYM at 2 MT/ha + biofertilizers with mulch (Baswana and Rana, 2010).

### 17.15.5. Integrated disease management

#### i) Powdery mildew, *Erysiphe polygonum*

In organically grown pea, biopriming of pea seeds with *Trichoderma harzianum* + *Pseudomonas fluorescens* was most effective in improving the seedling stand. Foliar application of bioagents significantly reduced the incidence of powdery mildew disease. Application of *P. fluorescens* + *T. harzianum* through seed and foliar spray was most effective in increasing the yield of organically grown pea (Table 17.133).

**Table 17.133.** Effect of different biocontrol agents applied through seed, FYM and/or foliar application on powdery mildew disease and yield of pea

| Treatment | Seedling stand/m$^2$ | Powdery mildew incidence (%) | Yield (kg/ha) |
|---|---|---|---|
| Seed biopriming with *Trichoderma harzianum* + sprays with *Pseudomonas fluorescens* | 56 | 13 | 355 |
| FYM colonized with *Trichoderma harzianum* + sprays with *Pseudomonas fluorescens* | 63 | 16 | 352 |
| Seed biopriming with *Trichoderma harzianum* + FYM colonized with *T. harzianum* + sprays with *T. harzianum* | 69 | 12 | 388 |
| Control | 52 | 36 | 301 |
| CD (P = 0.05) | 6 | 10 | 53 |

## 17.16. COWPEA, *VIGNA UNGUICULATA*

### 17.16.1. High yielding varieties

**Table 17.134.** High yielding varieties of cowpea

| High yielding varieties | Yield (MT/ha) | Developed at |
|---|---|---|
| Pusa Komal | 10 | IARI, New Delhi |
| Arka Garima | 17-19 | IIHR, Bangalore |
| Arka Suman | 17-19 | IIHR, Bangalore |
| Arka Samrudhi | 18-19 | IIHR, Bangalore |
| Kashi Gauri | 11-13 | IIVR, Varanasi |
| Lola | 20 | KAU, Vellanikkara |
| Vyjayanthi | 20 | KAU, Vellanikkara |

| | | |
|---|---|---|
| Sel 263 | 10-12 | PAU, Ludhiana |
| Bidan Barbati 1 | 13.4 | BCKV, Nadia |
| Bidan Barbati 2 | 15.9 | BCKV, Nadia |
| SEL-2-1 | 12.5-15.0 | |
| VR-5 | 15.0-17.5 | |

**Table 17.135.** Cowpea varieties resistant to diseases

| High yielding varieties | Yield (MT/ha) | Resistance to disease | Developed at |
|---|---|---|---|
| Kashi Gauri | 11-13 | *Psudocercospora cruenta*, Cowpea golden mosaic virus | IIVR, Varanasi |
| Kashi Kanchan | 15-20 | *Psudocercospora cruenta*, Cowpea golden mosaic virus | IIVR, Varanasi |
| Kashi Shymal | 7-8 | Cowpea golden mosaic virus (T) | IIVR, Varanasi |
| Swarna Suphala | 21-15 | Rust, Cowpea golden mosaic virus | HARP. Ranchi |
| Pusa Komal | 10.0-11.5 | Bacterial wilt | IARI, New Delhi |
| Sel-263 | 20-22 | Cowpea mosaic virus, Cowpea golden mosaic virus | PAU, Ludhiana |
| Bidan Barbati 1 | 13.4 | Cowpea mosaic virus, Cowpea golden mosaic virus | BCKV, Nadia |

T - Tolerant

## 17.16.2. Integrated nutrient management

Maximum grain yield, bhusa yield and quality of cowpea were obtained in treatments receiving enriched vermicompost (with low grade carriers like rock phosphate) as an organic source (Table 17.136).

**Table 17.136.** Comparative efficacy of enriched vermicompost on cowpea.

| Treatment | Grain yield (kg/ha) | Bhusa yield (kg/ha) | Protein content (%) |
|---|---|---|---|
| Rock phosphate (RP) | 690.0 | 1324.5 | 20.13 |
| FYM | 817.5 | 1619.5 | 21.56 |
| Vermicompost | 877.5 | 1823.0 | 23.81 |
| Enriched vermicompost | 1072.5 | 2093.5 | 24.94 |
| FYM + Full RP | 837.5 | 1650.0 | 21.62 |
| FYM + Half RP | 831.5 | 1650.0 | 21.62 |
| Vermicompost + Full RP | 882.5 | 1839.5 | 23.87 |
| Vermicompost + Half RP | 879.0 | 1830.0 | 23.81 |
| Vermicompost + Half RP (primed) | 898.5 | 1810.5 | 24.19 |
| FYM + Full RP (primed) | 859.0 | 1678.0 | 22.18 |
| FYM + Half RP (primed) | 833.5 | 1623.5 | 22.06 |
| CD (P = 0.05) | 54.05 | 137.05 | 0.412 |

Application of vermicompost instead of FYM without inorganic fertilizer was equally effective for increasing the cowpea yield (21%) as the recommended manurial schedule (Jiji and Dale, 1999) (Table 17.137).

**Table 17.137.** Effect of *in situ* vermiculture on fruit yield of okra

| Treatment | Fruit yield (kg/4 $m^2$) |
|---|---|
| POP recommendations | 4.155 |
| Vermicompost substituting FYM | 3.415 |
| Vermicompost substituting FYM + Half inorganic fertilizers as per POP | 3.701 |
| Vermicompost substituting FYM + Three fourth inorganic fertilizers as per POP | 4.080 |
| Vermicompost substituting FYM + N requirement as per POP through vermicompost | 3.569 |
| Vermicompost substituting FYM + Full inorganic fertilizers as per POP | 4.944 |
| CD (P= 0.05) | 0.740 |

### 17.16.3. Intercropping

To improve the productivity of vegetable crops, several intercropping combinations have been developed at the Indian Institute of Horticultural Research, Bangalore (Mathura Rai, 1999). Cowpea grown as intercrop in amla gave maximum net returns (Table 17.138).

**Table 17.138.** Economics of comparative performance of various horticultural crop combinations

| Intercrop combinations | Yield (MT/ha) | | Net returns (Rs/ha) |
|---|---|---|---|
| | Main crop | Inter crop | |
| Mango | 2.60 | --- | 24,000 |
| Mango + Cowpea | 2.55 | 5.537 | 37,111 |
| Guava | 4.60 | --- | 21,000 |
| Guava + Cowpea | 4.60 | 5.066 | 33,198 |
| Ber | 10.85 | --- | 51,250 |
| Ber + Cowpea | 10.88 | 1.735 | 53,605 |
| Amla | 5.57 | --- | 53,700 |
| Amla + Cowpea | 5.52 | 5.242 | 65,926 |
| Maize + Pigeon pea (Traditional system) | 0.81 | 0.320 | 6,200 |

Intercropping of cowpea with okra gave 31-53% higher okra equivalent yield and significantly higher net returns compared to those of okra sole cropping (Table 17.139).

**Table 17.139.** Intercrop combinations and their productivity levels

| Intercrop combinations | Yield (MT/ha) | Net returns (Rs/ha) |
|---|---|---|
| Okra + Cowpea | 24.98 | 36824 |
| Okra | 16.31 | 26389 |
| Cowpea | 20.54 | 27917 |
| Finger millet + Dolichos bean (Rain fed) | --- | 5000 |

### 17.16.4. Integrated weed management

Smother crops grown as intercrops suppress weeds without reducing total production in cowpea. Many short duration pulses such as green gram and soybean; and sorghum for fodder effectively smother weeds and increase yields of cowpea (Masood Ali, 1986; Bhat, 1989; Mutnal, 1984) (Table 17.140).

**Table 17.140.** Effect of intercropping of cowpea with sorghum on weed suppression and yield

| Treatment | Green fodder yield (MT/ha) | Cowpea yield (MT/ha) |
|---|---|---|
| Sole sorghum (30 cm rows) | 31.8 | --- |
| Sorghum + Cowpea (1:1) | 31.5 | 16.7 |
| Sorghum + Cowpea (2:2) | 31.4 | 16.5 |
| Sole cowpea | --- | 22.4 |

### 17.16.5. Integrated nematode management

**(i) Reniform, *Rotylenchulus reniformis* and stunt nematodes, *Tylenchorhynchus* sp.**

Seed soaking in profenfos at 200 ppm for 1 hour and spray with profenfos at 200 ppm and foliar spray of ß-aminobutyric acid at 500 ppm reduced nematode population and improved the yield of cowpea cv. Pusa Komal by more than 100% (Sharma *et al.*, 2010).

## 17.17. FRENCH BEAN, *PHASEOLUS VULGARIS*

### 17.17.1. High yielding varieties

**Table 17.141.** High yielding varieties of French bean

| High yielding varieties | Yield (MT/ha) | Developed at |
|---|---|---|
| Contender | 20 | IARI, New Delhi |
| Pusa Parvati | 22.5 | IARI, New Delhi |
| Pusa Himalata | 20 | IARI, RS, Katrain |
| Arka Komal | 18.0-19.5 | IIHR, Bangalore |
| Arka Suvidha | 18.0-19.5 | IIHR, Bangalore |
| Arka Bold | 15 | IIHR, Bangalore |

| | | |
|---|---|---|
| Kashi Param | 12-14 | IIVR, Varanasi |
| Phule Suyash | 23.0-24.5 | MPKV, Rahuri |
| Ooty 1 | 33.68 | HRS, TNAU, Ooty |
| Ooty 2 | 14.30 | HRS, TNAU, Ooty |

**Table 17.142.** French bean varieties resistant to pests/diseases

| High yielding varieties | Yield (MT/ha) | Resistance to pest/disease | Developed at |
|---|---|---|---|
| Phule Suyash | 23.0-24.5 | Fusarium wilt (T), Bean common mosaic (T) | MPKV, Rahuri |
| Pant Anupam | 9-10 | Bean common mosaic (MR), Angular leaf spot (MR) | GBPUAT, Pantnagar |
| Pant Bean 2 | 9 | Bean common mosaic (MR) | GBPUAT, Pantnagar |
| Swarna Priya | 12-14 | Leaf miner, Powdery mildew (T), Downy mildew (T) | HARP, Ranchi |
| Swarna Lata | 12-14 | Leaf miner, Powdery mildew (T), Downy mildew (T) | HARP, Ranchi |
| Arka Bold | 15 | Rust | IIHR, Bangalore |
| Arka Anoop | 20 | Rust, Bacterial blight (T) | IIHR, Bangalore |

### 17.17.2. Integrated nutrient management

The yield obtained in the foliar nutrition applied through micronutrient formulation (0.3%) was 18.7 MT/ha as compared to 16.5 MT/ha in control plot in French bean cv. Arka Anoop. There was an increase of 13.3% in yield in treated plots (Shivananda *et al.*, 2009).

Prilled urea coated with neem oil was superior to prilled urea in French bean cv. Arka Komal and produced 9% higher yield on an overall basis (Kotur *et al.*, 2006).

## 17.18. PIGEON PEA, *CAJANUS CAJAN*

### 17.18.1. High yielding varieties

**Table 17.143.** Pigeon pea varieties resistant to diseases

| High yielding variety | Resistance to disease |
|---|---|
| Maruti | *Fusarium* wilt |

### 17.18.2. Integrated nutrient management

Combined application of *Rhizobium* and FYM along with recommended dose of fertilizer treatment gave maximum yield (Anon, 1996) (Table 17.144).

**Table 17.144.** Grain yield of pigeon pea as influenced by application of FYM, *Rhizobium* and sulphur

| Treatment | Grain yield (kg/ha) | | |
|---|---|---|---|
| | 1993 | 1992 | Mean |
| *Rhizobium* + FYM + RDF (dry sowing) | 1611 | 1526 | 1568 |
| *Rhizobium* + FYM + RDF | 1055 | 1296 | 1175 |
| *Rhizobium* + FYM + RDF + 20 kg S/ha | 1183 | 1428 | 1305 |
| *Rhizobium* + RDF + 500 kg well decomposed FYM (furrow application at sowing) + 20 kg S/ha | 1338 | 1453 | 1395 |

## 17.18.3. Integrated weed management

Smother crops grown as intercrops suppress weeds without reducing total production in pigeon pea. Many short duration pulses such as green gram, cowpea and soybean; and sorghum for fodder effectively smother weeds and increased yields of pigeon pea (Masood Ali, 1986; Bhat, 1989; Mutnal, 1984) (Table 17.145).

**Table 17.145.** Effect of intercropping of pigeon pea with pulses on weed suppression and yield

| Treatment | Weed suppression (%) | Pigeon pea yield (kg/ha) |
|---|---|---|
| Pigeon pea (weed-free) | --- | 2578 |
| Pigeon pea (weedy) | --- | 1911 |
| Pigeon pea + Black gram | 207 | 2377 |
| Pigeon pea + Green gram | 22.2 | 2518 |
| Pigeon pea + Soybean | 22.2 | 2266 |
| Pigeon pea + Cowpea | 33.3 | 2349 |

## 17.18.4. Integrated pest management

### (i) Pod Borer, *Helicoverpa armigera*

Integrated management of pigeon pea pod borer using *Ha*NPV at 2 x $10^6$ PIB's/ml + nematode DD-136 (*Steinernema feltiae*) at 3 x $10^3$ infective juveniles/ml was attempted. The least population of *H. armigera* was observed in virus + nematode treated plots (0.2) when compared to control (12.6). The percentage of pod damage was significantly lower in virus + nematode (1.07) when compared to control (40.31). Significant increase in pod yield was obtained in virus + nematode treated plots (10.7 kg) when compared to control (3.2 kg) (Narayanan and Gopalakrishnan, 1987).

### Validated IPM technology (Gulburga, Karnataka)

- Fall ploughing to expose pupae to hot sun and natural enemies.
- Mixing of sorghum or mesta seeds at 250 g/ha with pigeon pea seeds, which act as live bird perches.
- Installation of pheromone traps at 5/ha to monitor the pests.

- Erecting the branched twigs at 20/ha which act as bird perches.
- First spraying with ovicide indoxacarb or methomyl at 300 g/ha or profenophos at 2 litres/ha.
- Second spraying with 5% NSKE or commercial neem formulation (1500 ppm Azadirachtin) at 2 litres/ha.
- Third spraying with *Ha*NPV at 250 LE/ha.
- Fourth spraying with indoxacarb at 300 ml/ha or chlorpyriphos at 2.5 litres/ha or quinalphos or endosulfan at 2 litres/ha.
- If necessary, spraying of synthetic pyrethroids at 500 ml/ha.
- In case of scarcity of water, dust 4% endosulfan followed by 1.5% quinalphos or 0.4% fenvalerate at 25 kg/ha.

**Validated IPM technology (Nanded, Maharashtra)**

**Pre-sowing**

- Deep ploughing and exposure of soil to hot summer to kill pupating larvae and fungal propagules.
- Use of FYM enriched with *Trichoderma viride* at 50 kg/ha mixed along with neem cake to reduce disease as well as nematode population.
- In areas where termite and cut worm is a problem, seed treatment with chlorpyriphos at 8 ml/kg or pre-sowing mixing of soil with chlorpyriphos dust is recommended as they are relatively low cost and initially protect the seedlings from cut worms.
- Synchronized sowing of multiple pest resistant varieties such as Sharad, Asha, Maruti, Bahar, Abhaya, BSMR-736.
- Ridge planting to prevent incidence of blight caused by *Phytophthora drechsleri*.

**Sowing time**

- Intercrop sorghum to reduce wilt, conserve beneficial insects and serve as bird perches.
- Early planting around mid–June in North West Plain Zone to avoid *Helicoverpa armigera*.
- Use marigold as a trap crop on border or interspersed with crop for pod borer control.

**Post-sowing**

- Monitoring through regular field scouting along with pheromone traps (5-10/ha) to assess population build up of borers and their management. A threshold level of 5-6 moths/trap/day is indicative of its peak activity period and warrants initiation of management practices.

- Rogue out and destroy sterility mosaic virus affected plants.
- Mechanical shaking of plants and collection of larvae for preparation of *Ha*NPV locally to cut down the cost.
- Spray 5% NSKE or neem based formulation against *H. armigera* and other Lepidopterous pests.
- Spray with *Ha*NPV at 450 LE/ha along with UV retardant.
- Spray eco-friendly pesticide like endosulfan if pest complex exceeds ETL.

**Validated IPM technology (Varanasi, Uttar Pradesh)**

- Deep summer ploughing to destroy immature stages and fungal propagules.
- Soil application of *Trichoderma harzianum* before sowing at 10 g/kg of FYM for controlling wilt.
- Ridge planting to prevent incidence of blight caused by *Phytophthora drechsleri*.
- Planting high yielding varieties with pest/disease tolerance – Asha and Maruthi against pigeon pea sterility mosaic virus for Andhra Pradesh and Karnataka, and Bahar for Uttar Pradesh (Varanasi).
- Installation of pheromone traps at 10/ha in the month of September.
- Erection of bird perches at 25/ha for facilitating predation of *Helicoverpa* larvae.
- One spray with 2% neem oil.
- Two applications of 5% NSKE in September and October.
- Spray *Ha* NPV at 500 LE/ha (1.5 x $10^{12}$ POB)/ha in September and October – when larvae are small.
- Shaking of plants 5 times a day starting from October for short duration pigeon pea.
- Need based spray with endosulfan at 2 litres/ha.

**Table 17.146.** Economics of IPM on pigeon pea at different locations in Gulburga district of Karnataka

| Centre | | Yield (q/ha) | Profit over non-IPM (Rs/ha) |
|---|---|---|---|
| Sannur | – IPM | 9.64 | 1600 |
| | – Non-IPM | 9.25 | |
| Farthabad | – IPM | 10.81 | 2687 |
| | – Non-IPM | 9.75 | |
| Tadtegnoor | – IPM | 10.50 | 1400 |
| | – Non-IPM | 10.25 | |
| Kodla | – IPM | 15.62 | 9590 |
| | – Non-IPM | 9.50 | |

In IPM validation studies conducted at four locations in Gulburga district of Karnataka state (Sannur, Farthabad, Tadteganoor and Kodla), IPM fields recorded higher grain yields of 9.64, 10.81, 10.50 and 15.62 q/ha as compared to 9.25, 9.75, 10.25 and 9.50 q/ha, respectively in non-IPM fields (Amerika Singh *et al.*, 2004) (Table 17.146).

### 17.18.5. Integrated disease management

#### i) Wilt, *Fusarium udum*

Soil application of *T. harzianum* (PDBC TH 10 and 15) at 1 kg/m row (1 kg of FYM enriched with 10 g of *T. harzianum*) prior to sowing resulted in the disease incidence of 22.5 and 28.8%, respectively compared to 57.3% in carbendazim and 83% in control. The above treatment also gave better seedling emergence (90.3 and 88.0%, respectively) and high plant vigour index (1191.5 and 1122.8, respectively). Soil application of the above 2 isolates of *T. harzianum* gave higher yield (1120.5 and 1219.5 kg/ha, respectively) compared to 267.3 kg/ha in control (Table 17.147).

**Table 17.147.** Effect of *Trichoderma harzianum* isolates on Fusarial wilt incidence and yield of pigeon pea

| Treatment | % disease incidence | Yield (kg/ha) |
|---|---|---|
| *T. harzianum* (PDBC TH 15) seed treatment at 10g/kg | 40.3 | 712.5 |
| *T. harzianum* (PDBC TH 10) seed treatment at 10g/kg | 35.3 | 752.0 |
| *T. harzianum* (PDBC TH 15) soil application at 1 kg/m row (1 kg FYM + 10 g *Th*) | 28.8 | 1120.5 |
| *T. harzianum* (PDBC TH 10) soil application at 1 kg/m row (1 kg FYM + 10 g *Th*) | 22.5 | 1219.5 |
| Carbendazim seed treatment at 2g/kg | 57.3 | 683.8 |
| Control (Pathogen + 1 kg FYM/m row) | 83.0 | 267.3 |

**Table 17.148.** Field performance of bacterial antagonists against Fusarium wilt of pigeon pea

| Treatment | Formulation used and cfu/g | Disease control (%) | Increase in yield (%) |
|---|---|---|---|
| Seed treat. with *Pseudomonas fluorescens* at 10 g/kg + Soil appln. at 2.5 kg/ha mixed with FYM | Talc $10^7 - 10^8$ | 22 | 6 |
| Seed treat. with *Pseudomonas putida* at 10 g/kg + Soil appln. at 2.5 kg/ha mixed with FYM | Talc $10^6 - 10^7$ | 28 | 12 |
| Seed treat. with *Bacillus subtilis* at 10 g/kg + Soil appln. at 2.5 kg/ha mixed with FYM | Talc $10^6 - 10^7$ | 26 | 14 |

Pigeon pea seed treatment with bacterial antagonists at 10 g/kg seed + soil application at 2.5 kg/ha mixed with FYM gave 22 to 28% Fusarium wilt control and increased the yield by 6 to 14% (Ramanujam *et al.*, 2003) (Table 17.148).

### 17.18.6. Integrated nematode management

#### i) Pigeon pea cyst nematode, *Heterodera cajani*

Integration of *T. harzianum* at 5 kg/ha along with *P. chlamydosporia* at 2 kg/ha significantly increased the plant height and pigeon pea yield (0.56 kg/plant as compared to 0.40 kg/plant in control) and reduced the seedling mortality, eggs per cyst, parasitization of cysts by the bioagents and cyst nematode population (Table 17.149).

**Table 17.149.** Effect of bioagents on plant growth and management of *Heterodera cajani* infecting pigeon pea

| Treatment | Seedling mortality | Yield (kg/plant) | eggs/cyst |
|---|---|---|---|
| *T. harzianum* – 5 kg/ha | 13.21 | 0.529 | 36.8 |
| *P. chlamydosporia* – 2 kg/ha | 13.67 | 0.545 | 30.1 |
| *T. harzianum* – 5 kg/ha + *P. Chlamydosporia* – 2 kg/ha | 9.20 | 0.560 | 24.0 |
| Carbofuran – 2 kg a.i./ha | 2.16 | 0.634 | 18.4 |
| Control | 18.40 | 0.404 | 45.1 |
| CD (P = 0.05) | 1.14 | 0.0053 | 3.45 |

Combined treatment of 10% neem seed kernel powder + *Trichoderma viride* at 10 g/kg seeds was found to reduce *Heterodera cajani* in pigeon pea by 58% and increased yield by 32% (Patel *et al.*, 2010).

Application of neem cake at 100 kg/ha + *Trichoderma viride* at 2.5 kg/ha was found to reduce *Heterodera cajani* by 62% and increased pigeon pea yield by 34% (Patel *et al.*, 2010).

Seed treatment with neem seed kernel powder at 10% w/w + *Pseudomonas fluorescens* at 10 g/kg seed was highly effective which recorded 30.11% reduction in nematode population and 29.10% increase in yield over control. Soil application of neem cake at 10 g/m$^2$ + *Pseudomonas fluorescens* at 2.5 kg/ha was also highly effective which recorded 31.63% % reduction in nematode population and 17.88% increase in yield over control (Subramanian and Sivakumar, 2010).

### 17.18.7. Integrated management of disease complex

#### (i) Cyst nematode, *Heterodera cajani* and wilt, *Fusarium udum* disease complex

In a trial conducted in the field during 2002, no wilting was observed in plots treated with neem seed powder + dimethoate/*T. harzianum*/*P. lilacinus*/latex, and neem seed powder and *T. harzianum* alone. Neem seed powder + *T. harzianum* was found to be the most effective treatment in increasing the yield and suppressing the pathogens, followed by carbofuran, neem seed powder + *P. lilacinus*, neem seed powder + dimethoate and neem seed powder + latex. Data regarding *F. udum* infection in roots gathered at the pre-flowering stage (90 DAS) indicated that all the treatments maintained significant protection of the roots as compared to control (Haseeb and Shukla, 2005) (Table 17.150).

**Table 17.150.** Effect of seed treatment with bioagents, chemicals and botanicals on wilt and cyst nematodes infecting pigeon pea under field conditions during 2002.

| Treatment | Grain yield (kg/100 m$^2$) | % wilting of plants | Nematodes /g of root |
|---|---|---|---|
| Carbofuran | 23.47 | 5.0 | 9 |
| Neem seed powder (soil application) | 19.55 | 0.0 | 23 |
| Neem seed powder (seed treatment) | 18.33 | 0.0 | 30 |
| *Trichoderma harzianum* | 15.46 | 0.0 | 31 |
| *Paecilomyces lilacinus* | 12.24 | 5.0 | 36 |
| Dimethoate | 11.75 | 10.0 | 39 |
| Latex of Calotropis | 11.34 | 10.0 | 39 |
| NSP (soil) + Dimethoate | 21.05 | 0.0 | 16 |
| NSP (soil) + *T. harzianum* | 23.35 | 0.0 | 10 |
| NSP (soil) + *P. lilacinus* | 21.85 | 0.0 | 14 |
| NSP (soil) + Latex | 20.75 | 0.0 | 18 |
| Control | 5.35 | 45.0 | 51 |
| CD (P = 0.05) | 1.22 | 2.8 | 0.68 |

### ii) Fusarium wilt, *Fusarium udum* and nematodes (*Heterodera cajani, Meloidogyne incognita* and *Rotylenchulus reniformis*) disease complex

The wilt disease complex caused by *Fusarium udum* in association with *Heterodera cajani, Meloidogyne incognita* and *Rotylenchulus reniformis* has been reported as the most severe constraint in the cultivation of pigeon pea.

In a farmers' field trial conducted at Arania, Bulandshahr district of Uttar Pradesh during 2001; wilting of plants was not seen in plots treated with carbofuran, neem seed powder, neem jivan, *T. harzianum* and *Aspergillus niger*, 1-5% wilting was observed in plots treated with *P. lilacinus* (Haseeb and Shukla, 2005) (Table 17.151).

**Table 17.151.** Effect of seed treatment with biopesticides, chemicals and botanicals on wilt and parasitic nematodes infecting pigeon pea under field conditions during 2001.

| Treatment | Grain yield (kg/100 m$^2$) | % wilting of plants | % root infection by *F. udum* | Nematodes / g of root | Cysts & larvae/100 ml soil |
|---|---|---|---|---|---|
| Carbofuran | 27.57 | 0.0 | 10.0 | 15 | 20 |
| Neem seed powder (soil application) | 22.66 | 0.0 | 5.0 | 18 | 22 |
| Neem seed powder (seed treatment) | 19.66 | 0.0 | 5.0 | 17 | 26 |
| *Trichoderma harzianum* | 18.66 | 0.0 | 5.0 | 20 | 27 |
| Neem jivan | 17.33 | 0.0 | 10.0 | 24 | 27 |
| *Aspergillus niger* | 16.49 | 0.0 | 10.0 | 26 | 30 |
| *Paecilomyces lilacinus* | 15.83 | 2.0 | 15.0 | 30 | 36 |
| Dimethoate | 15.83 | 5.0 | 20.0 | 31 | 32 |
| Latex of Calotropis | 12.66 | 5.0 | 25.0 | 30 | 33 |
| Control | 8.83 | 33.0 | 65.0 | 44 | 37 |
| CD (P = 0.05) | 1.58 | 1.8 | 2.7 | 2.9 | 2.1 |

## 17.19. LAB LAB BEAN, *DOILICHOS LAB LAB*

### 17.19.1. High yielding varieties

**Table 17.152.** Increased productivity in high yielding varietiesof lab lab bean

| HYV | Yield (MT/ha) | Developed at |
|---|---|---|
| Pusa Sem-2 | 15 | IARI, New Delhi |
| Pusa Sem-3 | 17 | IARI, New Delhi |
| Arka Jay | 12 | IIHR, Bangalore |
| Arka Vijay | 12 | IIHR, Bangalore |
| Hebbal Aware | 18-20 | UAS, Bangalore |
| CO-1 | 18-20 | TNAU, Coimbatore |
| CO-2 | 18-20 | TNAU, Coimbatore |
| Konkan Bhushan | 13 | KKV, Akola |

**Table 17.153.** Lab lab bean varieties resistant to pests/diseases

| HYV | Yield MT/ha | Resistance to pest/disease | Developed at |
|---|---|---|---|
| Pusa Sem-2 | 15 | Anthracnose (T), aphid (T), jassid (T), pod borer (T) | IARI, New Delhi |
| Pusa Sem-3 | 17 | Anthracnose (T), aphid (T), jassid (T), pod borer (T) | IARI, New Delhi |

## 17.20. WATERMELON, *CITRULLUS LUNATUS*

### 17.20.1. High yielding varieties

**Table 17.154.** Increased productivity in high yielding varieties/F1 hybrids of water melon

| HYV/F1 hybrid | Yield (MT/ha) | Developed at |
|---|---|---|
| Sugar Baby | 30 | IARI, New Delhi |
| Arka Manik | 60-61 | IIHR, Bangalore |
| Arka Jyothi* | 70-72 | IIHR, Bangalore |
| Durgapura Kesar | 40-45 | RARS, Durgapaura |
| Durgapura Meetha | 48-50 | RARS, Durgapaura |
| Durgapura Lal* | 43-45 | RARS, Durgapaura |
| PKM 1 | 36-38 | TNAU, Coimbatore |

* Hybrids

**Table 17.155.** Water melon varieties resistant to diseases

| HYV/F1 hybrid | Yield (MT/ha) | Resistance to disease | Developed at |
|---|---|---|---|
| Arka Manik | 60-61 | Anthracnose, Powdery mildew, Downy mildew | IIHR, Bangalore |
| Durgapura Lal* | 43-45 | Blight (MR), Bud necrosis (MR) | RARS, Durgapaura |

* Hybrids

### 17.20.2. Micro irrigation

Drip irrigation increased yields of watermelon (Table 17.156).

**Table 17.156.** Yield and water use efficiency of cucurbits under drip and furrow irrigation

| Crop | Yield (MT/ha) | | Water use efficiency (kg/ha/mm) | |
|---|---|---|---|---|
| | Furrow | Drip | Furrow | Drip |
| Watermelon | 29.5 | 82.3 | 84 | 103 |

### 17.20.3. Fertigation

Application of the entire recommended quantity of NK or NPK fertilizer through fertigation increased marketable yield significantly (62.19 and 61.89 MT/ha) over control. Further, the use of polyethylene mulch resulted in 22.63% higher yield even with entire quantity of NPK supplied through fertigation. Water use efficiency was higher with fertigation treatments (147 to 180 kg/ha-mm) compared to soil application of fertilizers in watermelon (140 kg/ha-mm) (Prabhakar *et al.*, 2006).

### 17.20.4. Integrated nutrient management

The farmers achieved the additional yield of 14.4 MT/ha through the application of 150 kg $K_2O$/ha as compared with the zero K treatment. The percentage of fruit yield increase due to K was as high as 47%. Such quality parameters as vitamin C and sugar content were also improved as resulted from potash fertilizer application at the above-indicated rate. Importantly, experimental data indicated that the percentage of cracked watermelon fruits in the NP treatment was 44% vs. 25% in the treatment in which K was applied at 150 kg $K_2O$/ha.

### 17.20.5. Integrated pest and disease management

Seed treatment with thiamethoxam + Phule *Trichoderma* at 5 + 5 g/kg was the most promising treatment against serpentine leaf miner (*Liriomyza trifolii*), thrips (*Thrips palmi*), bud necrosis disease and root rot disease in watermelon and registered highest fruit yield (17.722 MT/ha) followed by imidacloprid + Phule *Trichoderma* at 5 + 5 g/kg (12.555MT/ha), imidacloprid at 5 g/kg (14.444 MT/ha), and acetamiprid + Phule *Trichoderma* at 5 + 5 g/kg (12.555 MT/ha); while untreated plots recorded significantly lower yield (8.333 MT/ha) (Kumar and Patil, 2006).

## 17.21. MUSK MELON, *CUCUMIS MELO*

### 17.21.1. High yielding varieties

**Table 17.157.** Increased productivity in high yielding varieties/F1 hybrids of musk melon

| HYV/F1 hybrid | Yield (MT/ha) | Developed at |
|---|---|---|
| Pusa Rasraj * | 22.5-24.0 | IARI, New Delhi |
| Pusa Sharabati | 15-17 | IARI, New Delhi |
| Arka Jeet | 24.1 | IIHR, Bangalore |
| Arka Rajahans | 28.5-30.0 | IIHR, Bangalore |
| Punjab Hybrid* | 20 | PAU, Ludhiana |
| Punjab Sunehri | 16-18 | PAU, Ludhiana |
| Punjab Hybrid-1* | 16-18 | PAU, Ludhiana |
| Punjab Raseela | 16.0-17.5 | PAU, Ludhiana |
| MH-10* | 24-25 | PAU, Ludhiana |
| RM-43 | 20-22 | RARS, Durgapaura |
| RM-50 | 17.5-20.0 | RARS, Durgapaura |
| MHY-3 | 18-20 | RARS, Durgapaura |
| MHY-5 | 18-20 | RARS, Durgapaura |
| Kashi Madhu | 25-27 | IIVR. Varanasi |

* Hybrids

**Table 17.158.** Musk melon varieties resistant to pests/diseases

| HYV/F1 hybrid | Yield (MT/ha) | Resistance to pest/disease | Developed at |
|---|---|---|---|
| Arka Rajhans | 28.5-30.0 | Powdery mildew | IIHR, Bangalore |
| Punjab Hybrid-1* | 16-18 | Powdery mildew (MR), Fruit fly (MR) | PAU, Ludhiana |
| Kashi Madhu | 25-27 | Powdery mildew (T), Downy mildew (T) | IIVR, Varanasi |

* Hybrids

### 17.21.2. Micro irrigation

Adoption of micro-irrigation in musk melon produced 20-40% higher yields as compared to furrow method and also performed well with irrigation scheduled at 0.5 Epan level (Table 17.159).

**Table 17.159.** Yield performance of hybrid cantaloupe under micro-irrigation

| Irrigation method | Yield (MT/ha) | |
|---|---|---|
| | 50% Epan | 100% Epan |
| Micro tube | 22.6 | 27.6 |
| Drip | 14.7 | 19.3 |
| Chapin | 13.8 | 16.9 |
| Furrow | 8.9 | 16.8 |

### 17.21.3. Fertigation

Significant increase in yield was observed with pre-plant soil application of N at 67 kg/ha followed by post-plant fertigation with 100 ppm of N (49.4 MT/ha) followed by pre-plant soil application of N at 67 kg/ha followed by post-plant fertigation with 50 ppm of N (47.2 MT/ha) (Bhella and Wilcox, 1986) (Table 17.160).

**Table 17.160.** Nitrogen fertigation in musk melon

| Treatment (Nitrogen) | | Yield (MT/ha) |
|---|---|---|
| Pre-plant via broadcast (kg/ha) | Post-plant via fertigation (ppm) | |
| 0 | 0 | 30.4 |
| | 50 | 40.4 |
| | 100 | 44.2 |
| 67 | 0 | 42.1 |
| | 50 | 47.2 |
| | 100 | 49.4 |
| 100 | 0 | 42.3 |
| | 50 | 42.8 |
| | 100 | 43.2 |

## 17.22. CUCUMBER, *CUCUMIS SATIVUS*

### 17.22.1. High yielding varieties

**Table 17.161.** Increased productivity in high yielding varieties/F1 hybrids of Cucumber

| HYV/F1 hybrid | Yield (MT/ha) | Developed at |
|---|---|---|
| Pusa Sanyog (F1 hybrid) | 20 | IARI, New Delhi |
| Swarna Poorna | 30-35 | HARP, Ranchi |
| Swarna Agethi | 30-32.5 | HARP, Ranchi |
| Swarna Swetha | 25-30 | HARP, Ranchi |
| Swarna Sheetal | 40-45 | HARP, Ranchi |
| Japanese Long Green | 15.0-27.5 | IARI, Katrain |
| Sheetal | 38-40 | MPKV, Rahuri |
| Phule Subhangi | 30.0-32.5 | MPKV, Rahuri |
| PH 9 | 49 | MPKV, Rahuri |
| Phule Priyanka | 28.25 | MPKV, Rahuri |
| Gujarat Cucumber 1 | 24-26 | GAU, Anand |
| CO 1 | 25.28 | TNAU, Coimbatore |

**Table 17.162.** Cucumber varieties resistant to pests/diseases

| HYV/F1 hybrid | Yield (MT/ha) | Resistance to pest/disease | Developed at |
|---|---|---|---|
| Phule Subhangi | 30.0-32.5 | Anthracnose (T), Leaf spot (T), Pod borer (T), Leaf miner (T) | MPKV, Rahuri |
| Phule Priyanka | 28.25 | Downy mildew (T) | MPKV, Rahuri |
| Swarna Poorna | 31.0-33.5 | Powdery mildew (T) | HARP, Ranchi |
| Swarna Ageti | 30.0-32.5 | Powdery mildew (T) | HARP, Ranchi |
| PCUCH-3* | 25-27 | Powdery mildew (MR), Downy mildew (MR) | |

* Hybrids, MR- Moderately Resistant, T - Tolerant

### 17.22.2. Fertigation

The highest yield of 49.03 MT/ha was recorded when 150: 90: 90 kg NPK/ha (liquid fertilizers) was applied through fertigation. The increase in yield was 81.2% higher than the control (Chaudhari and More, 2002) (Table 17.163)

**Table 17.163.** Fertigation in cucumber

| Treatment | Fruit yield (MT/ha) | % increase |
|---|---|---|
| 100% RDF + Surface irrigation + Solid fertilizers | 27.1 | --- |
| 100% RDF + Drip irrigation + Solid fertilizers | 32.6 | 30.6 |

| | | |
|---|---|---|
| 100% RDF + Drip irrigation + Liquid fertilizers | 35.2 | 39.9 |
| 125% RDF + Drip irrigation + Liquid fertilizers | 43.9 | 62.4 |
| 150% RDF + Drip irrigation + Liquid fertilizers | 49.0 | 81.2 |
| 175% RDF + Drip irrigation + Liquid fertilizers | 41.9 | 54.9 |
| 200% RDF + Drip irrigation + Liquid fertilizers | 40.7 | 50.5 |

### 17.22.3. Protected cultivation

Maximum fruit yield was recorded from Hilton (122.98 MT/ha), followed by NVH-2-26 (120.07 MT/ha) and Istis (119.00 MT/ha) which were at par with each other. However, minimum fruit yield was noticed from Mohini (55.00 MT/ha) and Pant Shankar Khira (56.12 MT/ha) under protected cultivation (Bhatia *et al.*, 2009).

## 17.23. BITTER GOURD, *MEMORDICA CHARANTIA*

### 17.23.1. High yielding varieties

**Table 17.164.** Increased productivity in high yielding varieties/F1 hybrids of bitter Gourd

| HYV/F1 hybrid | Yield (MT/ha) | Developed at |
|---|---|---|
| Pusa Vishesh | 15 | IARI, New Delhi |
| Pusa Hybrid-1* | 21.8 | IARI, New Delhi |
| Pusa Hybrid-2* | 21.5 | IARI, New Delhi |
| Coimbatore Long | 20-22 | TNAU, Coimbatore |
| CO-1 | 14-15 | TNAU, Coimbatore |
| Kalyanpur Baramasi | 20-22 | CSAUAT, Kanpur |
| MDU-1 | 30.0-32.5 | AC&RI, TNAU, Madurai |
| COBgoH 1 | 51.29 | TNAU, Coimbatore |
| Konkan Tara | 23-25 | KKV, Konkan |
| Priya | 30.0-32.5 | KAU, Thrissur |
| Phule Green Gold | 22.0-23.5 | MPKV, Rahuri |
| Phule Priyanka | 20-25 | MPKV, Rahuri |

* Hybrids

**Table 17.165.** Bitter gourd varieties resistant to diseases

| HYV/F1 hybrid | Yield (MT/ha) | Resistance to disease | Developed at |
|---|---|---|---|
| Phule Priyanka* | 20-25 | Downy mildew (T) | MPKV, Rahuri |
| Phule Green Gold | 22.0-23.5 | Downy mildew (T) | MPKV, Rahuri |

* Hybrids, T - Tolerant

### 17.23.2. Integrated nutrient management

Application of vermicompost instead of FYM without inorganic fertilizer was equally effective for increasing the bitter gourd yield (19%) as the recommended manurial schedule (Jiji and Dale, 1999) (Table 17.166).

**Table 17.166.** Effect of vermicompost on fruit yield of bitter gourd

| Treatment | Fruit yield ($kg/4m^2$) |
|---|---|
| POP recommendations | 3.222 |
| Vermicompost substituting FYM | 2.322 |
| Vermicompost substituting FYM + ½ fertilizers as per POP | 2.842 |
| Vermicompost substituting FYM + ¾ inorganic fertilizers as per POP | 3.254 |
| Vermicompost substituting FYM + N requirement as per POP through vermicompost | 2.965 |
| Vermicompost substituting FYM + Full inorganic fertilizers as per POP | 4.023 |
| CD (P = 0.05) | 0.702 |

Application of higher doses of nitrogen (100 kg/ha), phosphorous (75 kg/ha) and potassium (50 kg/ha) gave higher seed yield (799 kg/ha, 758 kg/ha and 722 kg/ha, respectively) of bitter gourd cv. Coimbatore Long (Green) (Tirakannanavar *et al.*, 2006).

### 17.23.3. Integrated pest management

Among the different food attractants (with dichlorvos as poison), the cue lure (40.18), banana + jaggery (34.18), banana + molasses (34.66) and banana (34.07) were found promising in attracting and killing the melon flies (*Bactrocera cucurbitae*). The highest yield was recorded in banana as attractant (8.695 MT/ha) and the lowest was recorded in untreated check (6.901 MT/ha) (Shivayya and Ashok Kumar, 2006).

## 17.24. BOTTLE GOURD, *LAGENARIA SICERARIA*

### 17.24.1. High yielding varieties

**Table 17.167.** Increased productivity in high yielding varieties/F1 hybrids of bottle gourd

| HYV/F1 hybrid | Yield (MT/ha) | Developed at |
|---|---|---|
| Pusa Meghdoot | 45 | IARI, New Delhi |
| Pusa Manjari | 40 | IARI, New Delhi |
| Pusa Summer Prolific Long | 25-30 | IARI, New Delhi |
| Pusa Sandesh | 32 | IARI, New Delhi |
| Pusa Summer Prolific Round | 30.0-32.5 | IARI, New Delhi |
| Pusa Naveen | 35-40 | IARI, New Delhi |
| Pusa Hybrid-3* | 47 | IARI, New Delhi |

| | | |
|---|---|---|
| Punjab Round | 35.0-37.5 | PAU, Ludhiana |
| CO-1 | 36-40 | TNAU, Coimbatore |
| Arka Bahar | 40-50 | IIHR, Bangalore |
| Narendra Sankar Lauki-4 | 40-45 | NDUAT, Faizabad |
| Narendra Dharidar | 32.5-35.0 | NDUAT, Faizabad |
| Narendra Rashmi | 27.5-31.0 | NDUAT, Faizabad |
| Narendra Jyoti | 30-35 | NDUAT, Faizabad |
| Kashi Ganga | 48-55 | IIVR, Varanasi |
| Kashi Bahar | 50-55 | IIVR, Varanasi |
| Pant Sankar Lauki-3* | 35-37 | GBPUAT, Pantnagar |
| Pant Sankar Lauki-2* | 40.0-42.5 | GBPUAT, Pantnagar |
| Pant Sankar Lauki-1 | 40.0-42.5 | GBPUAT, Pantnagar |
| Azad Sankar Lauki-1* | 58-60 | CSAUAT, Kanpur |
| Samrat | 40 | MPKV, Rahuri |
| NDBG-132 | 43-45 | |

* Hybrids

**Table 17.168.** Bottle gourd varieties resistant to diseases

| HYV | Yield (MT/ha) | Resistance to disease | Developed at |
|---|---|---|---|
| Kashi Bahar | 50-55 | Anthracnose (T), Downy mildew (T), *Cercospora* leaf spot (T) | IIVR, Varanasi |
| Kashi Ganga | 48-55 | Anthracnose (T) | IIVR, Varanasi |
| Narendra Rashmi | 27.5-31.0 | Powdery mildew (MR), Downy mildew (MR) | NDUAT, Faizabad |

MR – Moderately Resistant, T - Tolerant

### 17.24.2. Micro irrigation

Drip irrigation increased yields of bottle gourd (Singh and Singh, 1978) (Table 17.169).

**Table 17.169.** Yield and water use efficiency of cucurbits under drip and furrow irrigation

| Crop | Yield (MT/ha) | | Water use efficiency (kg/ha/mm) | |
|---|---|---|---|---|
| | Furrow | Drip | Furrow | Drip |
| Bottle gourd | 38.0 | 55.8 | 45 | 75 |

### 17.24.3. Integrated nematode management

#### (i) Root-knot nematode, *Meloidogyne* spp.

Spot application of mustard cake at 15 g/plant one week before sowing was effective in controlling root-knot nematodes and significantly improved the yield (8.30 MT/ha) followed by neem cake (8.23 MT/ha) (Paruthi *et al.*, 2010).

## 17.25. POINTED GOURD, *TRICHOSANTHES DIOIDA*

### 17.25.1. High yielding varieties

**Table 17.170.** Increased productivity in high yielding varieties of pointed gourd

| HYV/F1 hybrid | Yield (MT/ha) | Developed at |
|---|---|---|
| Swarna Rekha | 17.5-20.0 | HARP, Ranchi |
| Swarna Alaukik | 23-25 | HARP, Ranchi |
| Rajendra Parwal-1 | 17.5-19.0 | RAU, Pusa |
| Rajendra Parwal-2 | 16-18 | RAU, Pusa |

**Table 17.171.** Pointed gourd varieties resistant to pests/diseases

| HYV | Yield (MT/ha) | Resistance to pest/disease | Developed at |
|---|---|---|---|
| Rajendra Parwal-1 | 17.5-19.0 | Fruit fly (T) | RAU, Pusa |
| Rajendra Parwal-2 | 16-18 | Fruit fly (T), Vine & fruit rot (T) | RAU, Pusa |

### 17.25.2. Integrated nematode management

#### (i) Root-knot nematode, *Meloidogyne incognita*

Integration of *Paecilomyces lilacinus* (50 g) + *Trichoderma harzianum* (100g) + neem cake (250g) + marigold (3 plants/pit) increased plant growth parameters and yield (6.6 kg/plant compared to 1.4 kg/plant in control), and reduced root galling (14.3/5g roots compared to 75.0/5g roots in control), number of egg masses (40/5g roots compared to 150/5g roots in control) and final nematode population (66.6/100g soil compared to 86660/100g soil in control) (Verma *et al.*, 2005) (Table 17.172).

**Table 17.172.** Effect of integration of bioagents, neem cake and marigold as intercrop for the management of *Meloidogyne incognita* infecting pointed gourd

| Treatment | No. of galls / 5g root | No. of egg masses/ 5g root | Yield (kg/ plant) |
|---|---|---|---|
| Neem cake + *P. lilacinus* (50 g) + *T. harzianum* (100g) + (250 g) + Marigold (3 plants)/pit | 10.4 | 30 | 8.5 |
| *P. lilacinus* (50 g/pit) | 15.3 | 53 | 6.4 |
| Carbofuran (3 g/pit) | 22.5 | 72 | 5.8 |
| Neem cake (250 g/pit) | 22.4 | 75 | 5.0 |
| *T. harzianum* (100 g/pit) | 25.6 | 45 | 4.2 |
| Marigold (3 plants/pit) | 35.3 | 113 | 4.0 |
| Control | 80.4 | 186 | 1.3 |
| CD (P = 0.05) | 3.5 | 9.6 | 0.5 |

Vine dipping in monocrotophos at 1000 ppm for 6 hr + soil application of *T. viride* at 10 g/pit in 2 split doses (once at planting and another at 40 DAP) reduced root galling caused by *M. incognita*, nematode population in soil and gave fruit yield almost double of the untreated plots (Khan *et al.*, 2009) (Table 17.173).

**Table 17.173.** Effect of bioagents, botanicals and chemicals for the management of *Meloidogyne incognita* infecting pointed gourd

| Treatment | Gall index (0-5 scale) | Nema popn. $J_2$/200 ml soil | Yield (kg/12m$^2$) |
|---|---|---|---|
| Vine dipping in 1000 ppm monocrotophos for 6 hr + Organic matter at 20 MT/ha | 1.93 | 264 | 22.10 |
| Vine dipping in 1000 ppm monocrotophos for 6 hr + Vermicompost at 2 MT/ha | 1.78 | 232 | 25.53 |
| Vine dipping in 1000 ppm monocrotophos for 6 hr + *P. lilacinus* at 10 g/pit (2 splits) | 2.54 | 198 | 11.93 |
| Vine dipping in 1000 ppm monocrotophos for 6 hr + *T. viride* at 10 g/pit in 2 split doses | 1.75 | 162 | 33.80 |
| Vine dipping in 1000 ppm monocrotophos for 6 hr + Carbofuran at 1 kg a.i./ha | 3.60 | 386 | 16.90 |
| Vine dipping in 1000 ppm monocrotophos for 6 hr + Neem cake at 500 kg/ha | 2.16 | 280 | 17.43 |
| Vine dipping in 1000 ppm monocrotophos for 6 hr alone | 3.27 | 390 | 12.60 |
| Untreated control | 3.67 | 535 | 16.83 |
| CD (P = 0.05) | 0.93 | --- | 6.02 |

## 17.26. RIDGE GOURD, LAGENARIA ACUTANGULA

### 17.26.1. High yielding varieties

**Table 17.174.** Increased productivity in high yielding varieties of ridge gourd

| HYV | Yield (MT/ha) | Developed at |
|---|---|---|
| Arka Sujat | 62.5-64.0 | IIHR, Bangalore |
| Arka Sumeet | 50 | IIHR, Bangalore |
| PKM-1 | 25-28 | TNAU, Coimbatore |
| CO-2 | 25-27 | TNAU, Coimbatore |
| Swarna Uphar | 28-31 | HARP, Ranchi |
| Deepthi | 30.0-32.5 | KAU, Vellanikkara |

**Table 17.175.** Ridge gourd varieties resistant to pests/diseases

| HYV | Yield (MT/ha) | Resistance to pest/disease | Developed at |
|---|---|---|---|
| PKM-1 | 25-28 | Pumpkin beetle (T), Fruit fly (T), Leaf spot (T) | TNAU, Coimbatore |
| Deepthi | 30.0-32.5 | Powdery mildew, Mosaic | KAU, Vellanikkara |

### 17.26.2. Micro irrigation

Drip irrigation increased yields of ridge gourd (Singh and Singh, 1978) (Table 17.176).

**Table 17.176.** Yield and water use efficiency of cucurbits under drip and furrow irrigation

| Crop | Yield (MT/ha) | | Water use efficiency (kg/ha/mm) | |
|---|---|---|---|---|
| | Furrow | Drip | Furrow | Drip |
| Ridge gourd | 10.7 | 12.0 | 13 | 16 |

## 17.27. SPONGE GOURD, *LUFFA CYLINDRICA*

### 17.27.1. High yielding varieties

**Table 17.177.** Increased productivity in high yielding varietiesof sponge gourd

| HYV | Yield (MT/ha) | Developed at |
|---|---|---|
| Pusa Chikni | 20.0-22.5 | IARI, New Delhi |
| Kalyanpur Hari Chikni | 18-20 | CSAUAT, Kanpur |
| Pusa Sneha | 20-23 | IARI, New Delhi |
| Swarna Prabha | 20-25 | HARP, Ranchi |
| Azad Torai Chikni-1 | 18-20 | CSAUAT, Kanpur |

### 17.27.2. Integrated nematode management

#### i) Root-knot nematode, *Meloidogyne incognita*

Spot application of neem cake at 30 g/plant was most effective in reducing the root-knot nematode population up to 28.0% and increased the yield up to 13.9% in comparison to control (Verma *et al.*, 2010).

## 17.28. SNAKE GOURD, *TRICHSATHES CUCUMERINA*

### 17.28.1. High yielding varieties

**Table 17.178.** Increased productivity in high yielding varieties of snake gourd

| HYV | Yield (MT/ha) | Developed at |
|---|---|---|
| CO-1 | 18-19 | TNAU, Coimbatore |
| CO-2 | 35-37 | TNAU, Coimbatore |
| PKM-1 | 25.5-27.0 | TNAU, Coimbatore |

| | | |
|---|---|---|
| MDU 1 | 31.75 | AC&RI, TNAU, Madurai |
| PLR(SG) | 35-40 | TNAU, Coimbatore |
| Baby | 39-42 | KAU, Vellanikkara |
| Manushree | 57.5-62.0 | KAU, Vellanikkara |

### 17.28.2. Micro irrigation

Among two drip irrigation frequencies tried (5 mm CPE and 10 mm CPE), 5 mm CPE recorded highest yield of 25.95 MT/ha which was significantly higher over 10 mm CPE (23.25 MT/ha) (Elizabeth Syriac *et al.*, 1999) (Table 17.179).

**Table 17.179.** Effect of irrigation levels on fruit yield of snake gourd

| Irrigation levels | Fruit yield (MT/ha) |
|---|---|
| 5 mm CPE | 24.98 |
| 10 mm CPE | 23.12 |

### 17.28.3. Integrated nutrient management

Among the four doses of nitrogen tried (35, 70, 105 and 140 kg/ha), N at 105 kg/ha recorded the highest yield of 25.36 MT/ha which was on par with N at 140 kg/ha (24.68 MT/ha) (Elizabeth Syriac *et al.*, 1999) (Table 17.180).

**Table 17.180.** Effect of nitrogen levels on fruit yield of snake gourd

| Nitrogen levels | Fruit yield (MT/ha) |
|---|---|
| 35 kg N/ha | 21.94 |
| 70 kg N/ha | 23.28 |
| 105 kg N/ha | 25.86 |
| 140 kg N/ha | 25.15 |
| Control (no nitrogen) | 15.15 |

### 17.28.4. Bioregulators

Among the four levels of ethephon tried (0, 50, 100 and 200 ppm), 100 ppm recorded the highest fruit yield of 24.11 MT/ha which was on par with 200 pm and 50 ppm (23.00 MT/ha) (Elizabeth Syriac *et al.*, 1999) (Table 17.181).

**Table 17.181.** Effect of ethephon levels on fruit yield of snake gourd

| Ethephon levels | Fruit yield (MT/ha) |
|---|---|
| 0 ppm | 22.72 |
| 50 ppm | 23.86 |
| 100 ppm | 24.47 |
| 200 ppm | 25.14 |
| Control (no ethephon) | 14.19 |

For getting the maximum fruit yield in snake gourd, the suitable levels of inputs were 105 kg N/ha, 200 pm ethephon and drip irrigation at 5 mm CPE.

### 17.28.5. Integrated nematode management

#### i) Root-knot nematode, *Meloidogyne incognita*

Spot application of neem cake at 30 g/plant 10 days prior to sowing recorded the lowest final nematode population (282/200 ml soil and 59/5g root) and root-knot index (3.0) which were on par with carbofuran 3G at 10 g/plant at sowing (293/200 ml soil, 62/5g root and 3.0 respectively). The highest yield of 14.80 MT/ha was recorded in neem cake treatment which was on par with carbofuran (14.68 MT/ha) compared to control (11.60 MT/ha) (Devarajan *et al.*, 2010).

## 17.29. PUMPKIN, *CUCURBITA MOSCHATA*

### 17.29.1. High yielding varieties

**Table 17.182.** Increased productivity in high yielding varieties/F1 hybrids of pumpkin

| HYV/F1 hybrid | Yield (MT/ha) | Developed at |
|---|---|---|
| Pusa Vishwas | 40.0-42.5 | IARI, New Delhi |
| Pusa Hybrid-1* | 52 | IARI, New Delhi |
| CO-2 | 45.0-47.5 | TNAU, Coimbatore |
| Sooraj | 35-40 | KAU, Vellanikkara |
| Azad Pumpkin-1 | 42.5-45.0 | CSAUAT, Kanpur |
| Narendra Amrit | 35-40 | NDUAT, Faizabad |
| Narendra Abhooshan* | 65-70 | NDUAT, Faizabad |
| Saras | 39-41 | KAU, Vellanikkara |

* Hybrids

## 17.30. SQUASH, *CUCURBITA PEPEO*

### 17.30.1. High yielding varieties

**Table 17.183.** Increased productivity in high yielding varietiesof squash

| HYV | Yield (MT/ha) | Developed at |
|---|---|---|
| Pusa Alankar | 50 | IARI, Katrain |
| Early Yellow Prolific | 25.0-27.5 | IARI, Katrain |
| Punjab Chappan Kaddu-1 | 30-32 | PAU, Ludhiana |
| Patty Pan | 35-40 | IIHR, Bangalore |
| Arka Suryamukhi | 36.0-38.5 | IIHR, Bangalore |

**Table 17.184.** Squash varieties resistant to pests/diseases

| HYV | Yield (MT/ha) | Resistance to pest/disease | Developed at |
|---|---|---|---|
| Punjab Chappan Kaddu-1 | 30-32 | CMV (T), Pumpkin beetle (T) | PAU, Ludhiana |

## 17.31. LETTUCE, *LACTUCA SATIVA*

### 17.31.1. Integrated nutrient management

As compared with the NP treatment, the additional lettuce yield obtained by potash application at 68 kg $K_2O$/ha was 5.8 MT/ha that is equivalent to the yield increment by 13%. In this experiment, K application decreased the content of nitrates 1.3 and 2.5 times over the control NP treatment in lettuce leaves and stems, respectively. Hence, with balanced fertilizer practices, lesser nitrates are accumulated in crop production and crop biomass is safer for human consumption.

Integrated use of organic inputs (FYM, digested sludge, and biofertilizers- *Azotobacter*, P-solubilizers and arbuscular mycorrhizal fungi) gave highest head yield (27.195-28.664 MT/ha) of lettuce (cv. Great Lakes) compared to lone application of organic input. The above treatment was found on par with recommended dose of NPK (Bahadur *et al.*, 2006).

### 17.31.2. Integrated nematode management

#### i) Root-knot nematodes, *Meloidogyne incognita, M. hapla*

The number of root galls was significantly decreased in *Pasteuria penetrans* (PP) treated pots with 75 (77.2% control effect) and 150/root (54.4% control effect) in 200,000 and 100,000 *Pasteuria penetrans* endospores/g medium, respectively, compared to the 330/root in inoculated control when harvested at 10 weeks after planting (Table 17.185).

**Table 17.185.** Suppression effect of *Pasteuria penetrans* KW1 (PP) on reproduction of *Meloidogyne incognita* in lettuce

| Treatments | No. of root galls | | Wt. of leaves/ (g/plant) | Total no. of leaves/plant | Root wt.(g/plant) |
|---|---|---|---|---|---|
| | No. of galls/root | Control effect (%) | | | |
| *Meloidogyne incognita* J2 at 5,000/plant (N) | 330 c | --- | 54.6 a | 40.9 a | 11.9 ab |
| N + PP at 100,000 endospores/g medium | 150 b | 54.4 | 55.5 a | 38.4 a | 11.6 b |
| N + PP at 200,000 endospores/g medium | 75 a | 77.2 | 65.2 a | 40.6 a | 11.6 b |
| Uninoculated control | 0 a | --- | 57.8 a | 39.5 a | 14.2 a |

Biofumigation with mustard under black polythene for 30 days followed by soil application of *Pseudomonas fluorescens* at 2.5 kg/ha at the time of planting was on par with carbofuran at 1 kg a.i./ha in reducing the root-knot nematode population by 76.39% and increased the lettuce leaf yield by 23.29% (Anita and Selvaraj, 2010).

## 17.32. AMARANTHUS, *AMARANTHUS BLITUM, A. TRICOLOR*

### 17.32.1. High yielding varieties

**Table 17.186.** Increased productivity in high yielding varieties of leafy amaranthus

| High yielding varieties | Yield (MT/ha) | Developed at |
|---|---|---|
| Pusa Kirti | 50-55 | IARI, New Delhi |
| Pusa Lal Chaulai | 40-45 | IARI, New Delhi |
| Pusa Kiran | 35 | IARI, New Delhi |
| CO - 3 | 30.72 | TNAU, Coimbatore |
| CO - 5 | 40 | TNAU, Coimbatore |
| Arka Suguna | 27-28 | IIHR, Bangalore |
| Renusree | 15.5 | KAU, Thrissur |
| Krishnasree | 14.8 | KAU, Thrissur |
| Annapurna | 25.0-27.5 | --- |
| Arun | 93.65 | --- |
| Kannara Local | 101.30 | --- |

**Table 17.187.** Amaranthus varieties resistant to diseases

| HYV | Yield (MT/ha) | Resistance to disease | Developed at |
|---|---|---|---|
| Annapurna | 25.0-27.5 | *Alternaria* spp., Leaf spot | |
| Arka Arunima | | White rust | IIHR, Bangalore |

### 17.32.2. Integrated nutrient management

Integration of biofertilizers, AMF and 75% RDF application gave maximum marketable yield (Niranjana *et al.*, 1999) (Table 17.188).

**Table 17.188.** Effect of biofertilizers and chemical fertilizers on yield of amaranth

| Treatment | Total yield (MT/ha) | Marketable yield (MT/ha) |
|---|---|---|
| *Azospirillum* | 105.55 | 96.73 |
| AMF | 95.90 | 92.39 |
| *Azospirillum* + AMF | 109.26 | 103.32 |
| 25% RDF | 96.07 | 92.01 |
| 50% RDF | 113.54 | 105.01 |

| | | |
|---|---|---|
| 75% RDF | 124.38 | 116.08 |
| *Azospirillum* + 50% RDF | 144.80 | 127.00 |
| *Azospirillum* + AMF + 75% RDF | 153.29 | 144.31 |
| Control | 80.31 | 76.08 |

## 17.33. SPINACH BEET/PALAK, *BETA VULGARIS* VAR. *BENGALENSIS*

### 17.33.1. High yielding varieties

**Table 17.189.** Increased productivity in high yielding varieties of palak

| High yielding varieties | Yield (MT/ha) | Developed at |
|---|---|---|
| Ooty-1 | 33.8 | HRS, TNAU, Ooty |
| Arka Anupama | 41.0-42.5 | IIHR, Bangalore |
| Pusa Jyoti | 42.5 | IARI, New Delhi |
| Pusa Bharati | 50 | IARI, New Delhi |

**Table 17.190.** Palak varieties resistant to pests/diseases

| HYV | Yield (MT/ha) | Resistance to pest/disease | Developed at |
|---|---|---|---|
| Arka Anupama | 41 | *Cercospora* leaf spot | IIHR, Bangalore |
| Ooty-1 | 33.8 | *Cercospora* leaf spot, root-knot nematode, whiteflies, aphids | HRS, TNAU, Ooty |

### 17.33.2. Protected cultivation

**Table 17.191.** Production of beet leaf (Palak) in different protected structures

| Structure | Production (kg/m$^2$) |
|---|---|
| Glasshouse | 3.3 |
| Ladakhi polyhouse | 5.4 |
| Polyhouse | 2.7 |
| Trench | 6.3 |
| Tunnel | 2.2 |
| Unit polyhouse | 4.5 |

## 17.34. DRUMSTICK, *MORINGA OLEIFERA*

### 17.34.1. High yielding varieties

**Table 17.192.** High yielding varieties of drumstick

| High yielding varieties | Yield (Pods/tree) | Developed at |
|---|---|---|
| PKM-1 | 218 (52.8 MT/ha) | HC&RI, TNAU, Periyakulam |
| PKM-2 | 220 (98.0 MT/ha) | HC&RI, TNAU, Periyakulam |

### 17.34.2. Integrated pest management

#### (i) Fruit fly, *Gitona distigma*

Modules A to F were effective in reducing the fruit fly infestation and in increasing the fruit yield of drumstick (Raghumoorthi and Subba Rao, 1998) (Table 17.193).

**Table 17.193.** Impact of IPM components on fruit fly infestation and yield of drumstick

| IPM module* | % fruit fly infestation | % reduction over control | Fruit yield (MT/ha) | % increase over control |
|---|---|---|---|---|
| Components 1,2,3,4 | 4.67 | 93.06 | 39.390 | 93.06 |
| Components 2,3,4 | 7.33 | 89.11 | 38.291 | 89.11 |
| Components 1,6,3,4 | 5.67 | 91.58 | 38.977 | 91.59 |
| Components 5,2,7,4 | 7.33 | 89.11 | 38.291 | 89.11 |
| Components 1,6,7,4 | 6.00 | 91.09 | 38.840 | 91.09 |
| Components 5,2,3,4 | 5.00 | 92.57 | 39.254 | 92.58 |
| Control-without any treatment | 67.33 | 0.00 | 13.499 | 0.00 |

* Components: 1. Application of Fenthion 80 EC 0.04% at vegetative & flowering stage.
2. Application of Nuvan 76 WSC 0.04% at 50% fruit set & 35 days later.
3. Soil application of Endosulfan 4D (200 g/tree) at 50% fruit set.
4. Removal of affected fruits regularly at weekly intervals.
5. Application of Nimbecidine 0.03% at 60 ppm conc. during vegetative & flowering stage.
6. Application of Nimbecidine 0.03% at 150 ppm conc. during 50% fruit set & 35 days later.
7. Soil application of NSKE 4% at 2 litres/tree during 50% fruit set.

Chapter 18

# ORNAMENTAL CROPS

India is known for growing of traditional flowers such as jasmine, marigold, chrysanthemum, tuberose, crossandra and aster. Commercial cultivation of cut flowers- rose, orchids, gladiolus, carnation, gerbera, anthurium and lilies-has also become popular.

With the liberalization of Indian economy, the domestic floricultural trade, which hitherto remained an unorganized sector, became an organized sector with the active participation of corporate giants. The establishment of state-of-art hi-tech floricultural units in early 90's for intensive cultivation of cut flowers contributes about Rs. 500 crores of valuable foreign exchange today. Various government agencies are actively involved in the promotion of this sector and are highly committed to resolve the teething problems. The world demand for floricultural products is approximately US $ 40 billions and cut flowers contribute nearly 60% (US$ 25 billions) to the global trade. India's share in the world flower market is about 0.7% which is contributed mainly from open cultivation.

Floriculture is estimated to cover an area of 161,000 ha with production of 870,000 MT of loose flowers during 2008. The area under cut flowers has increased significantly in recent years and the production has reached 4341 million numbers (Tables 18.1 and 18.2). Another 500 ha of climatic control greenhouses are available for growing quality flowers for export. Many corporate houses (more than 100) have set up 100% export oriented units with an investment of more than Rs. 1000 crores. India is bestowed with natural advantages like favourable weather conditions, soil, cheap labour which are likely to make a real mark in international trade. With this, India is emerging as one of the powerful forces in world flower trade. India exported floriculture products worth US $ 84.12 million during 2007.

**Table 18.1.** Area and production of ornamental crops in India (2007-08) (Bijay Kumar, 2009)

| Ornamental Crops | Area ('000 ha) | Production ('000 MT) | Productivity (MT/ha) |
|---|---|---|---|
| Flowers Loose | 161 | 870 | 5.4 |
| Flowers Cut* | --- | 4341* | --- |

* Million numbers

**Table 18.2.** State-wise area and production of flowers (2007-08)

| State/UT's | Area ('000 ha) | Production | | | |
|---|---|---|---|---|---|
| | | Loose ('000 MT) | % of total production | Cut (million nos.) | % of total production |
| Karnataka | 22.34 | 169.12 | 19 | 555.000 | 13 |
| Maharashtra | 16.74 | 69.45 | 8 | 572.800 | 13 |
| Uttar Pradesh | 8.41 | 12.36 | --- | 375.200 | 9 |
| Uttarakhand | 0.90 | 0.70 | --- | 145.545 | 3 |
| Gujarat | 9.74 | 49.50 | 6 | 506.300 | 12 |
| Delhi | 5.50 | 5.70 | --- | 103.800 | 2 |
| West Bengal | 27.42 | 48.45 | 6 | 1968.000 | 45 |
| Tamil Nadu | 26.74 | 214.38 | 25 | --- | --- |
| Andhra Pradesh | 23.52 | 126.27 | 14 | 6.782 | --- |
| Punjab | 1.00 | 77.90 | 9 | --- | --- |
| Haryana | 6.11 | 61.76 | 7 | 1.053 | --- |
| Others | 12.30 | 34.78 | 6 | 7.266 | 3 |
| Total | 160.72 | 870.37 | --- | 4341.746 | --- |

Floriculture is an intensive type of agriculture wherein the income per unit area is much higher than any other agricultural product if it is done in a scientific way. The area used for production of ornamental plants is low, but the cash returns are high because of the intensive cropping. The flowers grown in large areas include jasmine, rose, marigold, chrysanthemum, crossandra, champak, tuberose, gladiolus and aster. New crops like lilies, tulips, anthurium, carnation, gerbera, etc., are also finding potential place in the foreign market. With the rapid development in urban planning and hotel industry, there has been an increasing demand for cut flowers and ornamentals. The cut flowers followed by live plants, cut foliage, seeds and bulbs have got an immense potential for export to Western countries where the cost of production is comparatively high. India exports flowers to France, England, Netherlands and Singapore. Our share in the export market has registered a sharp increase from Rs.14.45 crores in 1991-92 to Rs.378.44 crores in 2007-08.

## 18.1. ROSE, *ROSA* SPP.

### 18.1.1. High yielding varieties

**Table 18.3.** High yielding varieties of rose

| Rose cultivar | Flower yield | Developed at |
|---|---|---|
| Arjun | 13 blooms/bush | IARI, New Delhi |
| Pusa Komal | 60 flowers/bush | IARI, New Delhi |
| Pusa Shatabdi | 35-40 flowers/bush | IARI, New Delhi |

| | | |
|---|---|---|
| Pusa Urmil | 40-45 flowers/bush | IARI, New Delhi |
| YCD-1 | 100-120 flowers/bush | HRS, TNAU, Yercaud |
| YCD-2 | 200 flowers/bush | HRS, TNAU, Yercaud |
| YCD-3 | 150 flowers/bush | HRS, TNAU, Yercaud |

**Table 18.4.** Rose varieties resistant to pests/diseases

| Rose cultivar | Flower yield | Resistance to pest/disease | Developed at |
|---|---|---|---|
| Pusa Mohit | | Black spot (T) | IARI, New Delhi |
| Pusa Komal | 60 flowers/bush | Thrips (T), Powdery mildew (T), Black spot (T) | IARI, New Delhi |
| Pusa Barahmasi | | Powdery mildew (T), Black spot (T), Dieback (T) | IARI, New Delhi |
| Pusa Manhar | | Powdery mildew (T), Black spot (T) | IARI, New Delhi |
| Pusa Shatabdi | 35-40 flowers/bush | Powdery mildew (T), Black spot (T) | IARI, New Delhi |

T – Tolerant

### 18.1.2. Micro irrigation

Hazan *et al.* (1994) reported that in rose cv. Mercedes grown in 5 cm thick rock wool and supplied with 180 ppm N through drip irrigation resulted in greater stalk length and highest cumulative flower yield (1421 flowers/3 $m^2$).

### 18.1.3. Fertigation

Ammonium nitrate at 150 ppm applied through fertigation gave highest flower yield (153/$m^2$) compared to the other treatments under protected conditions (Ashok *et al.*, 1999) (Table 18.5).

**Table 18.5.** Nitrogen fertigation in rose cv. First Red

| Treatment (Nitrogen source) | Level (ppm) | Flower yield (No./$m^2$) |
|---|---|---|
| Aqueous ammonia | 50 | 69 |
| | 100 | 103 |
| | 150 | 135 |
| Nitric acid | 50 | 68 |
| | 100 | 104 |
| | 150 | 134 |
| Ammonium nitrate | 50 | 79 |
| | 100 | 112 |
| | 150 | 153 |
| Urea | 50 | 67 |
| | 100 | 99 |
| | 150 | 130 |

### 18.1.4. Bioregulators

Foliar spray of BA at 200 ppm gave maximum number of cut flowers (18.81 lakh/ha) of rose cv. Lather under protected conditions (Sonwalker and Barad, 2009).

### 18.1.5. Integrated nematode management

#### (i) Root-knot nematodes, *Meloidogyne* spp.

Pre-plant treatment of soil with dazomet (at 25 g/m$^2$) followed by soil amendment with neem cake (at 1 kg/m$^2$) along with *Pochonia chlamydosporia* (at 2 x $10^{10}$ spores/m$^2$) recorded reduced root galling and increased flower yield (Nagesh and Janakiram, 2004) (Table 18.6).

**Table 18.6.** Integrated management of root-knot nematode on rose

| Treatment | Root gall index (1-5 scale) | Plant mortality (%) | % increase in flower yield |
|---|---|---|---|
| Dazomet-40 g | 1.8 | 18 | 14.0 |
| Neem cake-1kg + *P. chlamydosporia* -2 x $10^{12}$ spores | 1.7 | 14 | 15.5 |
| Dazomet-40 g + *P. chlamydosporia* -2 x $10^{12}$ spores | 0.6 | 10 | 18.0 |
| Neem cake-1kg + Dazomet-40 g + *P. chlamydosporia* -2 x $10^{12}$ spores | 0.2 | 8 | 24.5 |
| Control | 3.0 | 40 | --- |
| CD (P = 0.05) | 0.19 | 3.11 | --- |

## 18.2. CARNATION, *DIANTHUS CARYOPHYLLUS*

### 18.2.1. High yielding varieties

Carnation cvs. Madame Collette, West Pretty, Desio and Acardi were found superior in terms of flower yield (Shiragur *et al.*, 2004) (Table 18.7).

**Table 18.7.** High yielding varieties of carnation under protected conditions

| Carnation variety | Yield (No. of flowers/m$^2$) |
|---|---|
| Aicardi | 310 |
| Alma | 291 |
| Candy | 295 |
| Desio | 317 |
| Madame Collette | 297 |

| | |
|---|---|
| Pirandello | 257 |
| Sorisso | 273 |
| Sugar Baby | 233 |
| West Pretty | 347 |
| CD (P = 0.05) | 48.17 |

### 18.2.2. Fertigation

Application of 80% RDF (120: 80: 250 g NPK/$m^2$/year) in the form of water soluble fertilizers through fertigation at fortnightly intervals is recommended to get maximum flower yield (26838 flowers/100$m^2$/annum) and net returns from carnation (Khan *et al.*, 2001) (Table 18.8).

**Table 18.8.** Productivity of carnation and levels of fertigation as influenced by sources and levels of fertilizers

| Treatment | Flower yield/ 100 $m^2$ | Net profit (Rs/100 $m^2$) | Benefit: cost ratio |
|---|---|---|---|
| Normal fertilizers 60% fertigation | 14970 | 25,053 | 0.74 |
| Normal fertilizers 80% fertigation | 18850 | 40,629 | 1.20 |
| Normal fertilizers 100% fertigation | 16475 | 30,833 | 0.91 |
| Water soluble fertilizers 60% fertigation | 22693 | 55,576 | 1.62 |
| Water soluble fertilizers 80% fertigation | 26838 | 72,213 | 2.09 |
| Water soluble fertilizers 100% fertigation | 18900 | 40,202 | 1.16 |

Carnation cv. Master gave maximum flower yield with 80% RDF supplied through fertigation (Table 18.9).

**Table 18.9.** Yield of carnation cultivars as influenced by fertigation

| Carnation cv. | Fertigation regime* | | |
|---|---|---|---|
| | 80% RDF | 100% RDF | 120% RDF |
| Krishna | 1209 | 1223 | 1232 |
| Aleda | 1181 | 1190 | 1204 |
| Master | 1253 | 1253 | 1253 |
| Vienna | 987 | 997 | 994 |

* Recommended dose of fertilizer: from planting to end of pinching – N/4.64, P/1.62, K/3.62, Ca/2.72, Mg/0.55 g/plant; from end of pinching to start of harvesting – N/4.86, P/1.50, K/7.03, Ca/3.09, Mg/0.50 g/plant.

Application of 80% RDF (200: 48: 160g N, $P_2O_5$ and $K_2O$/$m^2$/annum) in water soluble form not only gave higher cumulative yield of first grade (256.36/$m^2$) and low cumulative yield of second grade (3.02/$m^2$) flowers (Gopinath *et al.*, 2010).

### 18.2.3. Integrated nematode management

#### (i) Root-knot nematode, *Meloidogyne incognita*

Soil application of *P. chlamydosporia* at 50 g/m$^2$ ($10^6$ chlamydospores/g) gave significant reduction in root galling, nematode population (*M. incognita*) in soil and roots, and increased egg parasitization, root colonization of the bioagent and flower yield of carnation under polyhouse conditions (Rao and Shylaja, 2003) (Table 18.10).

**Table 18.10.** Effect of *Pochonia chlamydosporia* for the management of *Meloidogyne incognita* infecting carnation under polyhouse conditions

| Treatment/ Dose (g/m$^2$) | Nema popn. in 100 ml soil | Nema popn. in 5 g roots | % increase in flower yield | % egg parasitization | Root colonization (cfu/g) |
|---|---|---|---|---|---|
| *P. chlamydosporia* – 25 | 78 | 14.6 | 12.89 | 40.56 | 32368 |
| *P. chlamydosporia* – 50 | 55 | 10.4 | 19.35 | 52.58 | 44278 |
| *P. chlamydosporia* – 100 | 53 | 11.2 | 21.37 | 54.79 | 45679 |
| Control | 135 | 29.0 | --- | --- | --- |
| CD (P = 0.05) | 12.46 | 2.96 | 4.11 | 7.13 | 4356.89 |

Pre-plant treatment of beds with dazomet (at 40 g/m$^2$) followed by the application of neem cake (1 kg/m$^2$ 15 days later) along with antagonistic fungi, *P. chlamydosporia*/*P. lilacinus* (at 2 x $10^{12}$ spores/m$^2$) significantly reduced root-knot nematode population (*M. incognita*), mortality of plants and suppressed the nematode infection for 2 years in carnation. The antagonistic fungi established better in the beds treated with dazomet. The above treatment also reduced root galling, nematode multiplication rate and plant mortality, and increased spike/stem length, flower yield and root colonization with the bioagents (Nagesh and Parvatha Reddy, 2005) (Table 18.11).

**Table 18.11.** Integrated management of root-knot nematode on carnation

| Treatment Dose/m$^2$ | Root gall index | Plant mortality (%) | Stem length (cm) | No. of flowers/m$^2$ |
|---|---|---|---|---|
| Dazomet-40 g | 2.0 | 32 | 96.0 | 42.5 |
| Carbofuran-10 g | 3.2 | 30 | 92.0 | 40.0 |
| Carbosulfan-1 litre of 0.03% | 2.2 | 33 | 96.5 | 47.5 |
| Chlorpyriphos-1 litre of 0.03% | 2.4 | 38 | 95.0 | 45.5 |
| Neem cake-1kg + *P. lilacinus*-2 x $10^{12}$ spores | 2.8 | 31 | 95.5 | 43.0 |
| Neem cake-1kg + *P. chlamydosporia* - 2 x $10^{12}$ spores | 2.8 | 29 | 98.0 | 46.0 |
| Dazomet-40 g + Neem cake-1kg + *P. lilacinus*-2 x $10^{12}$ spores | 1.6 | 11 | 104.8 | 63.0 |

| | | | | |
|---|---|---|---|---|
| Dazomet-40 g + *P. chlamydosporia* -2 x $10^{12}$ spores | 0.4 | 12 | 116.0 | 68.5 |
| Control | 4.2 | 52 | 75.5 | 37.5 |
| CD (P = 0.05) | 0.18 | 1.84 | 3.43 | 1.66 |

## 18.3. GERBERA, *GERBERA JAMESONII*

### 18.3.1. High yielding varieties

**Table 18.12.** High yielding varieties of gerbera

| Gerbera cultivar | No. of flowers/$m^2$ | Developed at |
|---|---|---|
| YCD-1 | 60 flowers/plant | HRS, TNAU, Yercaud |
| YCD-2 | 80 flowers/plant | HRS, TNAU, Yercaud |
| Tornado | 68.58 | --- |
| Goldspot | 72.67 | --- |
| Ibiza | 94.92 | --- |
| Rosabella | 69.42 | --- |
| Rendezvous | 56.25 | --- |

### 18.3.2. Fertigation

Application of 80% RDF (200: 80: 160 g NPK/$m^2$/year) in the form of water soluble fertilizers through fertigation at weekly intervals is recommended to get maximum flower yield (17880 flowers/100$m^2$/annum) and net returns from gerbera (Khan *et al.*, 2001) (Table 18.13).

**Table 18.13.** Productivity of gerbera and levels of fertigation as influenced by sources and levels of fertilizers

| Treatment | Flower yield/ 100 $m^2$ | Net profit (Rs/100 $m^2$) | Benefit: cost ratio |
|---|---|---|---|
| Normal fertilizers 60% fertigation | 11540 | 13,170 | 0.71 |
| Normal fertilizers 80% fertigation | 14960 | 22,569 | 1.21 |
| Normal fertilizers 100% fertigation | 13900 | 19,478 | 1.04 |
| Water soluble fertilizers 60% fertigation | 14970 | 9,881 | 0.88 |
| Water soluble fertilizers 80% fertigation | 17880 | 29,150 | 1.46 |
| Water soluble fertilizers 100% fertigation | 14320 | 19,089 | 0.94 |

The cumulative flower yield was highest with 80% recommended dose of water soluble fertilizers applied through fertigation under protected conditions (Sujatha *et al.*, 2002) (Table 18.14).

**Table 18.14.** Fertigation in gerbera

| Treatment | Flower yield (No./m$^2$) |
|---|---|
| 100% straight fertilizers through fertigation | 69.50 |
| 80% straight fertilizers through fertigation | 79.90 |
| 60% straight fertilizers through fertigation | 57.80 |
| 100% water soluble fertilizers through fertigation | 71.60 |
| 80% water soluble fertilizers through fertigation | 89.40 |
| 60% water soluble fertilizers through fertigation | 66.00 |

### 18.3.3. Integrated nutrient management

Blomme and Dambre (1984) obtained higher yield of 34 flowers/m$^2$ when supplied with N and $K_2O$ at 20 and 10 g/m$^2$, respectively in the form of slow release compounds than with nutrient solution containing N and $K_2O$ at 150 mg/litre applied in irrigation water (23 flowers/m$^2$) (Fig. 18.1).

### 18.3.4. Integrated weed management

The highest yield of gerbera flowers (2.782 million/ha) was recorded in weed-free check which was on par with pendamethalin at 1.0 kg a.i./ha (2.465 million/ha). This was due to the fact that the gerbera plants in these treatments put forth good vegetative growth right from the early stages of growth period because of less competition of weeds for nutrients, water, space and sunlight which might have resulted in higher photosynthetic activity and higher number of flowers per plant (Shalini and Patil, 2004) (Table 18.15).

**Table 18.15.** Effect of weed control treatments on flower yield in gerbera

| Treatment | Flower yield (million/ha) |
|---|---|
| Weed free | 2.782 |
| Unweeded control | 1.125 |
| 2 hoeings + 2 hand weedings at 30 and 60 DAP | 1.860 |
| Pendimethalin at 1.0 kg a.i./ha | 2.465 |
| Alachlor at 1.5 kg a.i./ha | 2.310 |
| Metolachlor at 1.0 kg a.i./ha | 2.122 |
| Butachlor at 1.0 kg a.i./ha | 1.730 |
| Atrazine at 1.0 kg a.i./ha | 1.370 |
| Oxyfluorfen at 0.15 kg a.i./ha | 1.832 |
| Oxadiargyl at 90 g a.i./ha | 1.960 |
| Paddy straw mulch | 2.219 |
| Black polythene sheet | 2.270 |
| CD (P = 0.05) | 0.431 |

### 18.3.5. Integrated nematode management

#### i) Root-knot nematode, *Meloidogyne incognita*

Pre-plant treatment of beds with dazomet (at 40 g/m$^2$) followed by the application of neem cake (at 1 kg/m$^2$ 15 days later) along with antagonistic fungi, *P. chlamydosporia* / *P. lilacinus* (at 2 x $10^{12}$ spores/m$^2$) significantly reduced root-knot nematode population (*M. incognita*), mortality of plants and suppressed the nematode infection for 2 years in gerbera. The antagonistic fungi established better in the beds treated with dazomet. The above treatment also reduced root galling, nematode multiplication rate and plant mortality, and increased spike/stem length, flower yield and root colonization with the bioagents (Nagesh and Parvatha Reddy, 2005) (Table 18.16).

**Table 18.16.** Integrated management of root-knot nematode on gerbera

| Treatment Dose/m$^2$ | Root gall index | Plant mortality (%) | Spike length (cm) | No. of flowers/m$^2$ |
|---|---|---|---|---|
| Dazomet-40 g | 1.8 | 28 | 66.5 | 44.0 |
| Carbofuran-10 g | 2.9 | 32 | 60.8 | 42.5 |
| Carbosulfan-1 litre of 0.03% | 1.8 | 30 | 65.5 | 44.0 |
| Chlorpyriphos-1 litre of 0.03% | 2.0 | 25 | 63.5 | 42.5 |
| Neem cake-1kg + *P. lilacinus*-2 x $10^{12}$ spores | 2.6 | 25 | 62.0 | 43.0 |
| Neem cake-1kg + *P. chlamydosporia* -2 x $10^{12}$ spores | 2.6 | 28 | 64.2 | 44.0 |
| Dazomet-40 g + Neem cake-1kg + *P. lilacinus*-2 x $10^{12}$ spores | 1.4 | 15 | 72.5 | 55.5 |
| Dazomet-40 g + *P. chlamydosporia* -2 x $10^{12}$ spores | 0.4 | 12 | 75.9 | 58.0 |
| Control | 3.8 | 40 | 48.5 | 40.5 |
| CD (P = 0.05) | 0.11 | 4.11 | 4.41 | 5.11 |

### 18.3.6. Integrated management of disease complex

#### i) Root-knot nematode, *Meloidogyne incognita* and foot rot, *Phytophthora parasitica* disease complex

Sustainable production of gerbera is seriously hampered by the disease complex caused by *M. incognita* and *P. parasitica*. These two pathogens reduce the productivity of gerbera significantly to the tune of 40-60%.

Combined application of neem cake enriched with either *Trichoderma harzianum* or *Pseudomonas fluorescens* [mixing 50 g of *T. harzianum* (2 x $10^6$ cfu/g) or *P. fluorescens* (2 x $10^8$ cfu/g) in 1 kg of neem cake] applied at 25 g/m$^2$ was found effective for the management of disease complex and increased the flower yield by 26% in gerbera cv. Debora (Manoj Kumar *et al.*, 2010).

## 18.4. TUBEROSE, *POLIANTHES TUBEROSA*

### 18.4.1. High yielding varieties

**Table 18.17.** High yielding varieties of tuberose

| Tuberose variety | Yield | % increase over standard variety |
|---|---|---|
| Shringar (single) | Flower yield – 15 MT/ha<br>No. of spikes/m² – 28.89<br>No. of bulbs/m² – 379.22 | 40% higher than Calcutta or Mexican single |
| Suvasini (double) | No. of spikes/m² – 20.00<br>No. of bulbs/m² – 315.55 | 25% more than Double |
| Prajwal (single) | --- | 20% more than Shringar |
| Vaibhav (double) | --- | 50% more than Suvasini |
| Mexican Single | No. of spikes/m² – 23.67<br>No. of bulbs/m² – 193.33 | --- |
| Variegated | No. of spikes/m² – 15.56<br>No. of bulbs/m² – 362.44 | --- |
| Pearl Double | No. of spikes/m² – 15.56<br>No. of bulbs/m² – 328.89 | --- |

**Table 18.18.** Performance of tuberose germplasm at BCKV, Mohanpur

| Name of the variety | No. of spikes/m² | No. of spikes/plant/year |
|---|---|---|
| Calcutta Double | 15.88 | 12.66 |
| Calcutta Single | 101.45 | 40.66 |
| Rajat Rekha | 50.90 | 9.66 |
| Swrna Rekha | 30.65 | 1.00 |
| Shringar | 55.73 | 63.33 |
| CD (P = 0.05) | 5.98 | 16.98 |

### 18.4.2. Fertigation

Application of 80: 40: 40 kg NPK/ha in the form of water soluble fertilizers along with 0.1% boron through sprinkler irrigation at monthly intervals is recommended to get maximum yield of flowers (1165 kg/ha), concrete (24.96 kg/ha) and net returns from tuberose (Khan *et al.*, 2001) (Tables 18.19 and 18.20).

**Table 18.19.** Productivity and economics of tuberose cv. Single as influenced by sources and level of fertilizers and micronutrients

| Treatment | Flower yield (MT/ha) | Net income (Rs.) | Benefit: cost ratio |
|---|---|---|---|
| No fertilizers + Furrow irrigation | 2.632 | 28,486 | 2.18 |
| Normal fertilizers 80% + Furrow irrigation | 5.683 | 86,429 | 4.17 |
| Normal fertilizers 100% + Furrow irrigation | 6.646 | 104,921 | 4.75 |
| No fertilizers + Sprinkler irrigation | 2.750 | 32,139 | 2.41 |
| Water soluble fertilizers 80% + Sprinkler irrigation | 6.730 | 105,420 | 4.61 |
| Water soluble fertilizers 100% + Sprinkler irrigation | 8.560 | 140,455 | 5.57 |
| Water soluble fertilizers 80% + 0.1% Zn + Sprinkler irrigation | 7.980 | 125,085 | 4.62 |
| Water soluble fertilizers 80% + 0.05% Zn + Sprinkler irrigation | 10.834 | 184,830 | 6.80 |
| Water soluble fertilizers 80% + 0.1% B + Sprinkler irrigation | 11.650 | 202,865 | 7.73 |
| Water soluble fertilizers 80% + 0.05% B + Sprinkler irrigation | 9.623 | 189,517 | 6.49 |

**Table 18.20.** Concrete yield and economics of tuberose cv. Single as influenced by sources and level of fertilizers and micronutrients

| Treatment | Concrete yield (kg/ha) | Net income (Rs.) | Benefit: cost ratio |
|---|---|---|---|
| No fertilizers + Furrow irrigation | 3.73 | 24,142 | 1.48 |
| Normal fertilizers 80% + Furrow irrigation | 8.51 | 86,002 | 2.02 |
| Normal fertilizers 100% + Furrow irrigation | 11.00 | 124,980 | 2.32 |
| No fertilizers + Sprinkler irrigation | 4.08 | 31,156 | 1.62 |
| Water soluble fertilizers 80% + Sprinkler irrigation | 10.63 | 115,962 | 2.20 |
| Water soluble fertilizers 100%+Sprinkler irrigation | 15.00 | 183,043 | 2.70 |
| Water soluble fertilizers 80% + 0.1% Zn + Sprinkler irrigation | 14.79 | 181,278 | 2.58 |
| Water soluble fertilizers 80% + 0.05% Zn + Sprinkler irrigation | 22.00 | 298,352 | 3.12 |
| Water soluble fertilizers 80% + 0.1% B + Sprinkler irrigation | 24.69 | 346,887 | 3.36 |
| Water soluble fertilizers 80% + 0.05% B + Sprinkler irrigation | 18.96 | 253,080 | 3.00 |

### 18.4.3. Integrated nematode management

#### i) Root-knot nematode, *Meloidogyne incognita*

Application of *P. chlamydosporia* at 100 g/4m$^2$ significantly reduced root galling, nematode population both in soil and roots, and increased root colonization by the bioagent, egg parasitization and flower yield of tuberose under field conditions (Rao *et al.*, 2004) (Table 18.21).

**Table 18.21.** Effect of *P. chlamydosporia* for the management of *Meloidogyne incognita* infecting tuberose

| Treatment/Dose/4m$^2$ | Root-knot index | No. of spikes/ 4m$^2$ | Spike weight (g) | Root colonization (cfu/g) | % Egg parasitization by the bioagent |
|---|---|---|---|---|---|
| *P. chlamydosporia* - 50 g | 6.5 | 39 | 41.2 | 39,353 | 42.7 |
| *P. chlamydosporia* - 100 g | 4.9 | 54 | 45.5 | 47,489 | 56.2 |
| *P. chlamydosporia* - 200 g | 4.2 | 56 | 46.9 | 51,986 | 59.7 |
| Control | 8.3 | 27 | 33.4 | --- | --- |
| CD (P = 0.05) | 1.94 | 9.38 | 8.37 | 6347.86 | 9.48 |

Split application (at planting and 45 DAP) of neem cake at 1kg/1.5m$^2$ in combination with *P. lilacinus* (at 25g/1.5m$^2$ containing 18 x 10$^8$ spores/g) significantly reduced root galling. The above treatment also significantly increased number of spikes, and parasitization of egg masses by *P. lilacinus* (Nagesh *et al.*, 1998) (Table 18.22).

**Table 18.22.** Effect of split application of oil cakes and *Paecilomyces lilacinus* on root galling, spike yield and propagule density on tuberose infected with *Meloidogyne incognita*

| Treatment | | Root-knot index | No. of Spikes/ 1.5 m$^2$ | % egg mass parasitization by *P. lilacinus* |
|---|---|---|---|---|
| At planting | 45 DAP | | | |
| Control | --- | 4.2 | 34 | --- |
| *P. lilacinus*-25 g + Castor cake-1 kg | *P. lilacinus*-25 g + Castor cake-1 kg | 3.0 | 55 | 32.3 |
| *P. lilacinus*-25 g + Karanj cake -1 kg | *P. lilacinus*-25 g + Karanj cake-1 kg | 2.4 | 56 | 40.3 |
| *P. lilacinus*-25 g + Neem cake-1 kg | *P. lilacinus*-25 g + Neem cake-1 kg | 1.8 | 69 | 48.9 |
| Carbofuran-2 kg a.i./ha | --- | 3.2 | 45 | --- |
| CD (P = 0.05) | | 0.26 | 4.82 | 5.22 |

Integration of AMF (*Glomus mosseae, G. fasciculatum*), a botanical (neem cake) and a nematicide (aldicarb) gave effective management of *M. incognita* and increased plant growth and yield of bulbs (Khan and Parvatha Reddy, 1994) (Table 18.23).

**Table 18.23.** Effect of integration of *Glomus mosseae, G. fasciculatum*, neem cake and aldicarb on root galling, plant growth and bulb weight of tuberose infected with *Meloidogyne incognita*

| Treatment | Dose | Plant weight (g) | Root-knot index | Bulb wt. (g) |
|---|---|---|---|---|
| *Glomus mosseae* + Aldicarb | 100 spores/100 g soil + 0.5 kg a.i./ha | 7.99 | 2.0 | 2.3 |
| *G. fasciculatum* + Aldicarb | 100 spores/100 g soil + 0.5 kg a.i./ha | 9.53 | 2.6 | 1.7 |
| *G. mosseae* + Neem cake | 100 spores/100 g soil + 0.5 kg a.i./ha | 5.46 | 2.5 | 2.8 |
| *G. fasciculatum* + Neem cake | 100 spores/100 g soil + 0.5 kg a.i./ha | 9.96 | 2.4 | 2.5 |
| *G. mosseae* + Aldicarb + Neem cake | 100 spores/100 g soil + 0.5 kg a.i./ha + 0.5 MT/ha | 11.80 | 1.5 | 2.8 |
| *G. fasciculatum* + Aldicarb + Neem cake | 100 spores/100 g soil + 0.5 kg a.i./ha + 0.5 MT/ha | 7.43 | 2.2 | 0.9 |
| Control | --- | 3.26 | 4.5 | 1.1 |
| CD (P = 0.05) | --- | 1.35 | 0.18 | 0.50 |

### 18.4.4. Integrated management of disease complex

#### i) Root-knot nematode, *Meloidogyne incognita* and wilt, *Fusarium oxysporum* f. sp. *dianthi* disease complex

The root-knot nematode, *M. incognita* reduced flower yield considerably and the tuberose plants became highly susceptible to the attack by *F. oxysporum* f. sp. *dianthi* (Rao *et al.*, 2002).

In course of the experiments carried out to evaluate combination of bio-agents for the control of wilt and root-knot nematode disease complex in tuberose, the best result was obtained in plants treated with *P. chlamydosporia* + *T. harzianum* (Rao *et al.*, 2003). The above treatment gave significant reduction in root galling, nematode population both in soil and roots, disease index, and increased egg parasitization and root colonization by bioagents and flower yield (Table 18.24).

**Table 18.24.** Effect of integration of bioagents for the management of disease complex and flower yield in tuberose

| Treatment/Dose (g/4m$^2$) | Root-knot index (1-10 scale) | Disease index (1-5 scale) | No. of Spikes/4m$^2$ | Spike length (cm) |
|---|---|---|---|---|
| *P. chlamydosporia* – 20 | 6.21 | 3.76 | 24 | 60.56 |
| *P. chlamydosporia* – 40 | 4.26 | 3.11 | 26 | 65.32 |
| *T. harzianum* – 20 | 6.56 | 2.82 | 20 | 63.37 |
| *T. harzianum* – 40 | 5.21 | 2.33 | 23 | 66.87 |

| | | | | |
|---|---|---|---|---|
| *P. chlamydosporia* – 20 + *T. harzianum* – 20 | 4.79 | 2.18 | 29 | 71.79 |
| *P. chlamydosporia* – 40 + *T. harzianum* – 40 | 4.20 | 1.59 | 24 | 75.73 |
| Control | 8.49 | 4.42 | 18 | 45.69 |
| CD (P = 0.05) | 1.76 | 0.76 | 2.33 | 7.52 |

## 18.5. GLADIOLUS, *GLADIOLUS* SPP.

### 18.5.1. High yielding varieties

**Table 18.25.** High yielding varieties of gladiolus

| Gladiolus variety | Yield (No. of spikes/ha) | Developed at |
|---|---|---|
| KKL-1 | 211100 | HRS, TNAU, Kodaikanal |
| Vedanapoli | 488333 | --- |
| Sylvia | 445000 | --- |
| Melody | 266667 | --- |
| American Beauty | 233333 | --- |
| Magic | 211667 | --- |
| Snow White | 211667 | --- |

### 18.5.2. Integrated disease management

#### i) Wilt, *Fusarium oxysporum* f. sp. *gladioli*

Successful control of corm rot of gladiolus was achieved by using several strains of *T. harzianum* isolated from soil (Table 18.26).

**Table 18.26.** Effect of different fungal antagonists on corm rot disease of gladiolus.

| Antagonists | Germination (%) | Mortality (%) | Plant height (cm) | Spike length (cm) | No. of cormels/ plant |
|---|---|---|---|---|---|
| *T. harzianum* MTCC-792 | 100.00 | 2.33 | 72.33 | 85.33 | 26.33 |
| *T. virens* MTCC-794 | 95.60 | 3.33 | 70.87 | 82.10 | 21.67 |
| *T. harzianum* MTCC-795 | 96.77 | 3.50 | 70.67 | 82.56 | 22.33 |
| *T. koningii* MTCC-796 | 98.25 | 2.64 | 70.67 | 80.50 | 21.00 |
| *T. viride* MTCC-800 | 96.00 | 6.66 | 67.50 | 79.53 | 15.25 |
| *T. harzianum* MTCC-801 | 100.00 | 6.67 | 66.00 | 75.30 | 16.69 |
| *T. harzianum* (UP) | 96.67 | 8.77 | 65.00 | 75.66 | 16.00 |
| Control | 79.00 | 28.67 | 50.80 | 45.80 | 12.67 |

### 18.5.3. Integrated nematode management

#### i) Root-knot nematode, *Meloidogyne incognita*

Corms of gladiolus treated with the suspension of neem enriched with *Pseudomonas fluorescens* strain 1 and 2 (5 g/litre of water) and application of neem cake enriched with *P. fluorescens* strains to beds at 20g/m$^2$ was found to be effective in reducing the root population of *M. incognita* by 68% and increased the flower yield by 28% (Sowmya and Rao, 2010).

## 18.6. CHRYSANTHEMUM, *DANDRATHEMA GRANDIFLORA*

### 18.6.1. High yielding varieties

**Table 18.27.** High yielding varieties of chrysanthemum

| Chrysanthemum variety | Yield (g/plant) | Chrysanthemum variety | Yield (g/plant) |
|---|---|---|---|
| Chandrakant | 272 | Harvest Home | 14.87 MT/ha |
| Nilima | 281 | Mutant No. 9 | 14.68 MT/ha |
| Pankaj | 363 | Karnool | 14.24 MT/ha |
| Red Gold | 400 | Selection-5 | 14.20 MT/ha |
| Yellow Gold | 244 | Saraval | 14.13 MT/ha |
| Yellow Star | 272 | Chandrika | 13.70 MT/ha |
| Baggi | 14.30 MT/ha | CO 1 | 16.70 MT/ha |
| Ratlam Selection | 17.50 MT/ha | MDU 1 | 30.59 MT/ha |

### 18.6.2. Integrated nutrient management

Experiments conducted with integrated nutrient management revealed that the plants supplied with 50% recommended dose of fertilizers (RDF = 125: 120: 120 kg/ha) + vermicompost (5 MT/ha) + 3% Panchagavya was superior in respect of all the economic floral parameters (ICAR Annual report, 2006-07) (Table 18.28).

**Table 18.28.** Integrated nutrient management studies in chrysanthemum cv. CO 1

| Treatment | No. of flowers/ spray/plant |
|---|---|
| 100% of Recommended dose of FYM and inorganic fertilizers (RDF) (control) | 71.54 |
| 50% (RDF) | 73.27 |
| 50% RDF + 3% Manchurian tea | 81.49 |
| 50% of RDF + 3% Panchagavya | 83.53 |

| | |
|---|---|
| 50% of RDF + 3% Panchagavya + 3% Manchurian tea | 89.44 |
| 50% RDF + Vermicompost (5 MT/ha) | 104.16 |
| 50% of fertilizers + Vermicompost (5 MT/ha)+ 3% Manchurian tea | 95.96 |
| 50% of fertilizers (500 g/m2-twice a year) + Vermicompost (5 MT/ha) + 3% Panchagavya | 110.36 |
| CD (P = 0.05) | 2.98 |

Application of vermicompost at 300 kg/ha + FYM at 10 MT/ha + 50% of RDF (100: 150: 100 kg NPK/ha) gave maximum flower yield of chrysanthemum (15.5 MT/ha) (Hilli *et al.*, 2006).

### 18.6.3. Bioregulators

The treatment of IBA at 2000 ppm as a quick dip method to tip cutting along with two sprays of IBA at 10 ppm after 30 and 60 DAP cuttings in main field recorded maximum survival (96.57%) along with highest marketable yield (Ranpise *et al.*, 2004) (Table 18.29).

**Table 18.29.** Effect of different levels of IBA on survival and flower yield of chrysanthemum

| Treatment | Survival (%) | Flower yield (MT/ha) |
|---|---|---|
| IBA – 1000 ppm | 90.14 | 64.579 |
| IBA – 2000 ppm | 96.57 | 76.849 |
| IBA – 3000 ppm | 79.28 | 40.505 |
| Control | 65.42 | 29.081 |
| CD (P = 0.05) | 3.44 | 0.525 |

### 18.6.4. Protected cultivation

The growing medium, irrigation regime, fertilizer application and photoperiod regulation for cut chrysanthemum under poly house conditions were standardized (Table 18.30).

**Table 18.30.** Protected cultivation of chrysanthemum

| Variety | Low Regan |
|---|---|
| Area of polyhouse | 1000 $m^2$ |
| Growing media | Soil + Compost + Coco peat at 1: 1: 2 ratio |
| Irrigation | 40 kPa |
| Spacing | 20 x 15 cm |
| Nutrition | 10: 20: 10 g/$m^2$ NPK through soil application as basal, i.e. 50% N as basal + 50% N i.e. 10 g/$m^2$ as top dressing at flower bud opening stage. |
| Photoperiod | 14 hours dark and 10 hours light |

| | |
|---|---|
| Light intensity | 3000 Lux |
| Temperature | 25-35°C |
| Relative humidity | 60-70% |
| Yield | 240 sprays/m$^2$ (239.3 during summer and 240.3 in winter) with 3.54 flowers/spray in summer and 3.56 flowers /spray in winter |
| B: C Ratio | 3.44 |

### 18.6.5. Integrated nematode management

Field application of two formulations of *P. lilacinus* (talc and pesta granules) in combination with neem cake reduced root gall index, nematode populations and enhanced flower yield of chrysanthemum by 23 to 28% and increased the *P. lilacinus* spore viability in the rhizosphere for longer time.

## 18.7. CHINA ASTER, *CALLISTEPHUS CHINENSIS*

### 18.7.1. High yielding varieties

**Table 18.31.** High yielding varieties of China aster

| China aster variety | Yield (MT/ha) | Developed at |
|---|---|---|
| Phule Ganesh Pink | 35.08 | MPKV, Rahuri |
| Phule Ganesh White | 32.05 | MPKV, Rahuri |
| Phule Ganesh Violet | 27.31 | MPKV, Rahuri |
| Phule Ganesh Purple | 24.07 | MPKV, Rahuri |
| Namdhari | 15.14 | Namdhari Seeds, Bangalore |
| Kamini | 9.44 | IIHR, Bangalore |

### 18.7.2. Fertigation

The maximum number of flowers (12.70 MT/ha) were recorded in treatment with water soluble fertilizer at 120% of RDF through fertigation as compared to 7.19 MT/ha in control (120% of RDF with flood irrigation). Additional yield of 5.51 MT/ha were obtained over traditional application of fertilizers and irrigation system (Kumar *et al.*, 2009).

### 18.7.3. Integrated nutrient management

Foliar application of Zn $SO_4$ at 0.5% recorded significantly higher number of flowers (63.24/plant), weight of flowers (164.42 g/plant) as well as flower yield (18.269 MT/ha) (Rajput *et al.*, 2006).

## 18.8. MARIGOLD, *TAGETES ERECTA, T. PATULA*

### 18.8.1. High yielding varieties

**Table 18.32.** High yielding varieties of marigold

| High yielding varieties | Flower Yield | Developed at |
|---|---|---|
| Pusa Narangi Gainda | 34.9 MT/ha | IARI, New Delhi |
| Pusa Basanti Gainda | 58 flowers/plant | IARI, New Delhi |
| MDU-1 | 41.54 MT/ha | AC&RI, TNAU, Madurai |
| Giant African Yellow | 24.0 MT/ha | --- |

### 18.8.2. Integrated nutrient management

The flower yield was highest (21.40 MT/ha) with the treatment combination of N at 225 kg and P at 120 kg/ha. The above treatment also recorded the highest petal meal (1.855 MT/ha) and xanthophylls yield (33.17 kg/ha) (Hemla Naik *et al.*, 2009).

### 18.8.3. Bioregulators

The maximum marigold flower yield was recorded with gibberellic acid at 200 ppm (9617.48 kg/ha) which was on par with gibberellic acid at 150 ppm (8614.86 kg/ha) and ethrel at 250 ppm (8344.92 kg/ha) (Verma and Arha, 2004) (Table 18.33).

**Table 18.33.** Influence of gibberellic acid, ethrel and MH on flower yield of marigold

| Treatment | Flower yield (kg/plant) |
|---|---|
| Giberellic acid - 100 ppm | 7164.91 |
| Giberellic acid -150 ppm | 8614.86 |
| Giberellic acid - 200 ppm | 9617.48 |
| Ethrel - 250 ppm | 8344.92 |
| Ethrel - 500 ppm | 7828.18 |
| Ethrel - 750 ppm | 6231.70 |
| MH - 250 ppm | 7550.53 |
| MH - 500 ppm | 6385.95 |
| MH - 750 ppm | 5792.08 |
| Control | 7018.37 |
| CD (P = 0.05) | 1531.14 |

Among pinching and chemical sprays, the pinched crop of African marigold sprayed with 2% DAP recorded highest xanthophylls yield (16.53 kg/ha) followed by 1000 ppm cycocel (14.84 kg/ha) and 1% DAP (14.41 kg/ha). However, the maximum flower yield was obtained by pinching + 1000 ppm cycocel (16.44 MT/ha) followed by pinching + 2% DAP (15.78 MT/ha) (Hemla Naik *et al.*, 2004) (Table 18.34).

**Table 18.34.** Effect of pinching and chemicals on flower and xanthophylls yield of marigold

| Treatment | Flower yield (MT/ha) | Xanthophyll yield (kg/ha) |
|---|---|---|
| CCC - 500 ppm | 14.95 | 13.57 |
| CCC - 1000 ppm | 16.44 | 14.84 |
| Mepiquat chloride - 500 ppm | 13.89 | 12.24 |
| Mepiquat chloride - 1000 ppm | 13.98 | 10.71 |
| DAP - 1% | 14.56 | 14.41 |
| DAP - 2% | 15.78 | 16.53 |
| MOP - 1% | 13.87 | 12.56 |
| MOP - 2% | 13.99 | 12.77 |
| CD (P = 0.05) | 0.31 | 0.51 |

## 18.9. GAILLARDIA

### 18.9.1. Integrated nutrient management

Application of nitrogen at 150 kg/ha in two splits (50% as basal dose and 50% at initiation of flowering) recorded the highest flower yield (93.04 and 33.16 MT/ha) on black and red soils, respectively. The increase in flower yield was mainly due to improved yield attributes like number of flowers per plant and flower size (Patil *et al.*, 2004) (Table 18.35).

**Table 18.35.** Effect of split application of nitrogen on flower yield of gaillardia

| Treatment | Flower yield (MT/ha) | |
|---|---|---|
| | Red soil | Black soil |
| Recommended dose of nitrogen - 150 kg/ha | 27.17 | 30.33 |
| 75% N basal + 25% N at initiation of flowering | --- | 28.30 |
| 50% N basal + 25% N at tillering + 25% N at initiation of flowering | 26.23 | 29.16 |
| 50% N basal + 50% N at initiation of flowering | 29.04 | 33.16 |
| 50% N basal + 50% N at 50% flowering | 25.64 | 27.40 |
| 25% N basal + 50% N at tillering + 25% N at initiation of flowering | 24.30 | 26.27 |
| 25% N basal + 25% N at tillering + 50% N at initiation of flowering | 24.76 | 23.61 |
| CD (P = 0.05) | 0.98 | 2.20 |

## 18.10. CROSSANDRA, *CROSSANDRA UNDULAEFOLIA*

### 18.10.1. Integrated nematode management

#### i) Root-knot nematode, *Meloidogyne incognita*

Combinations of *Verticillium lecanii* and *Paecilomyces lilacinus* (2 x $10^4$ spores/ml each) with 5% neem leaf extract resulted in significantly higher plant growth parameters and crossandra flower yield (Nagesh and Parvatha Reddy, 1995). Root galling, nematode multiplication factor were least and the parasitization of eggs and egg masses was highest in *V. lecanii* + 5% neem leaf extract (Table 18.36).

**Table 18.36.** Effect of integration of *Paecilomyces lilacinus, Verticillium lecanii* and leaf extracts on root galling, root colonization, egg parasitization and flower yield of crossandra infected with *Meloidogyne incognita*

| Treatment | Dose/plant | Rot-knot index | Yield (g/plant) |
|---|---|---|---|
| Control | --- | 4.21 | 18 |
| Castor leaf extract | 5% | 3.8 | 24 |
| Neem leaf extract | 5% | 3.0 | 28 |
| *P. lilacinus* | $10^4$ spores/ml | 3.0 | 25 |
| *V. lecanii* | $10^4$ spores/ml | 3.4 | 24 |
| *P. lilacinus* + Castor leaf extract | $10^4$ spores/ml + 5% | 2.8 | 32 |
| *P. lilacinus* + Neem leaf extract | $10^4$ spores/ml + 5% | 2.8 | 33 |
| *V. lecanii* + Castor leaf extract | $10^4$ spores/ml + 5% | 2.5 | 28 |
| *V. lecanii* + Neem leaf extract | $10^4$ spores/ml + 5% | 2.0 | 38 |

## 18.11. GOLDEN ROD, *SOLIDAGO CANADENSIS*

### 18.11.1. Bioregulators

The maximum number of golden rod flower stalks were obtained from the plants treated with 1000 ppm cycocel (768,450/ha) which was on par with Gibberellic acid at 50 ppm (657,120/ha), 100 ppm (732,600/ha) and 200 ppm (682,650/ha). Flower stalk yield in terms of tonnage was highest in plants treated with gibberellic acid at 100 ppm (41.20 MT/ha) which was on par with Gibberellic acid at 200 ppm (38.55 MT/ha) (Patil *et al.*, 2004) (Table 18.37).

**Table 18.37.** Effect of growth substances on flower stalk yield of golden rod

| Growth substances (ppm) | No. of flower stalks '000/ha | Weight of flower stalks (MT/ha) |
|---|---|---|
| Gibberellic acid – 50 | 657.12 | 26.640 |

| | | |
|---|---|---|
| Gibberellic acid – 100 | 732.60 | 38.550 |
| Gibberellic acid – 200 | 682.65 | 41.200 |
| Cycocel – 500 | 647.50 | 25.340 |
| Cycocel – 1000 | 768.45 | 32.710 |
| Cycocel – 2000 | 638.25 | 25.510 |
| Paclobutrazol – 500 | 371.85 | 15.460 |
| Paclobutrazol – 1000 | 438.45 | 21.210 |
| Paclobutrazol – 2000 | 421.80 | 12.080 |
| Control (water) | 441.78 | 13.770 |
| CD (P = 0.05) | 117.050 | 7.090 |

## 18.12. ANTHURIUM, *ANTHURIUM* SPP.

### 18.12.1. Integrated nutrient management

In TNAU, application of NPK at 30: 10: 10 at 0.2 per cent spray + *Azospirillum* + phosphobacteria + VAM + $GA_3$ at 200 ppm in anthurium cv. Temptation has been found to influence the number of flowers per plant, besides improving the floral attributes (Table 18.38) (Padmadevi *et al*., 2004).

**Table 18.38.** Effect of nutrients, biofertilizers, and growth regulator on flower production in *Anthurium andreanum* cv. Temptation

| Treatment | Number of flowers/plant |
|---|---|
| T 1 - NPK at 30: 10: 10 at 0.2 % spray and $GA_3$ at 200 ppm | 1.33 |
| T 2 - T 1 + *Azospirillum* | 2.33 |
| T 3 - T 1 + Phosphobacteria | 3.67 |
| T 4 - T 1 + AMF | 3.67 |
| T 5 - T 1 + *Azospirillum* + Phosphobacteria | 3.33 |
| T 6 - T 1 + *Azospirillum* + AMF | 3.33 |
| T 7 - T 1 + Phosphobacteria + AMF | 4.00 |
| T 8 - T 1 + *Azospirillum* + Phosphobacteria + AMF | 5.67 |
| T 9 - NPK at 15: 10: 10 at 0.2 % spray and $GA_3$ 200 ppm + *Azospirillum* | 3.33 |
| T 10 - NPK at 30: 5: 10 at 0.2 % spray and $GA_3$ 200 ppm + Phosphobacteria | 4.00 |
| T 11 - NPK at 30: 5: 10 at 0.2 % spray and $GA_3$ 200 ppm + AMF | 3.00 |
| T 12 - NPK at 15: 5: 10 at 0.2 % spray and $GA_3$ 200 ppm + *Azospirillum* + Phosphobacteria | 4.00 |
| T 13 - NPK at 15: 5: 10 at 0.2 % spray and $GA_3$ 200 ppm + *Azospirillum* + AMF | 3.33 |
| T 14 - NPK at 30: 0: 10 at 0.2 % spray and $GA_3$ 200 ppm + Phosphobacteria + AMF | 4.00 |

| | |
|---|---|
| T 15 - NPK at 15: 0: 10 at 0.2% spray and $GA_3$ 200 ppm + *Azospirillum* + Phosphobacteria + AMF | 4.33 |
| T 16 - Control (no fertilizer + no biofertilizer) | 0.33 |
| CD (P = 0.05) | 0.79 |

## 18.13. HIBISCUS, *HIBISCUS ROSA-SINENSIS*

### 18.13.1. High yielding varieties

**Table 18.39.** High yielding varieties of hibiscus

| High yielding varieties | Flowers/plant | Developed at |
|---|---|---|
| CO 1 (Thilagan) | 3055 | TNAU, Coimbatore |
| CO 2 (Punnagai) | 988 | TNAU, Coimbatore |
| CO 3 | 1309 | TNAU, Coimbatore |

## 18.14. BARLERIA

### 18.14.1. High yielding varieties

**Table 18.40.** High yielding varieties of barleria

| High yielding varieties | Flower yield (kg/plant) | Developed at |
|---|---|---|
| CO 1 (Thilagan) | 2.11 | TNAU, Coimbatore |

**Chapter 19**

# MEDICINAL CROPS

In recent times, due to increasing realization of health hazards and toxicity caused by synthetic drugs (side effects), there has been a renewal of interest in the use of plant based drugs throughout the world. A campaign is going on worldwide to utilize more and more plant derived chemicals in the human health care/personal care system. As a result, the demand for plant-based raw materials has increased enormously in both national and international markets. Herbal plants are finding diverse uses as raw materials, not only for medicines but also as biopesticides.

In medicinal plant sector, the WHO has estimated that about 80% of the population in developing countries rely on traditional medicines mostly plant drugs for primary health-care needs. Even modern medicines contain about 25% drugs derived from plants. A large number of people are earning their livelihood from different activities of medicinal plants such as collection, cultivation, marketing, processing, etc. The area under cultivation of the above crops is low, but the cash returns are high because of intensive cropping.

Medicinal crops form one of the important groups due to their demand in various pharmaceutical industries and also to earn foreign exchange by way of export. The potential for earning foreign exchange by India from the exports of medicinal and aromatic plants is estimated over US$ 3,000 million per annum. The export of these plants and their products has a tremendous potential in advanced countries like Europe, USA and Japan. The international market for medicinal plants related trade is estimated at US$ 60 billion/year having a growth rate of 7% per annum. Destination of foreign tourists is now aimed at India for health treatment. Kerala and Gujarat have become their favourite destinations since Ayurvedic treatment is very popular in these two states.

Unless we start cultivating the important medicinal plants, it will not be possible to meet the increasing demands. However, cultivation of medicinal plants has to be taken up in polyculture models rather than in monoculture models. There is a huge task to be carried out at every level to make the

Indian system of medicine a viable, sustainable and modern. For this task, urgency is felt which can only be established through regular and systematic scientific research on collection, conservation, evaluation and development of good agricultural practices to compete in the world market.

## 19.1. COLEUS, *COLEUS FORSKOHLII*

### 19.1.1. High yielding varieties

**Table 19.1.** High yielding varieties of coleus

| High yielding varieties | Yield (MT/ha) | Developed at |
|---|---|---|
| CO 1 | 31.93 | TNAU, Coimbatore |

### 19.1.2. Integrated nematode management

#### i) Root-knot nematode, *Meloidogyne incognita*

The root-knot nematode infestation was reported on coleus from Kerala and Orissa. The dry weight of the tubers was reduced by 20% due to root-knot nematodes. The percentage of starch on fresh weight basis showed drastic reduction (16%) in the infested tubers. *M. incognita* was responsible for 70.2% loss in tuber yield of coleus.

Soil application of *T. viride* ($10^8$ cfu/g) and *P. fluorescens* ($10^8$ cfu/g) each at 2.5 kg/ha recorded increased plant growth, tuber yield and propagule density in soil; and reduced root galling, egg mass production and *M. incognita* population in soil (Senthamarai *et al.*, 2006) (Table 19.2).

**Table 19.2.** Effect of bioagents for the management of *Meloidogyne incognita* infecting medicinal coleus

| Treatment | Tuber yield (g/plant) | No. of galls/ plant | No. of egg masses/g root | Spore density (cfu/g soil) | |
|---|---|---|---|---|---|
| | | | | *P. fluorescens* | *T. viride* |
| Control | 22.71 | 1680 | 98 | --- | --- |
| *P. fluorescens* ($10^8$ cfu/g) at 2.5 kg/ha | 61.57 | 678 | 31 | 72.05 | --- |
| *T. viride* ($10^8$ cfu/g) at 2.5 kg/ha | 36.29 | 729 | 71 | --- | 12.95 |
| CD (P = 0.05) | 13.84 | 168.29 | 19.83 | --- | --- |

Integration of dipping stem cutting in 0.1% *Pseudomonas fluorescens* + soil application of neem cake at 400 kg/ha + growing marigold as an intercrop followed by their biomass incorporation during earthing up increased the root tuber yield by 22.7 to 30.0% and reduced the root-knot nematode (*M. incognita*) population by 71.2 to 73.8%. However, integration of *P. fluorescens* + marigold intercrop proved to be more economical (with benefit: cost ratio of 6.4 to 8.8) and effective management practice for the management of root-knot nematodes in medicinal coleus (Seenivasan and Devarajan, 2008) (Table 19.3).

**Table 19.3.** Effect of integration of bioagents, neem cake and marigold intercrop for the management of *Meloidogyne incognita* infecting Coleus

| Treatment | Gall index (1-5 scale) | | Yield (MT/ha) | | Benefit: cost ratio | |
|---|---|---|---|---|---|---|
| | *Kharif* | *Rabi* | *Kharif* | *Rabi* | *Kharif* | *Rabi* |
| Cutting dip in 0.1% *P. fluorescens* + neem cake at 400 kg/ha | 4.3 | 3.6 | 10.41 | 9.86 | 1.8 | 1.3 |
| *P. lilacinus* at 2.5 kg/ha + neem cake at 400 kg/ha | 4.3 | 3.6 | 10.21 | 9.81 | 1.4 | 1.2 |
| Cutting dip in 0.1% *P. fluorescens* + marigold intercrop | 2.3 | 2.1 | 12.08 | 11.07 | 8.8 | 6.4 |
| *P. lilacinus* at 2.5 kg/ha + marigold intercrop | 3.6 | 3.3 | 10.92 | 10.20 | 4.6 | 3.2 |
| *P. fluorescens* + neem cake + marigold | 2.3 | 2.1 | 12.11 | 11.12 | 3.4 | 2.5 |
| *P. lilacinus* + neem cake + marigold | 3.6 | 3.3 | 11.06 | 10.26 | 2.0 | 1.4 |
| Carbofuran at 1 kg a.i./ha | 4.3 | 3.6 | 10.36 | 9.85 | 2.1 | 1.6 |
| Control | 4.6 | 4.3 | 9.31 | 9.06 | --- | --- |
| CD (P = 0.05) | 0.36 | 0.36 | 0.83 | 0.80 | --- | --- |

## 19.2. ASHWAGANDHA, *WITHANIA SOMNIFERA*

### 19.2.1. High yielding varieties

**Table 19.4.** High yielding varieties of ashwagandha

| High yielding varieties | Developed at |
|---|---|
| Jawahar Asgand 20 | JNKVV, Jabalpur |
| Jawahar Asgand 134 | JNKVV, Jabalpur |

### 19.2.2. Integrated nutrient management

Application of vermicompost at 2.5 MT/ha + *Azotobacter* at 2 kg/ha + phosphate solubilizing bacteria at 2 kg/ha gave maximum fresh root (4.620 MT/ha) and herb yield (20.653 MT/ha) as well as dry root (2.256 MT/ha) and dry herb yield (7.195 MT/ha) (Singh *et al.*, 2009).

### 19.2.3. Integrated nematode management

#### (i) Root-knot nematode, *Meloidogyne incognita*

Integration of neem cake + *T. harzianum,* vermicompost + *T. harzianum,* and cow urine + *T. harzianum* considerably reduced the root-knot nematode development and enhanced plant growth and yield. Maximum root-knot suppression was noticed in vermicompost + *T. harzianum* followed by menthe distillate + *Glomus aggregatum*. Highest increase in plant yield was recorded when the soil was amended with menthe/curry leaf distillates along with *T. harzianum* / *G. aggregatum* (Pandey and Kalra, 2005) (Table 19.5). Vermicompost apart from providing a complete nutrition to the plant may also support the growth *T. harzianum* thus helping it proliferate in the rhizosphere and improve the soil health and thus sustainability to restrict the nematode population.

**Table 19.5.** Effect of organic materials and bioagents on plant growth of Ashwagandha infected with *Meloidogyne incognita*

| Treatment | Dry weight of plant (g) | Root-knot index | Total nema popn (soil + roots) |
|---|---|---|---|
| Untreated – Uninoculated | 8.5 | --- | --- |
| Untreated – Inoculated | 5.2 (-38.8) | 4.00 | 6460 |
| Carbofuran | 8.0 (-5.9) | 1.66 | 3012 |
| Neem cake + Davana | 23.8 (+62.4) | 1.66 | 2820 |
| Neem cake + Curry leaf | 7.4 (-12.9) | 1.33 | 2261 |
| Neem cake + Vermicompost | 12.0 (+41.2) | 1.66 | 2649 |
| Neem cake + *T. harzianum* | 12.8 (+50.6) | 1.66 | 2674 |
| Neem cake + *G. aggregatum* | 14.3 (+68.2) | 3.00 | 2964 |
| Neem cake + *Mentha* distillate | 14.2 (+67.1) | 1.66 | 2409 |
| Davana + Curry leaf | 10.8(+27.1) | 1.33 | 2030 |
| Davana + Vermicompost | 14.1 (+65.9) | 3.00 | 2594 |
| Davana + *T. harzianum* | 9.6 (+12.9) | 1.66 | 2110 |
| Davana + *G. aggregatum* | 10.3 (+21.2) | 1.33 | 2050 |
| Davana + *Mentha* distillate | 13.2 (+55.3) | 1.33 | 2150 |
| Curry leaf + Vermicompost | 14.0 (+64.7) | 1.66 | 2440 |
| Curry leaf + *T. harzianum* | 15.0 (+76.5) | 3.33 | 3252 |
| Curry leaf + *G. aggregatum* | 13.4 (+57.6) | 3.00 | 2635 |
| Curry leaf + *Mentha* distillate | 15.3 (+80.0) | 1.66 | 2404 |
| Vermicompost + *T. harzianum* | 14.4 (+69.4) | 0.66 | 1400 |
| Vermicompost + *G. aggregatum* | 13.6 (+60.0) | 1.99 | 2480 |
| Vermicompost + *Mentha* distillate | 13.9 (+63.5) | 1.33 | 2000 |
| *G. aggregatum* + *Mentha* distillate | 15.6 (+83.5) | 1.00 | 1784 |
| CD (P = 0.05) | 0.71 | 0.01 | 465.70 |

Root-knot nematode infection was drastically impaired and plant growth biomass improved when plants were inoculated with three AMF simultaneously as compared to their inoculation alone (Table 19.6).

**Table 19.6.** Effect of integration of AMF for the management of *Meloidogyne incognita* infecting ashwagandha

| Treatment | Fresh wt (g) | Dry wt. (g) | Oil yield (%) | % AMF colonization | Root-knot indices |
|---|---|---|---|---|---|
| Untreated-uninoculated | 395.2 | 107.1 | 0.51 | --- | --- |
| Untreated-inoculated | 285.0 | 77.9 | 0.38 | --- | 3.6 |
| *Glomus aggregatum* (*Ga*) | 322.0 | 91.6 | 0.40 | 42.3 | 2.3 |
| *G. fasciculatum* (*Gf*) | 333.0 | 93.8 | 0.41 | 58.2 | 2.0 |
| *G. mosseae* (*Gm*) | 388.0 | 105.9 | 0.48 | 63.5 | 1.6 |
| *Ga* + *Gf* + *Gm* | 360.0 | 99.8 | 0.46 | 78.5 | 1.3 |

Soil application of super *Pseudomonas* (consortia of *Pseudomonas fluorescens* strains with chitinase, ACC deaminase and neem mixture) at 2.5 kg/ha was effective to suppress the root-knot nematode incidence by 78% and to enhance the biomass of economic parts of aswagandha plants by 26.87% (dry root) and 43.19% (seed) over untreated control (Ramakrishnan, 2010).

## 19.3. ISABGOL, *PLANTAGO OVATA*

### 19.3.1. High yielding varieties

**Table 19.7.** High yielding varieties of isabgol

| High yielding varieties | Seed yield | Developed at |
|---|---|---|
| Gujarat Isabgol-1 | 11% higher yield over local check | AAU, Anand |
| Gujarat Isabgol-2 | 20% higher yield over Gujarat Isabgol-1 | AAU, Anand |
| Jawahar Isabgol-3 | High yielding | JNKVV, Jabalpur |
| Haryana Isabgol-5 | Resistant to downy mildew | CCSHAU, Hisar |

### 19.3.2. Integrated nutrient management

Application of FYM at 10 MT/ha + Biovita at 20 kg/ha + Biovita at 500 ml/ha was found to be the best treatment in increasing the seed yield (1.272 MT/ha) and husk yield (422 kg/ha) and gave maximum benefit-cost ratio of 3.55 (Prasad *et al.*, 2009).

## 19.4. OPIUM POPPY, *PAPAVER SOMNIFERUM*

### 19.4.1. High yielding varieties

**Table 19.8.** High yielding varieties of opium poppy

| High yielding varieties | Latex yield (kg/ha) | Developed at |
|---|---|---|
| Jawahar Opium -16 | 60-65 | JNKVV, Mandasur |
| Jawahar Opium-539 | 65 | JNKVV, Mandasur |
| Jawahar Opium-540 | 75 | JNKVV, Mandasur |
| Kirtiman | 45-50 | NDUAT, Faizabad |
| Trishna | 49-53 | NBPGR, New Delhi |

### 19.4.2. Integrated disease management

#### (i) Sclerotinia rot and blight, *Sclerotinia sclerotiorum*

*Sclerotinia* rot and blight has been successfully controlled by using several strains of *Trichoderma harzianum* isolated from soil and procured from IMTECH, Chandigarh (Table 19.9).

**Table 19.9.** Effect of different fungal antagonists on *Sclerotinia* rot and blight disease of opium poppy

| Antagonists | % plants infected after 16 weeks | Plant height (cm) | Opium yield (mg) |
|---|---|---|---|
| *T. harzianum* – 792 | 9.88 | 97.70 | 161.90 |
| *T. harzianum* – 795 | 9.96 | 98.10 | 172.50 |
| *T. harzianum* – 796 | 8.72 | 98.40 | 122.90 |
| *T. harzianum* – 801 | 8.01 | 98.80 | 119.60 |
| NBRI *T. harzianum* – I | 7.52 | 97.70 | 188.70 |
| NBRI *T. harzianum* – III | 9.01 | 99.10 | 124.70 |
| NBRI *T. harzianum* – X | 9.99 | 100.50 | 127.90 |
| NBRI *T. harzianum* – XIV | 10.01 | 97.03 | 121.80 |
| NBRI *T. harzianum* – XV | 6.89 | 98.83 | 184.30 |
| NBRI *T. harzianum* – XVII | 6.54 | 98.13 | 132.90 |
| Control | 39.96 | 88.12 | 92.40 |

## 19.5. *MACUNA PRURIENS*

### 19.5.1. Integrated nutrient management

The integrated nutrient combination involving organic form of manures (coco peat at 5 MT/ha + FYM at 12.5 MT/ha) and inorganic fertilizers (40: 30: 30 kg NPK/ha) and harvesting at dry pod stage gave highest seed yield (112.06 g/plant). The highest amount of L DOPA (7.43%) was also recorded in the above treatment when harvesting was done at green pod stage (Kavitha *et al.*, 2006).

## 19.6. HENBANE, *HYOCYAMUS MUTICUS, H. NIGER, H. ALBUS*

### 19.6.1. High yielding varieties

**Table 19.10.** High yielding varieties of henbane

| High yielding varieties | Yield | Developed at |
|---|---|---|
| HMI-80-1 | 0.4-0.7% alkaloid content | JNKVV, Jabalpur |

### 19.6.2. Integrated nematode management

#### i) Root-knot nematode, *Meloidogyne incognita*

**(a) Cultural Methods:** It was observed that essential oils of palmarosa, citronella, basil and menthol mint were quite effective in reducing *M. incognita* population and improving plant growth. The oil of palmarosa at 2ml/pot was most effective (Pandey *et al.*, 2003) (Table 19.11).

**Table 19.11.** Effect of essential oils of aromatic crops on growth and root galling of henbane

| Treatment | Plant dry weight (g) | No. of leaves/plant | Root-knot index |
|---|---|---|---|
| Control | 4.4 | 39.0 | 4.00 |
| Menthol mint - 0.5 ml | 5.3 | 50.6 | 2.33 |
| Menthol mint - 1.0 ml | 8.4 | 63.6 | 1.66 |
| Menthol mint - 2.0 ml | 10.4 | 72.6 | 1.33 |
| Basil - 0.5 ml | 9.6 | 86.6 | 2.00 |
| Basil - 1.0 ml | 11.4 | 123.3 | 1.33 |
| Basil - 2.0 ml | 9.0 | 116.0 | 1.33 |
| Citronella - 0.5 ml | 10.0 | 112.0 | 2.33 |
| Citronella - 1.0 ml | 11.3 | 94.3 | 1.66 |
| Citronella - 2.0 ml | 9.8 | 88.0 | 1.33 |
| Palmarosa - 0.5 ml | 9.6 | 79.7 | 2.00 |
| Palmarosa - 1.0 ml | 12.5 | 83.6 | 1.66 |
| Palmarosa - 2.0 ml | 12.9 | 106.3 | 1.00 |
| CD (P=0.05) | 0.555 | 12.645 | 0.030 |

**b) Integrated methods:** Experimental findings indicated that the application of bio-inoculants have not only enhanced the total biomass yield of *Hyoscyamus niger* but also significantly decreased the multiplication of nematodes. However, a significantly higher reduction was recorded in the treatment where all bio-inoculants are combined. This may be attributed to the fact that these bioagents are secreting potential chemicals which are either non-favourable for multiplication of *M. incognita* or inducing tolerance in the plant against the attack of root-knot nematodes. Hence, mixed inoculation of rhizobacterium with AMF could be considered for biomanagement for reducing the deleterious effect of root-knot nematodes in black henbane. Similar results were also obtained with different bioagents in *H. muticus* plants against *M. incognita* (Pandey *et al.*, 2000) (Table 19.12).

**Table 19.12.** Effect of bioagents for the management of *Meloidogyne incognita* infecting henbane

| Treatment | Fresh biomass wt. (g) | Reproduction factor | Root-knot index |
|---|---|---|---|
| Control | 120.6 | 8.14 | 4.00 |
| *Pseudomonas fluorescens* (*Pf*) | 171.0 | 4.24 | 1.33 |
| *Glomus aggregatum* (*Ga*) | 172.7 | 4.52 | 1.66 |

| | | | |
|---|---|---|---|
| *G. fasciculatum (Gf)* | 144.6 | 5.78 | 2.66 |
| *G. mosseae (Gm)* | 142.0 | 6.05 | 3.00 |
| *Pf + Ga + Gf + Gm* | 202.0 | 4.03 | 1.00 |
| CD (P = 0.05) | 12.01 | --- | 1.16 |

## 19.7. KACHOLAM, *KAEMPFERIA GALANGA*

### 19.7.1. Integrated nematode management

#### i) Root-knot nematode, *Meloidogyne incognita*

Mulching of kacholam with green leaves of neem and chromolaena at 5 kg/m$^2$ 15 days before planting rhizomes, increased the rhizome yield by 160.7% and 143.3% over control, respectively. The above treatments also reduced root galling (1 as compared to 5 in control), the nematode population both in soil (81.8% and 80.4%, respectively) and roots (89.60% and 86.14%, respectively) (Table 19.13).

**Table 19.13.** Effect of mulching of green leaves on root-knot nematodes and yield of kacholam.

| Treatment | Root-knot index | Nema popn. (100g soil) | % redn. over control | Nema popn. (5g roots) | % redn. over control | Rhizome yield (kg/4m$^2$) | % increase over control |
|---|---|---|---|---|---|---|---|
| Neem leaf | 1 | 45.44 (6.74)* | 81.80 | 7.00 (2.76) | 89.60 | 5.6 | 160.7 |
| Glyricidia | 2 | 52.48 (7.24) | 79.05 | 14.67 (2.84) | 78.21 | 3.9 | 79.6 |
| Mangium | 5 | 178.26 (13.35) | 28.50 | 33.67 (5.87) | 49.99 | 2.6 | 21.6 |
| Clerodendron | 2 | 95.35 (9.92) | 60.60 | 22.70 (4.82) | 6.33 | 3.4 | 56.4 |
| Calotropis | 2 | 85.86 (9.27) | 65.60 | 16.00 (4.03) | 76.24 | 3.6 | 69.8 |
| Cromolaena | 1 | 48.96 (6.99) | 80.40 | 9.33 (3.18) | 86.14 | 5.3 | 143.3 |
| Control | 5 | 249.37 (15.79) | --- | 67.33 (8.26) | --- | 2.2 | --- |
| CD (P=0.05) | --- | (1.02) | --- | (1.45) | --- | 1.28 | --- |

*Figures in parentheses are values after square root transformation.

## 19.8. BRAHMI, *BACOPA MONNIERI*

### 19.8.1. Integrated nematode management

#### i) Root-knot nematode, *Meloidogyne incognita*

Neem compound, distillation waste of menthol mint, curry leaf, davana and vermicompost and bioinoculants such as *Glomus aggregatum* and

*Trichoderma harzianum* alone or in combination were root-knot nematode suppressive and enhanced the growth and yield of brahmi (Pandey *et al.*, 2002) (Table 19.14).

**Table 19.14.** Effect of organic materials and bioagents on plant growth, yield and root galling in brahmi infected with *Meloidogyne incognita*

| Treatment | Total dry weight (g/pot) | Root-knot index |
|---|---|---|
| Control | 47.1 | 3.66 |
| Carbofuran | 68.2 | 2.33 |
| Davana | 96.0 | 0.66 |
| Neem compound | 102.6 | 1.00 |
| Neem cake | 100.8 | 2.33 |
| Linseed cake | 64.0 | 2.66 |
| Sesamum cake | 68.3 | 2.00 |
| *Glomus aggregatum* | 69.5 | 3.33 |
| *G. fasciculatum* | 69.0 | 2.66 |
| *Trichoderma harzianum* | 77.3 | 2.66 |
| CD (P=0.05) | 3.271 | 0.041 |

## 19.9. SENNA, *CASSIA ANGUSTIFOLIA*

### 19.9.1. High yielding varieties

**Table 19.15.** High yielding varieties of senna

| High yielding varieties | Dry leaf yield (kg/ha) | Developed at |
|---|---|---|
| Anand Late Selection | 75% higher than local check | AAU, Anand |
| KKM Sel. 1 | 712 (pod yield-266 kg/ha) 47.7 and 53.8% higher than local check | AC&RI, TNAU, Killikulam |

## 19.10. DIOSCOREA, *DIOSCOREA FLORIBUNDA*

### 19.10.1. High yielding varieties

**Table 19.16.** High yielding varieties of dioscorea

| High yielding varieties | Tuber yield | Developed at |
|---|---|---|
| FBI-1 | 2.5% diosgenin content | IIHR, Bangalore |
| Arka Upakar | 3.5% diosgenin content | IIHR, Bangalore |

## 19.11. KHASI KATERI, *SOLANUM VIARUM*

### 19.11.1. High yielding varieties

**Table 19.17.** High yielding varieties of khasi kateri

| High yielding varieties | Dry berry yield | Developed at |
|---|---|---|
| Arka Sanjeevini | High alkaloid content | IIHR, Bangalore |
| Arka Mahima | 2% solasodine content | IIHR, Bangalore |

## 19.12. *SOLANUM LACIATUM*

### 19.12.1. High yielding varieties

**Table 19.18.** High yielding varieties of *Solanum laciatum*

| High yielding varieties | Dry berry yield (MT/ha) | Developed at |
|---|---|---|
| NH 88-12 | 3.138 | |

## 19.13. THYME

### 19.13.1. High yielding varieties

**Table 19.19.** High yielding varieties of thyme

| High yielding varieties | Green leaf yield (MT/ha) | Resistance to | Developed at |
|---|---|---|---|
| Ooty (Tv) 1 | 10.7 | Root rot, root-knot nematode, white fly, aphid | TNAU, Coimbatore |

Chapter 20

# AROMATIC CROPS

In recent years, there has been an increased interest in the cultivation of aromatic plants to meet the requirements of cosmetic, flavouring and perfumery industries. Aromatic crops are estimated to cover an area of 386,100 ha with production of 325,000 MT during 2008 (Bijay Kumar, 2009) (Table 20.1). The aromatic plants provide raw material for the production of flavours, herbal cosmetics, perfumery etc. Aromatic plants and their products, particularly the essential oils, are now becoming one of the more important export items from many developing countries of Asia. Chemically essential oils are terpenes, which act as carriers of the aromatic substances. Most essential oils also contain camphors, and the more odiferous compounds present in them consist of oxygen derivatives of terpenes, alcohols, esters, aldehydes and ketones. India has enjoyed a prominent position in the manufacture of superior perfumes and aromatics by using essential oils. The important aromatic plants grown are lemon grass, vetiver, patchouli, palmarosa, citronella, mints, geranium, lavender, basil, jasmine etc.

**Table 20.1.** State-wise area and production of aromatic crops (2007-08) (Bijay Kumar, 2009)

| State | Area ('000 ha) | Production ('000 MT) | Productivity (MT/ha) |
|---|---|---|---|
| Madhya Pradesh | 19.6 | 117.6 | 6.0 |
| Rajasthan | 198.2 | 94.1 | 2.1 |
| Chhattisgarh | 11.5 | 65.6 | 5.6 |
| Andhra Pradesh | 9.6 | 13.8 | 1.4 |
| Uttar Pradesh | 133.7 | 13.4 | 0.1 |
| Tamil Nadu | 7.6 | 11.2 | 1.4 |
| Karnataka | 2.0 | 7.2 | 3.6 |
| Punjab | 2.4 | 1.1 | 0.5 |
| Haryana | 1.4 | 1.0 | 0.7 |
| Jharkhand | 0.1 | 0.0 | --- |
| Total | 386.1 | 325.0 | 0.8 |

The world market, India stands in second position. Most of the essential oils produced though marketed within the country, but a sizeable amount has also been exported. Among the essential oils exported from India are Japanese mint oil, ambrette seed essence, sandal wood oil, citronella oil, lemon grass oil, palmarosa oil, ginger oil, clove oil, eucalyptus oil, vetiver oil, pepper oil, etc. The annual exports of the derivatives of aromatic plants are to the tune of Rs. 600-700 million.

## 20.1. JASMINE, *JASMINUM* SPP.

### 20.1.1. High yielding varieties

**Table 20.2.** High yielding varieties of jasmine

| Jasmine species | High yielding varieties | Flower bud Yield (kg/ha) | Developed at |
|---|---|---|---|
| *Jasminum auriculatum* | Parimullai | 7800 | TNAU, Coimbatore |
| | CO-1 | 8800 | TNAU, Coimbatore |
| | CO-2 | 11,198 | TNAU, Coimbatore |
| *J. grandiflorum* | Arka Surabhi | 10.000 MT/ha | IIHR, Bangalore |
| | CO-1 | 10.144 MT/ha | TNAU, Coimbatore |
| | CO-2 | 11.680 MT/ha | TNAU, Coimbatore |
| *J. sambac* | Arka Aradhana | 8.000 MT/ha | IIHR, Bangalore |
| | Arka Arpan | 6.000 MT/ha | IIHR, Bangalore |

### 20.1.2. Integrated nutrient management

The plants supplied with 50% recommended dose of fertilizers + 8 kg vermicompost per plant produced highest flower yield (1.268 kg/plant) followed by 50% recommended dose of fertilizers + 4 kg vermicompost per plant (1.188 kg/plant), 50% RDF + RDFYM (1.147 kg/plant) and 50% recommended dose of fertilizers + 2 kg vermicompost per plant (1.135 kg/plant) (Patil *et al.*, 2004) (Table 20.3).

**Table 20.3.** Effect of organic and inorganic fertilizers on yield of jasmine (*Jasminum sambac*)

| Treatment | Yield (kg/plant) |
|---|---|
| Control | 0.708 |
| RDF + RDFYM | 1.147 |
| RDF | 0.992 |
| RDFYM | 0.852 |
| 50% RDFYM | 0.796 |
| 2 kg vermicompost/plant | 0.909 |
| 4 kg vermicompost/plant | 0.971 |

| | |
|---|---|
| 8 kg vermicompost/plant | 1.040 |
| 50% RDF + 2 kg vermicompost/plant | 1.135 |
| 50% RDF + 4 kg vermicompost/plant | 1.188 |
| 50% RDF + 8 kg vermicompost/plant | 1.268 |
| *In situ* vermiculture (50 worms/plant) | 1.021 |
| *In situ* vermiculture (100 worms/plant) | 1.135 |
| Only organic residue | 0.897 |

### 20.1.3. Integrated nematode management

Application of phorate at 4 g a.i. per plant during May and September months and incorporation of 20 kg FYM per plant increased flower yield of jasmine by 50% and reduced the root-knot nematode population by 70% (Sundarababu, 1992).

## 20.2. MINT, *MENTHA* SPP.

### 20.2.1. High yielding varieties

**Table 20.4.** High yielding varieties of mint

| High yielding varieties | Yield | Developed at |
|---|---|---|
| Punjab Spearmint-1 | 0.57% essential oil (60-70% carvone) | YSPUH&F, Solan |

### 20.2.2. Integrated nutrient management

Maniram *et al.* (1987) found an increase of 31.6% in the yield of Japanese mint with neem coated urea as compared to uncoated urea (Table 20.5).

**Table 20.5.** Relative efficiency of urea and neem coated urea on productivity of mint

| Location | N (kg/ha) | Yield of essential oil (kg/ha) | | % increase |
|---|---|---|---|---|
| | | Urea | Neem coated urea | |
| CIMAP, Lucknow | 120 | 130 | 170 | 31.6 |

### 20.2.3. Integrated nematode management

#### i) Root-knot nematodes, *Meloidogyne incognita, M. javanica, M. hapla*

Root-knot disease of Menthol/Japanese mint, Spearmint, Scotch Spearmint, Peppermint and Bergamot mint caused by *M. incognita* and *M. javanica* was observed for the first time in Lucknow and Terai region of Uttar Pradesh which reduces the herbage yield and oil content. *M. incognita* dominates over *M. javanica* in mixed infection. The quality of mint oil was also adversely affected due to nematode infection. The root-knot nematode, *M. incognita* caused 40.2% loss in herbage yield and 46.6% loss in oil yield in menthol mint.

Maximum reduction in *M. incognita* population in Menthol mint was recorded in neem cake treated soil followed by mustard cake and *Trichoderma harzianum*. Higher herbage yield (21 MT/ha) was observed in the soil treated with oil cakes in comparison to control (14.66 MT/ha). The crop produced had significantly higher oil content (2.7%) and yield (151 kg/ha) in neem cake treatment followed by mustard cake (2.7% oil and 143.6 kg/ha yield), *T. harzianum* (2.5% oil and 116.4 kg/ha yield) and *Glomus aggregatum* (2.4% oil and 112 kg/ha yield). The untreated crop gave 1.8% oil and 70.4 kg/ha yield (Pandey *et al.*, 1998) (Table 20.6).

**Table 20.6.** Effect of bio-organics and a pesticide on herb and oil yield of menthol mint and root-knot indices of *Meloidogyne incognita*.

| Treatment | Herbage yield (MT/ha) | % oil content | Oil yield (kg/ha) | Root-knot indices |
|---|---|---|---|---|
| Untreated control | 1.466 | 1.8 | 70.4 | 3.66 |
| Carbofuran | 1.880 (+28.4) | 2.3 | 114.2 (+62.2) | 2.33 (-34.7) |
| Mustard cake | 1.998 (+36.3) | 2.7 | 143.6 (+104.0) | 2.00 (-45.4) |
| Neem cake | 2.100 (+43.2) | 2.7 | 151.0 (+114.5) | 2.00 (-45.4) |
| *Glomus aggregatum* | 1.758 (+19.9) | 2.4 | 112.2 (+59.4) | 2.66 (-27.3) |
| *Trichoderma harzianum* | 1.750 (+19.4) | 2.5 | 116.4 (+65.3) | 2.66 (-27.3) |
| CD (P = 0.05) | 0.284 | 0.5 | 40.4 | 0.1 |

Application of arbuscular mycorrhizal fungi such as *G. aggregatum, G. fasciculatum* and *G. mosseae* improved plant growth, enhanced herbage and oil yield and effectively inhibited root-knot nematode infection (Pandey *et al.*, 1997) (Table 20.7).

**Table 20.7.** Effect of arbuscular mycorrhizal fungi on the productivity of mint infected with *Meloidogyne incognita*

| Treatment | Fresh herbage weight (g) | % oil yield | % mycorrhizal colonization | Root-knot index |
|---|---|---|---|---|
| Untreated – Inoculated | 285.0 | 0.38 | --- | 3.6 |
| *Glomus aggregatum* (Ga) | 322.0 | 0.40 | 42.3 | 2.3 |
| *G. fasciculatum* (Gf) | 333.0 | 0.41 | 58.2 | 2.0 |
| *G. mosseae* (Gm) | 388.0 | 0.48 | 63.5 | 1.6 |
| Ga + Gf + Gm | 360.0 | 0.46 | 78.5 | 1.3 |

## 20.3. GERANIUM, *PELARGONIUM GRAVEOLENS*

### 20.3.1. High yielding varieties

**Table 20.8.** High yielding varieties of geranium

| High yielding varieties | Green leaf yield (MT/ha) | Essential oil yield (kg/ha) | Developed at |
|---|---|---|---|
| KKL-1 | 45.2 | 54.4 | AC&RI, TNAU, Killikulam |

### 20.3.2. Fertigation

Soil application of 200 : 35 : 35 kg NPK/ha in the form of normal fertilizers at monthly intervals through sprinkler irrigation is recommended to get maximum essential oil yield (91 liters/ha) and net returns from geranium (Khan *et al.*, 2001) (Table 20.9).

**Table 20.9.** Yield of essential oil and economics of geranium cv. Reunion as influenced by sources and level of fertilizers and micronutrients

| Treatment | Oil yield (litres/ha) | Net income (Rs.) | Benefit: cost ratio |
|---|---|---|---|
| No fertilizers + Sprinkler irrigation | 62 | 104,360 | 5.04 |
| No fertilizers + Surface irrigation | 62 | 124,035 | 5.85 |
| Normal fertilizers 50% + Sprinkler irrigation | 72 | 136,793 | 6.37 |
| Normal fertilizers 75% + Sprinkler irrigation | 79 | 154,751 | 6.91 |
| Normal fertilizers 100% + Sprinkler irrigation | 88 | 157,371 | 6.35 |
| Normal fertilizers 100% soil appln.+ Sprinkler irrigation | 91 | 103,780 | 5.13 |
| Normal fertilizers 100% soil appln.+Surface irrigation | 83 | 126,523 | 4.05 |
| Water soluble fertilizers 50% + Sprinkler irrigation | 76 | 145,145 | 5.09 |
| Water soluble fertilizers 75% + Sprinkler irrigation | 87 | 150,067 | 4.97 |
| Water soluble fertilizers 100% + Sprinkler irrigation | 90 | 142,811 | 6.07 |

The cumulative herbage yield was the highest with all the 100% RDF irrespective of the method of irrigation (70 to 76 MT/ha) which were on par with 75% WSF fertigation (71.1 MT/ha) compared to 51 and 54 MT/ha in controls. The oil yields were also higher with all 100% RDF (88.5 to 91.1 kg/ha) and also with 75% WSF fertigation (87.4 kg/ha) thus indicating a saving of 25% fertilizers. The controls gave only around 62 kg/ha of geranium oil (Sabina Ayyanna and Farooqi, 1995).

## 20.4. VETIVER, *VETIVERIA ZIZANIOIDES*

### 20.4.1. Micro irrigation

In vetiver, moisture regime of 1.00 IW/CPE ratio recorded maximum root yield (2.851 MT/ha), root dry matter (55.14 kg/ha) and essential oil yield (46.31 kg/ha) (Nataraja *et al.*, 2009).

## 20.5. ROSEMARY, *ROSAMARINUS OFFICINALIS*

### 20.5.1. High yielding varieties

**Table 20.10.** High yielding varieties of rosemary

| High yielding varieties | Green leaf yield (MT/ha) | Resistance to pest/disease | Developed at |
|---|---|---|---|
| Ooty (RM) 1 | 12.4 | Leaf blight (*Rhizoctonia solani*), whiteflies, aphids | TNAU, Coimbatore |

### 20.5.2. Fertigation

Application of 80: 32: 32 kg NPK/ha in the form of water soluble fertilizers along with 0.5% Zinc and 0.5% Boron through fertigation at monthly intervals is recommended to get maximum essential oil yield (408 liters/ha), concrete (24.96 kg/ha) and net returns from rosemary (Khan *et al.*, 2001) (Table 20.11).

**Table 20.11.** Yield of essential oil and economics of rosemary as influenced by sources and level of fertilizers and micronutrients

| Treatment | Oil yield (litres/ha) | Net income (Rs.) | Benefit: cost ratio |
|---|---|---|---|
| Normal fertilizers 100% soil appln. + Drip irrigation | 232 | 36,407 | 0.64 |
| Normal fertilizers 80% soil appln. + Drip irrigation | 195 | 22,876 | 0.41 |
| Normal fertilizers 60% soil appln. + Drip irrigation | 152 | 6,909 | 0.12 |
| Water soluble fertilizers 100% fertigation | 333 | 77,572 | 1.39 |
| Water soluble fertilizers 80% fertigation | 255 | 47,146 | 0.85 |
| Water soluble fertilizers 60% fertigation | 231 | 38,304 | 0.70 |
| Normal fertilizers 100% soil appln. + 0.5% Zn + 0.5% B + Drip irrigation | 338 | 77,827 | 1.35 |
| Normal fertilizers 80% soil appln. + 0.5% Zn + 0.5% B + Drip irrigation | 342 | 80,696 | 1.43 |
| Normal fertilizers 60% soil appln. + 0.5% Zn + 0.5% B + Drip irrigation | 215 | 31,159 | 0.56 |
| Water soluble fertilizers 100% + 0.5% Zn + 0.5% B fertigation | 390 | 99,392 | 1.75 |
| Water soluble fertilizers 80% + 0.5% Zn + 0.5% B fertigation | 408 | 106,967 | 1.91 |
| Water soluble fertilizers 60% + 0.5% Zn + 0.5% B fertigation | 346 | 83,324 | 5.10 |
| Normal fertilizers 100% soil appln. + Furrow irrigation | 185 | 64,011 | 6.40 |
| Normal fertilizers 100% soil appln. + 0.5% Zn + 0.5% B + Furrow irrigation | 135 | 81,186 | 6.33 |

## 20.6. PATCHOULI, *POGOSTEMON PATCHOULI*

### 20.6.1. Micro irrigation

Irrigation at 0.9 IW/CPE ratio increased dry herbage yield from 668 to 902 kg/ha in the first harvest, from 583 to 765 kg/ha in the second harvest and from 360 to 515 kg/ha in the third harvest. The total oil yield increased by 42% with 0.9 IW/CPE ratio over 0.3 IW/CPE ratio (Bhaskar *et al.*, 2009).

### 20.6.1. Integrated nutrient management

The total dry herbage yield and oil yield was higher when K was applied at 50 kg/ha/year (1891.9 kg/ha) over 100 kg/ha/year (1868.0 kg/ha) (Bhaskar *et al.*, 2009).

## 20.7. DAVANA, *ARTIMISIA PALLENS*

### 20.7.1. Integrated nematode management

#### i) Root-knot nematode, *Meloidogyne incognita*

**(a) Cultural Methods:** Soil application of neem cake at 1g N/kg soil significantly suppressed root-knot nematode population and increased oil yield (Haseeb and Butool, 1991). Better efficacy of neem cake over the other organic materials viz., mahua cake, castor cake and nematicides like aldicarb and carbofuran has been demonstrated on *M. incognita* infecting *A. pallens* (Pandey, 1994b) (Table 20.12). Increase in nematicidal efficacy of neem cake and FYM for 90 to 120 days in comparison to 60 days for fenamiphos has also been noticed (Anita and Vadivelu, 1997).

**Table 20.12.** Effect of oil cakes and pesticides on plant growth and yield of davana infected with *Meloidogyne incognita*

| Treatment | Fresh weight (g) | % oil content | Root-knot index |
|---|---|---|---|
| Control | 52.5 | 0.25 | 3.6 |
| Mahua cake | 54.9 | 0.27 | 1.4 |
| Castor cake | 73.4 | 0.33 | 2.4 |
| Neem cake | 87.5 | 0.36 | 1.4 |
| Carbofuran | 75.0 | 0.33 | 2.4 |
| Aldicarb | 83.3 | 0.35 | 1.4 |
| Bavistin | 53.0 | 0.27 | 2.6 |

A considerable enhancement of herbage biomass, flower buds and oil yields were noticed when davana plants were treated with the distillation waste of palmarosa, citronella, curry leaf, menthol mint and vermicompost of pyrethrum, marigold and menthol mint. Least nematode infections were recorded in distillation waste of curry leaf, palmarosa, lemon grass and vermicompost of pyrethrum, marigold and menthol mint (Pandey, 2005b) (Table 20.13).

**Table 20.13.** Effect of distillation waste and vermicompost on growth and root-knot development of davana infected with *Meloidogyne incognita*

| Distillation waste/ vermicompost | Fresh wt. of herbage & flower buds (g) | Total nematode population (soil + root) | Reproduction factor | Root-knot index |
|---|---|---|---|---|
| Untreated-Uninoculated control | 40.0 | --- | --- | --- |
| Untreated-Inoculated control | 24.7 | 3480 | 3.48 | 4.00 |

| | | | | |
|---|---|---|---|---|
| Carbofuran control | 36.8 | 1800 | 1.80 | 2.66 |
| Lemon grass (Distillation waste) | 41.8 | 1260 | 1.26 | 2.00 |
| Palmarosa (Distillation waste) | 48.8 | 1100 | 1.10 | 1.66 |
| Citronella (Distillation waste) | 44.2 | 1040 | 1.04 | 1.66 |
| Menthol mint (Distillation waste) | 42.7 | 1640 | 1.64 | 2.33 |
| Curry leaf (Distillation waste) | 58.5 | 1020 | 1.02 | 1.00 |
| Geranium (Distillation waste) | 32.3 | 1680 | 1.68 | 2.66 |
| Patchouli (Distillation waste) | 31.3 | 1800 | 1.80 | 2.66 |
| Marigold (Distillation waste) | 41.6 | 1600 | 1.60 | 2.33 |
| Menthol mint (Vermicompost) | 51.3 | 1260 | 1.26 | 2.00 |
| Citronella (Vermicompost) | 51.0 | 1380 | 1.38 | 2.00 |
| Chrysanthemum (Vermicompost) | 44.0 | 1240 | 1.24 | 1.66 |
| Marigold (Vermicompost) | 37.9 | 1040 | 1.04 | 1.66 |
| CD (P = 0.05) | 2.176 | --- | --- | 0.118 |

**(b) Biological methods:** A considerable enhancement of herbage biomass, flower bud and oil yields were noticed when davana plants were treated with the bioagent, *Trichoderma harzianum* (Pandey, 2005b) (Table 20.14).

**Table 20.14.** Effect of bio-agents on growth and root-knot development of davana infected with *Meloidogyne incognita*

| Bio-agents | Fresh wt. of herbage & flower buds (g) | Total nematode population (soil + root) | Reproduction factor | Root-knot index |
|---|---|---|---|---|
| Untreated-Uninoculated control | 40.5 | --- | --- | --- |
| Untreated-Inoculated control | 27.3 | 3480 | 3.48 | 4.00 |
| Carbofuran control | 39.0 | 1760 | 1.76 | 2.66 |
| *Trichoderma harzianum* | 56.7 | 1480 | 1.48 | 2.00 |
| *Glomus aggregatum* | 29.5 | 2080 | 2.08 | 2.66 |
| *G. fasciculatum* | 30.4 | 1880 | 1.88 | 2.66 |
| CD (P=0.05) | 2.174 | --- | --- | 0.137 |

## 20.8. CHAMOMILE, *MATRICARIA CHAMOMILLA*

### 20.8.1. Integrated nematode management

#### (i) Root-knot nematode, *Meloidogyne incognita*

**(a) Chemical methods:** Maximum reduction in root-knot severity and nematode population occurred with 4-chlorosalicylic acid, O-acetylsalicylic acid and 2-chloronicotinic acid. The result envisages that these chemical activators may be used in integrated management system for protection against root-knot nematodes (Pandey and Kalra, 2005) (Table 20.15).

**Table 20.15.** Effect of chemical activators on root-knot nematodes, plant growth and flower yield

| Treatment | Total nema popn. (soil + roots) | Reproduction factor (Rf = Pf/Pi) | Root-knot index | Dry weight of plant (g) | Flower yield (g) |
|---|---|---|---|---|---|
| Control | 3018 | 3.01 | 4.00 | 4.1 | 3.3 |
| Isonicotinamide | 2498 | 2.49 | 3.00 | 3.0 | 3.3 |
| 2-Chloronicotinic acid | 1710 | 1.71 | 2.00 | 6.3 | 6.1 |
| 5-Nitrosalicylic acid | 1818 | 1.81 | 2.00 | 7.7 | 6.3 |
| 4-Chlorosalicylic acid | 1412 | 1.41 | 1.33 | 9.6 | 6.7 |
| DL-2 Aminobutyric acid | 1777 | 1.77 | 2.00 | 6.7 | 6.1 |
| 2-Aminobutyric acid | 2020 | 2.02 | 2.33 | 4.3 | 3.7 |
| 0-Acetyl salicylic acid | 1640 | 1.64 | 1.66 | 10.6 | 7.4 |
| 4-Amino salicylic acid | 2176 | 2.17 | 2.66 | 6.7 | 6.2 |
| Salicylic acid | 2010 | 2.01 | 2.00 | 8.6 | 6.5 |
| Carbofuran | 1920 | 1.92 | 2.00 | 4.9 | 6.0 |
| CD (P = 0.05) | 321.83 | --- | 0.021 | --- | 0.01 |

**(b) Integrated methods:** Integration of *G. mosseae* and neem cake reduced the severity of root-knot disease and enhanced the growth, biomass and flower yield by 12 to 37% of chamomile.

Chapter 21

# TUBER CROPS

Tuber crops form an important part of the diet in many areas of the tropics and subtropics. They are most important after cereals and grain legumes. They are mostly grown for the carbohydrates stored in underground storage organs viz., roots or rhizomes. In addition, the leaves or leaf stalks of some of the tuber crops are rich in protein, vitamins and minerals and are used for either human consumption or cattle feed. Tuber crops are a major source of food and calories in many tropical countries. The annual tropical root crop production (sweet potato and cassava) is in the range of 10.200 million MT in 0.396 million ha with productivity of 20.5 MT/ha (Table 21.1). Root and tuber crops are predominantly grown in the upland or rainfed areas, mainly by resource poor small and marginal farmers. The major tuber crops include cassava (tapioca), sweet potato, colocasia, amorphophyllus, yams and coleus. Cassava and sweet potato are grown to a larger extent in southern and eastern India. Their cultivation has spread to Maharashtra, Gujarat and North-Eastern states as well. The Salem belt in Tamil Nadu and the Samalkot belt in Andhra Pradesh are known for cassava as an industrial crop. Yams and Colocasia are popular in Andhra Pradesh, Tamil Nadu, West Bengal, Uttar Pradesh and Orissa states.

**Table 21.1.** Area, production and productivity of tuber crops in India (2007-08) (Bijay Kumar, 2009)

| Tuber crop | Area ('000 ha) | Production ('000 MT) | Productivity (MT/ha) |
|---|---|---|---|
| Sweet potato | 126 | 1146 | 9.1 |
| Cassava (Tapioca) | 270 | 9054 | 33.5 |
| Total | 396 | 10200 | 20.5 |

Tuber crops, especially cassava can be successfully substituted as an alternative source of feed for livestock in both tropical and subtropical countries. Cassava has also been used as a substrate for the production of ethyl alcohol. The tropical root crops, in general, have a great potential in meeting basic food and energy needs of the developing world. Tuber crops are very efficient in solar energy transfer, having a clear superiority over cereals in biological efficiency and are comparatively free from severe pests and diseases. Cassava and sweet potatoes account for about 30% of the total production of root crops from developing countries. The potential of root crops as nutritionally rich sources of ß-carotene, anti-oxidants, dietary fibre and minerals like calcium have begun to be recognized as a result of the multifarious research programmes worldwide.

## 21.1. SWEET POTATO, *IPOMEA BATATAS*

### 21.1.1. High yielding varieties

**Table 21.2.** High yielding varieties of sweet potato

| High yielding varieties | Yield (MT/ha) | Developed at |
|---|---|---|
| H-41 (hybrid) | 20-25 | CTCRI, Trivandrum |
| H-42 (hybrid) | 22-25 | CTCRI, Trivandrum |
| H-268 (Varsha) (hybrid) | 25 | CTCRI, Trivandrum |
| Sree Nandini | 20-25 | CTCRI, Trivandrum |
| Sree Vardhini | 25 | CTCRI, Trivandrum |
| Sree Rethna (hybrid) | 20-26 | CTCRI, Trivandrum |
| Sree Bhadra | 20-27 | CTCRI, Trivandrum |
| Sree Arun (hybrid) | 20-28 | CTCRI, Trivandrum |
| Sree Varun (hybrid) | 20-28 | CTCRI, Trivandrum |
| VL Sakarkand-6 | 20 | VPKAS, Almora |
| VL Sakarkand-19 | 20-25 | VPKAS, Almora |
| CO-1 | 28 | TNAU, Coimbatore |
| CO-2 | 32 | TNAU, Coimbatore |
| CO-3 | 43.68 | TNAU, Coimbatore |
| CO.CIP-1 | 31.76 | TNAU, Coimbatore |
| Rajendra Sakarkand-5 (hybrid) | 20 | RAU, Dholi |
| Samrat | 20 | |
| Rajendra Sakarkand-35 | 25 | RAU, Dholi |
| Rajendra Sakarkand-43 | 20 | RAU, Dholi |
| Rajendra Sakarkand-47 | 30 | RAU, Dholi |
| Rajendra Sakarkand-92 | 25 | RAU, Dholi |
| Kiran | 22 | ANGRAU, Hyderabad |
| RNSP-1 | 22 | ANGRAU, Hyderabad |
| Bidhan Jagannath | 21-24 | BCKV, Kalyani |
| Birsa Sakarkand | 28 | BAU, Ranchi |
| Indira Madhur | 22 | IGAU, Raipur |
| IGSP-4 | 25 | IGAU, Raipur |
| IGSP-17 | 28 | IGAU, Raipur |
| Narendra Sakarkand-9 | 26 | NDUAT, Faizabad |
| Narendra Sakarkand-10 | 22 | NDUAT, Faizabad |
| Pusa Safed | 30 | IARI, New Delhi |

### 21.1.2. Integrated nutrient management

Potash fertilizer application increased the tuber yield of sweet potato over the control NP treatment by 5.4-6.7 MT/ha when organic manure was applied. Without organic manure application, the yield increment due to K addition was more significant and reached 7.6-8.6 MT/ha. Moreover, the share of large-sized tubers was tended to be higher as resulted from potash fertilizer application. At the same time, the share of small-sized tubers noticeably decreased when using K. The yield of sweet potato aerial parts mostly composed of vines (vines may be utilized as animal feed) was also considerably improved as a result of K application: the increment in vine yield due to K application was 3.3-5.3 MT/ha.

Application of 2/3 of the recommended dose of N + *Azospirillum* biofertilizer as vine dipping at 2 kg/ha could produce the maximum tuber yield of sweet potato (17.83 MT/ha) followed by 1/3 of the recommended dose of N (16.79 MT/ha). The net return and cost-benefit ratio are higher for the treatment receiving 2/3 of the recommended dose of N + *Azospirillum* biofertilizer as vine dipping at 2 kg/ha, which has fetched an additional income of Rs. 16,508/ha over that from control (Dutta *et al.*, 2002).

### 21.1.3. Integrated disease management

#### (i) Wilt/stem rot, *Fusarium oxysporum* f. sp. *batatas*

Stem rot is widespread in USA and 10-50% losses in susceptible varieties may occur. Losses up to 95% have been recorded.

Fusarium wilt of sweet potato has been reported to be successfully controlled by preinoculation of the cuttings with non-pathogenic strains of *Fusarium oxysporum* immediately before the cuttings are transplanted into the field (Ogawa and Komada, 1984, 1986). It is believed that the non-pathogenic strains not only compete with the pathogenic ones in the rhizosphere and for infection sites, but they also enhance the resistance of the host towards the pathogenic strains. The level of control of Fusarium wilt of sweet potato using non-pathogenic strains of *Fusarium* was equivalent to that obtained with benomyl (Table 21.3). This method has been applied in commercial fields in Japan.

**Table 21.3.** Effect of preinoculation with non-pathogenic *Fusarium oxysporum* on the incidence of Fusarium wilt and on yield of sweet potato

| Treatment | Field A | | | Field B | | |
|---|---|---|---|---|---|---|
| | No. of plants treated | % of plants diseased | Yield (kg/100 plants) | No. of plants treated | % of plants diseased | Yield (kg/100 plants) |
| Non-pathogenic *Fusarium* | 97 | 6.2 | 60.0 | 300 | 2.0 | 89.2 |
| Benomyl at 0.1% | 97 | 5.2 | 60.8 | 500 | 1.5 | 95.6 |
| Control | 89 | 58.3 | 33.5 | 150 | 39.3 | 45.6 |

## 21.2. CASSAVA (TAPIOCA), *MANIHOT ESCULENTA*

### 21.2.1. High yielding varieties

**Table 21.4.** High yielding varieties/hybrids of cassava (Edison, 2005)

| Variety/hybrid | Average yield (MT/ha) | Developed at |
|---|---|---|
| H-97 (hybrid) | 25-35 | CTCRI, Trivandrum |
| H-165 (hybrid) | 33-38 | CTCRI, Trivandrum |
| H-226 (hybrid) | 30-35 | CTCRI, Trivandrum |
| Sree Sahya (hybrid) | 35-40 | CTCRI, Trivandrum |
| Sree Visakham (hybrid) | 35-38 | CTCRI, Trivandrum |
| Sree Prakash | 30-35 | CTCRI, Trivandrum |
| Sree Harsha (hybrid) | 35-40 | CTCRI, Trivandrum |
| Sree Jaya | 26-30 | CTCRI, Trivandrum |
| Sree Vijaya | 25-28 | CTCRI, Trivandrum |
| Sree Rekha (hybrid) | 45-48 | CTCRI, Trivandrum |
| Sree Prabha (hybrid) | 40-45 | CTCRI, Trivandrum |
| Sree Padmanabha | 38 | CTCRI, Trivandrum |
| CO -1 | 30 | TNAU, Coimbatore |
| CO-2 | 35-37 | TNAU, Coimbatore |
| CO-3 | 42.58 (irrigated)<br>27.31 (rainfed) | TNAU, Coimbatore |
| CO-4 | 50.6 | TNAU, Coimbatore |
| CO(Tp) 5 | 38 | TNAU, Coimbatore |
| MVD 1 | 34.5 | Tapioca Expt. Stn., Mulluvadi, Tamil Nadu (State Dept. of Hort.) |

### 21.2.2. Integrated nutrient management

The research on nutrient management in cassava revealed that application of recommended dose of NPK (100: 50: 100 kg/ha) + FYM at12.5 MT/ha was on par with the application of half the recommended dose of 'P', with full dose of N and K and combined application of AMF and phosphobacteria (Saraswathi, 2005) (Table 21.5).

**Table 21.5.** Effect of biofertilizer in cassava

| Treatment | Tuber yield (MT/ha) |
|---|---|
| Recommendation dose of NPK (100:50:100 kg NPK/ha) + FYM (12.5 MT/ha) | 29.70 |
| Half the recommendation dose of P + AMF + Full recommendation dose of N&K | 27.10 |
| CD (P = 0.05) | 0.74 |

At N rates of 40 and 80 kg/ha, the highest K application rate of 160 kg $K_2O$/ha increased the cassava tuber yield over the control (zero K) treatment by 32% and 46%, respectively.

## 21.3. COLOCASIA, *COLOCASIA ESCULENTA*

### 21.3.1. High yielding varieties

**Table 21.6.** High yielding varieties of colocasia

| High yielding varieties | Yield (MT/ha) | Developed at |
|---|---|---|
| Narendra Bunda-1 | 35-40 | NDUAT, Faizabad |
| NDB-21 | 45-50 | NDUAT, Faizabad |
| NDB-03 | 35-40 | NDUAT, Faizabad |
| CO 1 | 24.30 | TNAU, Coimbatore |

### 21.3.2. Integrated disease management

#### (i) Leaf blight, *Phytophthora colocasiae*

Two sprays of Ridomil MZ at 20 days interval starting just after the appearance of the disease was found significantly superior to other treatments in terms of % disease incidence (12.06%) and % disease index (8.13%) yielding 20.3 MT/ha of corms which was at par with Indofil M-45 and Captaf (Ojha and Yadav, 2000) (Table 21.7) .

**Table 21.7.** Field efficacy of fungicides for the control of leaf blight disease and yield of colocasia

| Fungicide | Conc. (%) | Disease incidence (%) | Disease index (%) | Corm yield (MT/ha) |
|---|---|---|---|---|
| Indofil M-45 | 0.25 | 21.88 | 21.25 | 19.028 |
| Blitox-50 | 0.30 | 20.31 | 18.12 | 17.500 |
| Captaf | 0.20 | 18.75 | 17.75 | 19.514 |
| Ridomil MZ | 0.20 | 12.06 | 8.13 | 20.347 |
| Bordeaux mixture | 0.50 | 23.44 | 20.00 | 17.222 |
| Control | --- | 46.88 | 32.50 | 14.790 |
| CD (P = 0.05) | --- | 5.505 | 4.487 | 0.9465 |

## 21.4. YAMS, *DIOSCOREA* SPP.

### 21.4.1. High yielding varieties

**Table 21.8.** High yielding varieties/hybrids of yams

| Yams | Variety/hybrid | Average yield (MT/ha) | Developed at |
|---|---|---|---|
| Greater yam | Sree Keerthi | 25-30 | CTCRI, Trivandrum |
| | Sree Roopa | 25-30 | CTCRI, Trivandrum |
| | Sree Shilpa (hybrid) | 28 | CTCRI, Trivandrum |
| | Sree Karthika | 30 | CTCRI, Trivandrum |
| | Orissa Elite | 25 | CTCRI, Trivandrum |
| | CO 1 | 44.8 | TNAU, Coimbatore |
| White yam | Sree Priya | 35-40 | CTCRI, Trivandrum |
| | Sree Subhra | 35-40 | CTCRI, Trivandrum |
| | Sree Dhanya | 20 | CTCRI, Trivandrum |
| Lesser yam | Sree Latha | 25-30 | CTCRI, Trivandrum |
| | Sree Kala | 20 | CTCRI, Trivandrum |
| | Konkan Kanchan | 18 | Dr. BSKVV, Dapoli |
| Taro | Sree Pallavi | 15-18 | CTCRI, Trivandrum |
| | Sree Rashmi | 15-20 | CTCRI, Trivandrum |
| | Sree Kiran | 17.5 | CTCRI, Trivandrum |
| | Satamukhi | 15 | |
| | Muktakeshi | 20 | CTCRI, Trivandrum |
| | PKS-1 | 20-31 | NDUAT, Faizabad |
| | Indira Arvi-1 | 22 | IGAU, Jagdalpur |
| | CO 1 | 24.3 | TNAU, Coimbatore |
| Elephant foot yam | Sree Padma | 42 | CTCRI, Trivandrum |
| | Sree Athira | 41 | CTCRI, Trivandrum |
| | Gajendra | 50 | ANGRAU, Hyderabad |
| | Bidan Kusum | 50 | BCKV, Kalyani |
| | Narendra Asha | 65 | NDUAT, Faizabad |
| Chinese potato | Sree Dhara | 25 | CTCRI, Trivandrum |
| | Suphala | 15.93 | KAU, Thrissur |
| | CO 1 | 31.93 | TNAU, Coimbatore |
| Yam bean | Rajendra Mishrikand-1 | 35 | RAU, Dholi |

## Chapter 22

# PLANTATION CROPS

Plantation crops serve a variety of human needs such as food, oil, industrial raw materials, beverages and confectionary items. They generate huge employment opportunities directly or indirectly to several million people in their production, processing, marketing and international trade sectors. Plantation crops are a large group of crops grown over an area of 3.226 million ha with production of 12.045 million MT in India (Bijay Kumar, 2009) (Tables 22.1 and 22.2). The plantation crops contributed 5.82% of total horticulture production, in 2007-2008. India is the largest producer and consumer of areca nut, cashew nut, tea, third largest producer of coconut, fourth largest producer and consumer of rubber and sixth largest producer of coffee in the world. While the major plantation crops include coconut, areca nut, oil palm, cashew, coffee, tea and rubber, minor plantation crops are cocoa and betel vine. They play an important role in export as well as domestic requirement, employment generation and poverty alleviation particularly in rural sector. The annual income generated is more than Rs. 5,52,590 million and export earning is approximately Rs. 93,350 millions.

**Table 22.1.** Crop-wise area and production of plantation crops in India (2007-08) (Bijay Kumar, 2009)

| Plantation crops | Area ('000 ha) | Production ('000 MT) | Productivity (MT/ha) | % share of India production |
|---|---|---|---|---|
| Coconut | 1939.9 | 19893.9 | 5.6 | 90.4 |
| Cashew nut | 868.0 | 665.0 | 0.8 | 5.5 |
| Areca nut | 386.6 | 476.0 | 1.2 | 4.0 |
| Cocoa | 31.8 | 10.6 | 0.3 | 0.1 |
| Total | 3226.4 | 12045.5 | 3.7 | |

**Table 22.2.** State-wise area, production and productivity of plantation crops in India (2007-08) (Bijay Kumar, 2009)

| State | Area in '000 ha | Production in '000 MT | Productivity in MT/ha | % share of India production |
|---|---|---|---|---|
| Kerala | 1073.7 | 43.59.9 | 4.0 | 36.2 |
| Tamil Nadu | 503.9 | 3810.6 | 7.5 | 31.6 |
| Karnataka | 679.2 | 1401.0 | 2.0 | 11.6 |
| Andhra Pradesh | 289.0 | 1021.2 | 3.5 | 8.5 |
| Maharashtra | 190.2 | 334.1 | 1.7 | 2.8 |
| Orissa | 182.0 | 279.8 | 1.5 | 2.3 |
| West Bengal | 44.1 | 279.1 | 6.3 | 2.3 |
| Assam | 104.3 | 186.1 | 1.7 | 1.5 |
| Goa | 82.2 | 120.8 | 1.4 | 1.0 |
| Gujarat | 20.4 | 99.2 | 4.8 | 0.8 |
| Andaman & Nicobar | 25.5 | 66.9 | 2.6 | 0.6 |
| Others | 71.8 | 86.7 | 1.2 | 0.7 |
| Total | 3226.3 | 12045.5 | --- | |

There is a good scope for switching over the production of plantation crops viz., coffee, tea, coconut and areca nut to organic farming, as it is possible to realize higher returns from the unit quantity exported when grown organically. The plantation crops are generally grown in ecologically fragile hilly tracts, adopting organic farming methods would protect environment and also prevent contamination thereby producing products of high nutritional quality. Further, the plantation crops are amenable for organic farming as they produce huge amount of waste biomass for recycling, which can meet major portion of nutrients required.

## Export of plantation produce

India is the largest producer in the world of cashew nuts, coconuts and tea. India is largest producer, importer, processor, exporter and second largest consumer of cashew. Indian exports during 2006-2007 were US $ 866 million in tea and coffee (Table 22.3).

**Table 22.3.** Export of plantation crops (in US $)

| Commodity | Apr-Mar 2006 | Apr-Mar 2007 | % Growth | % Share |
|---|---|---|---|---|
| Tea | 390.92 | 432.00 | 10.51 | 0.34 |
| Coffee | 358.83 | 434.77 | 21.16 | 0.34 |
| Cashew | 585.76 | 553.41 | -5.52 | 0.44 |

India is the largest producer and consumer of tea in the world. During 2006-07, India produced 0.95 million MT of tea of which 0.218 million MT was exported. Tea accounts for Rs. 80 billion in turnover in India.

India produced 0.29 million MT of coffee accounting for 4% of the global production. Over 80% of the coffee produced in India is exported.

India is the fourth largest producer of natural rubber with a share of 8.8% in world production in 2006. The yield/ha in India at 1879 kg/ha, was amongst the highest in the world. India produced 0.852 million MT of natural rubber during 2006-07 and exported 56545 MT.

## 22.1. COCONUT, *COCOS NUCIFERA*

### 22.1.1. High yielding varieties

**Table 22.4.** High yielding varieties/F1 hybrids of coconut (Ragagopal *et al.*, 2005)

| Coconut variety/hybrid | Yield (nuts/ palm/year) | Developed at |
|---|---|---|
| West Coast Tall (WCT) (standard variety) | 80 | |
| Lakshadweep Ordinary | 97 | |
| Benawali Green Round | 151 | |
| Philippines Ordinary | 110 | |
| Chandra Sankara (COD x WCT) | 116 | CPCRI, Kasargod |
| Chandra Laksha (COD x WCT) | 109 | CPCRI, Kasargod |
| Chandra Kalpa | 97 | CPCRI, Kasargod |
| Kera Sankara (WCT x COD) | 108 | CPCRI, Kasargod |
| Laksha Ganga (LCT x GBGD) | 108 | CPCRI, Kasargod |
| Ananda Ganga (ADOT x GBGD) | 95 | CPCRI, Kasargod |
| Kalpa Pratibha | 91 | CPCRI, Kasargod |
| Kalpa Mitra | 80 | CPCRI, Kasargod |
| Kalpa Dhenu | 86 | CPCRI, Kasargod |
| Kera Sagara | 99 | KAU, Vellanikkara |
| Kera Ganga (WCT x GBGD) | 100 | KAU, Vellanikkara |
| Kera Sankara | 106 | KAU, Vellanikkara |
| Kera Chandra | 110 | KAU, Vellanikkara |
| Kera Sree (WCT x MYD) | 130 | KAU, Vellanikkara |
| Kera Soubhagya (WCT x SSA) | 130 | KAU, Vellanikkara |
| VHC-1 (ECT x CGD) | 98 | CRS, TNAU, Veppankulam |
| VHC-2 (ECT x MYD) | 107 | CRS, TNAU, Veppankulam |
| VHC-3 (ECT x MOD) | 156 | CRS, TNAU, Veppankulam |
| Godavari Ganga (ECT x GBGD) | 140 | ANGRAU, Hyderabad |
| Double Century | 90-200 | ANGRAU, Hyderabad |
| Kalyani Coconut-1 | 80 | BCKVV, Kalyani |

**Table 22.5.** Coconut varieties resistant to diseases

| Coconut variety/hybrid | Yield (nuts/palm/ year) | Resistance to | Developed at |
|---|---|---|---|
| Kalpa Raksha | 65 (2.44 MT/ha of copra) | Root wilt disease (field resistance) | CPCRI, Kasargod |

### 22.1.2. Fertigation

Application of 500: 320: 1200 g NPK/palm/year through fertigation at monthly intervals in the form of water soluble fertilizers is recommended for coconut for getting high productivity and returns (Khan *et al.*, 2001) (Table 22.6).

**Table 22.6.** Productivity of coconut and levels of fertigation as influenced by sources and levels of fertilizers

| Treatment | No. of nuts/palm |
|---|---|
| Normal fertilizers 80% + Drip irrigation | 81 |
| Normal fertilizers 100% + Drip irrigation | 84 |
| Normal fertilizers 80% Fertigation | 95 |
| Normal fertilizers 100% Fertigation | 110 |
| Water soluble fertilizers 80% Fertigation | 103 |
| Water soluble fertilizers 100% Fertigation | 115 |
| CD (P = 0.05) | 24.27 |

### 22.1.3. Integrated nutrient management

In a long term fertilizer experiment in red loam soil, Wahid *et al.* (1988) recorded significantly higher nut yield with potassium application. Further, they observed that early bearing was also achieved with increased levels of K application. The yield was 19, 68 and 77 nuts palm-1 year-1 in the 21$^{st}$ year after planting under no fertilizer, 450 g $K_2O$ and 900 g $K_2O$/palm/year, respectively.

In a recent study at Veppankulam in Tamil Nadu, potash response to coconut in soils adequately supplied with potassium has been established (Table 22.7) (AICRPP, 2000).

**Table 22.7.**Response of coconut to application of potash (selected treatments)

| Treatment | Yield (Nuts/tree) |
|---|---|
| $N_0$ $P_0$ $K_0$ | 108 |
| $N_0$ $P_0$ $K_1$ (1000 g) | 126 |
| $N_1$ (500g) $P_0$ $K_1$ (1000 g) | 149 |
| $N_0$ $P_0$ $K_2$ (2000 g) | 134 |
| $N_1$ (500g) $P_0$ $K_2$ (2000 g) | 154 |

| | |
|---|---|
| $N_1$ (500g) $P_1$ (250 g) $K_1$ (1000 g) | 159 |
| $N_2$ (1000 g) $P_2$ (500 g) $K_2$ (2000 g) | 201 |

A field experiment with var. Chandra Sankara hybrid (Chawghat Orange Dwarf x West Coast Tall) was started in 1988 from the seedling stage with different combinations of N, P and K. Results of 14 years of fertilizer application in different combinations indicated that a combination of 500 g N with 250g $P_2O_5$ and 1000 g $K_2O$ per palm per year produced a yield of 155 nuts per palm compared to 49 nuts per palm under control treatment. No further increase in yield was observed when the fertilizer dose was doubled. Further an increase in 11 nuts was recorded when the N dose was increased to 1000g N per palm.

VHC-2 hybrid palms with recommended dose of fertilizer application (N-560g, $P_2O_5$-320g, $K_2O$-1200g/plant/year) when brought under drip irrigation equal to 100% Eo, yielded 130-140 and in situations of water scarcity 122 and 89 nuts palm/year with drip irrigation at 66% Eo and 33% Eo respectively (Table 22.8). The corresponding quantity of water applied to realize the yields were 13591, 8970 and 4484 litres palm/year. With drip irrigation at 66%, 100% Eo and basin irrigation, the palms were in sufficiency range of nutrition for N and P and in a slightly higher status for K as judged by leaf analysis. The benefit cost ratio for drip irrigation equal to 100% Eo was 2.11 (Venkitaswamy and Khan, 2004).

**Table 22.8.** Summary of yield parameters influenced by different levels of irrigation to coconut VHC-2 hybrid, Aliyarnagar (Mean values for four years)

| Treatment | Water applied/palm (litres) * | Nuts/palm | % increase |
|---|---|---|---|
| No irrigation | --- | 68 | --- |
| Drip 33 % Eo | 4485 | 89 (+21) | 31 |
| Drip 60 % Eo | 8790 | 122 (+54) | 79 |
| Drip 100 % Eo | 11359 | 147 (+79) | 116 |
| Basin irrigation | 13533 | 146 (+78) | 114 |

*Exclusive of rainfall contribution of 5848 litres

To evolve appropriate technology of INM to improve and sustain productivity of coconut groves utilizing the resources available in different areas, country wide experiments were started under the aegis of the AICRPP during 1996 and the results are presented in Table 22.9.

**Table 22.9.** Yield of coconut under integrated nutrient management

| Treatment | Yield nuts/palm | | | |
|---|---|---|---|---|
| | Aliyarnagar * Tamil Nadu | Arsikere* Karnataka | Veppankulam* Tamil Nadu | Ratnagiri* Maharashtra |
| Farmers practice** | 171 | 60 | 89 | 70 |
| Chemical fertilizers(CF) | 180 | 81 | 113 | 91 |

| | | | | |
|---|---|---|---|---|
| Composted coir pith (CCP) | 178 | --- | 101 | 71** |
| 50 % CCP + 50 % CF | 174 | --- | 99 | 88** |
| Neem cake + Bone meal + Ash | 159 | 93 | 99 | 88 |
| 50 % CCP | --- | --- | 96 | --- |
| CD (P = 0.05) | NS | 24.4 | 21.76 | 6.99 |

***Aliyarnager:** Red Sandy loam, **Arsikere:** Red sandy loam (semi arid), **Veppankulam:** Alluvial soil, **Ratnagiri:** Coastal alluvium (sandy loam).
** : Coconut wastes cut and spread instead of CCP.

### 22.1.4. Intercropping

In a trial conducted to evaluate economic feasibility of raising turmeric, ginger and elephant foot yam along with WCT coconut garden, the maximum profit was received from turmeric (Table 22.10).

**Table 22.10.** Yield performance of intercrops in coconut garden

| Intercrop | Nut yield (MT/ha) | Net profit (Rs./ha) |
|---|---|---|
| Elephant foot yam | 5.233 | 458 |
| Turmeric | 4.441 | 2597 |
| Zinger | 2.426 | 611 |

CO-3 grass grown as intercrop in coconut garden, not only conserved the soil and water resources but also produced fodder at 100 MT/ha/year in seven harvests. Coconut yield also increased three times after imposing the treatments (Dhanapal *et al.*, 2006).

### 22.1.5. Integrated disease management

#### (i) Basal stem rot or Thanjavur wilt, *Ganoderma lucidum, G. applanatum*

Field trials carried out at Veppankulam, Tamil Nadu by soil application of *Azospirillum*, phosphobacteria and *Gigaspora calospora*, along with 5 to 10 kg FYM/palm/year reduced the disease index and increased the nut yield (Table 22.11) (Nambiar, 1996).

**Table 22.11.** Effect of biofertilizers on basal stem rot disease and nut yield of coconut

| Treatment/Dose | Disease index | Nut yield/palm |
|---|---|---|
| *Azospirillum* at 200 g/palm | 56.5 | 90 |
| Phosphobacteria at 200 g/palm | 9.2 | 102 |
| *Gigaspora calospora* at 500 g/palm | 49.8 | 89 |

| | | |
|---|---|---|
| Control | 74.3 | 76 |
| CD (P = 0.05) | 4.5 | 3 |

## 22.2. ARECA NUT, *ARECA CATECHU*

### 22.2.1. High yielding varieties

**Table 22.12.** High yielding varieties of areca nut

| High yielding varieties | Yield (kg/palm) | Developed at |
|---|---|---|
| Mangala | 3.00 | CPCRI, Kasargod |
| Sumangala | 3.28 | CPCRI, Kasargod |
| Sreemangala | 3.18 | CPCRI, Kasargod |
| Mohitnagar | 3.67 | CPCRI, Kasargod |
| Samrudhi | 4.34 | CPCRI, Kasargod & CARI, Port Blair |
| Swarnamangala | 3.88 | CPCRI, Kasargod |
| VTLAH-1 | 2.54 | CPCRI, Kasargod |
| SAS-1 | 4.60 | UAS, Dharwad |

## 22.3. CASHEW NUT, *ANACARDIUM OCCIDENTALE*

### 22.3.1. High yielding varieties

**Table 22.13.** High yielding varieties of cashew

| Cashew variety | Yield (kg/tree) | Developed at |
|---|---|---|
| NRCC Sel-1 | 10.0 | NRC Cashew, Puttur |
| NRCC Sel-2 | 9.0 | NRC Cashew, Puttur |
| Bhaskara | 10.7 | NRC Cashew, Puttur |
| BPP-4 | 10.5 | AICRP Cashew, Bapatla |
| BPP-6 | 10.5 | AICRP Cashew, Bapatla |
| BPP-8 | 14.5 | AICRP Cashew, Bapatla |
| Bhubaneswar-1 | 10.5 | AICRP Cashew, Bhubaneshwar |
| Chintamani-1 | 7.2 | AICRP Cashew, Chintamani |
| Jhargram-1 | 8.5 | AICRP Cashew, Madakkathara |
| Madakkathara-1 | 13.8 | AICRP Cashew, Madakkathara |
| Madakkathara-2 | 17.0 | AICRP Cashew, Madakkathara |
| K-22-1 | 13.2 | AICRP Cashew, Madakkathara |
| Dhana | 17.5 | AICRP Cashew, Madakkathara |
| Kanaka | 19.0 | AICRP Cashew, Madakkathara |
| Priyanka | 16.9 | AICRP Cashew, Madakkathara |
| Amrutha | 18.4 | AICRP Cashew, Madakkathara |

| | | |
|---|---|---|
| Vengurla-1 | 19.0 | AICRP Cashew, Vengurla |
| Vengurla-4 | 17.2 | AICRP Cashew, Vengurla |
| Vengurla-6 | 13.8 | AICRP Cashew, Vengurla |
| Vengurla-7 | 18.5 | AICRP Cashew, Vengurla |
| VRI-2 | 9.24 | AICRP Cashew, Vridhachalam |
| VRI-3 | 14.19 | AICRP Cashew, Vridhachalam |
| VRI 4 | 18.10 | AICRP Cashew, Vridhachalam |
| Ullal-1 | 16.0 | ARS, Ullal (UAS, Dharwad) |
| Ullal-3 | 14.7 | ARS, Ullal (UAS, Dharwad) |

### 22.3.2. High density planting

It is normally planted at 7.5m x 7.5m or 8m x 8 m (100-177 plants/ha). In case of high density planting, 312-625 or higher grafts/ha are planted at a closer spacing of 4m x 4m or 6m x 6m or 5m x 5m in the initial stage followed by thinning out in the later years thereby maintaining a spacing of 8m x 8m in the 11$^{th}$ year. It gives rise to four-fold increase in yield up to 6 years and about 2.5 fold increases up to 12 years.

High density planting with spacing of 4 x 4m accommodating 625 plants/ha resulted in a yield increase by 2.5 times over normal spacing of 8 x 8m in the initial 10 years (Bhat and Venkata Kumar, 2005).

High density planting with cashew cv. Madakkathara showed a yield increase of 92 and 116% with 400 and 500 trees/ha as compared to 200 trees/ha. Land utilization efficiency was significantly higher under high density planting (Mathew and Mini, 2006).

### 22.3.3. Micro irrigation

Irrigation at 75% PE recorded the highest yield of nuts (1.44 MT/ha) which was on par with irrigation at 50% PE (1.43 MT/ha) through subsurface irrigation by diffuser system (Sawke *et al.*, 2006).

### 20.3.4. Fertigation

Application of 400: 100: 100 g NPK/tree/year in the form of water soluble fertilizers through fertigation in four equal splits is recommended for cashew to get high productivity and returns (Khan *et al.*, 2001) (Table 22.14).

**Table 22.14.** Productivity of cashew nut and levels of fertigation as influenced by sources and levels of fertilizers

| Treatment | Cashew nut yield (kg/tree) |
|---|---|
| Normal fertilizers 100% + Rain fed | 0.44 |
| Normal fertilizers 100% + Drip irrigation | 1.29 |
| Normal fertilizers 80% Fertigation | 1.70 |
| Normal fertilizers 100% Fertigation | 1.70 |

| | |
|---|---|
| Water soluble fertilizers 80% Fertigation | 2.34 |
| Water soluble fertilizers 100% Fertigation | 2.28 |

### 22.3.5. Integrated nutrient management

N applied at 200 kg/ha recorded the maximum cashew nut yield (1.91 MT/ha, 1290 nuts/tree) and was significantly superior to N applied at 100 kg/ha (1.44 MT/ha, 949 nuts/tree) (Sawke *et al.*, 2006).

As compared with the NP treatment, K-fertilizer application at 24.5 kg $K_2O$/ha increased the cumulative nut yield of cashew by 8%.

### 22.3.6. Bioregulators

Paclobutrazol, inhibitor of gibberellic acid synthesis, when applied at 8 g a.i./plant/year as soil applicant to 11 year-old plants of cashew nut cv. Vengurla-3 reduced the plant height (72.9%), canopy spread (65%) and increased the nut yield (73%) as compared to control in the second season of application (Nayak *et al.*, 2006).

## 22.4. COFFEE, *COFFEA ARABICA, C. CANEPHORA*

### 22.4.1. Micro irrigation

Experiments on sprinkler irrigation revealed that irrigation for blossom + backing rains would increase the yield by 27 to 57%, while continuous irrigation during the dry period would improve yields by 47 to 95% in drought susceptible coffee cv. Robusta. Application of 4 to 8 litres of water daily per plant would boost the growth of coffee bushes and improve the yields by 25 to 45% when compared to unirrigated control (Raghuramulu, 1996).

Providing irrigation with micro-sprinklers followed by two way drippers resulted in the highest crop water use efficiency of 4.43 and 2.86 kg/cm with an yield of 6.2 kg/plant and 4.0 kg/plant, respectively. The yield increase due to micro-sprinkler and two way emitter treatments was 196% and 59%, respectively when compared to unirrigated control (Raghuramulu, 1996).

### 22.4.2. Fertigation

Application of 240: 180: 240 kg NPK/ha in the form of water soluble fertilizers through fertigation in four equal splits is recommended for Robusta coffee to get high productivity and returns (Khan *et al.*, 2001) (Table 22.15).

**Table 22.15.** Productivity of coffee and levels of fertigation as influenced by sources and levels of fertilizers

| Treatment | Fruit yield (kg/plant) |
|---|---|
| No fertilizers – Control | 2.92 |
| Normal fertilizers 100% + Blossom shower | 3.04 |

| | |
|---|---|
| Normal fertilizers 100% + Drip irrigation | 3.26 |
| Water soluble fertilizers 50% Fertigation | 3.15 |
| Water soluble fertilizers 100% Fertigation | 3.49 |
| Water soluble fertilizers 150% Fertigation | 3.93 |
| Water soluble fertilizers 200% Fertigation | 4.46 |

Coffee plants when supplied with 60: 45: 60 kg NPK/ha applied through fertigation gave highest yield of clean coffee (806.5 kg/ha) as compared to higher dose of 90: 60: 90 kg NPK/ha supplied as soil application. The requirement of fertilizers was reduced by 33% and the yield was doubled (Table 22.16).

**Table 22.16.** Fertigation in coffee cv. Robusta

| Treatment | Cured coffee yield (kg/ha) |
|---|---|
| 90: 60: 90 kg NPK/ha (Soil application) | 378.6 |
| 90: 60: 90 kg NPK/ha (Soil application) + Drip irrigation | 594.7 |
| 90: 60: 90 kg NPK/ha (Fertigation) | 531.7 |
| 60: 45: 60 kg NPK/ha (Fertigation) | 806.5 |
| 45: 30: 45 kg NPK/ha (Fertigation) | 644.0 |
| CD (P = 0.05) | 136.25 |

Application of 120: 90: 120 kg of NPK/ha through drip irrigation resulted in yield of 1277 kg/ha of clean coffee, as compared to soil application of 160: 120: 160 kg of NPK/ha in four split doses per year which produced an yield of 976 kg/ha of clean coffee (Venkataraman *et al.*, 1998).

### 22.4.3. Integrated disease management

#### i) Berry blotch, *Colletotrichum kahawae*

Spraying tank mixtures of organic fungicides (captafol, chlorothalanil, dithianon, prochloraz, benomyl and carbendazim), at half the normal rates, and copper-based fungicides at a reduced rate of 5 kg/ha were effective and resulted in better yields than when using the organic fungicides by themselves. The tank mixtures are less costly due to the reduced rates and inclusion of the much cheaper copper fungicides and hence their use improves the cost benefit ratio (Table 22.17). In addition, the tank mixtures are able to control coffee leaf rust on the same spray programme, and bacterial blight of coffee with a few additional copper sprays.

**Table 22.17.** Comparative efficacy of straight fungicides and tank mixtures on coffee yield and cost benefit ratio.

| Chemical | Conc. % | Yield (kg/ha) | Yield benefit (kg/ha) | Cost of chemical/ha (Rs.)* | Cost: benefit ratio** |
|---|---|---|---|---|---|
| Captafol 80% WP | 0.4 | 1169 | 735 | 6209 | 1: 4.73 |
| Captafol + Cupric hydroxide | 0.2 + 0.5 | 1163 | 729 | 5567 | 1: 5.24 |
| Chlorothalanil | 0.4 | 741 | 307 | 7920 | 1: 1.55 |
| Chlorothalanil + Cupric hydroxide | 0.2 + 0.4 | 889 | 455 | 6209 | 1: 2.93 |
| Anilazine | 0.4 | 852 | 418 | 5632 | 1: 2.97 |
| Anilazine + Cupric hydroxide | 0.2 + 0.4 | 1011 | 577 | 5236 | 1: 4.41 |
| Unsprayed | --- | 434 | --- | --- | --- |

* Market prices for 1988

** Based on assumption that market price for coffee was £ 2000/tonne but excludes labour costs.

## 22.5. TEA, *CAMELLIA SINENSIS*

### 20.5.1. Integrated nutrient management

Compared to the estate practice (application of inorganic sources of fertilizers), the application of Digested Coir Compost (DCC) as well as biofertilizers alone or in combination was helpful in increasing the green leaf yield at a decreased level of recommended dose of estate practice (Table 22.18) (Rajalingam and Kumar, 2001).

**Table 22.18.** Green leaf yield per plot of 100 bushes

| Variety | Best treatment plot | Yield in best treatment plot (kg) | Yield in control plot (kg) |
|---|---|---|---|
| Assam Jat | 75% EP* + Biofertilizer | 787.3 | 587.1 |
| China Jat | 75% EP + DCC** (3 MT/ha) + Biofertilizer | 648.6 | 557.8 |
| ATK | 100% EP + DCC (3 MT/ha) | 317.9 | 294.1 |

* Estate practice, ** Digested coir compost

## 22.6. BETEL VINE, *PIPER BETEL*

### 22.6.1. High yielding varieties

**Table 22.19.** High yielding varieties of betel vine

| High yielding varieties | Leaf yield (million/ha) | Developed at |
|---|---|---|
| SGM-1 | 10.9 | ARS, TNAU, Sirugamani |
| SGM(BV) 2 | 14.9 | ARS, TNAU, Sirugamani |

**Table 22.20.** Betel vine varieties resistant to diseases

| High yielding varieties | Resistance to | Developed at |
|---|---|---|
| SGM 2 | Bacterial leaf spot (T) | ARS, TNAU, Sirugamani |
| Krishna Pan | Foot rot (MR) | MPKV, Sangli |

### 22.6.2. Integrated management of disease complex

#### (i) Root-knot nematode, *Meloidogyne incognita* and foot rot, *Phytophthora capsici* disease complex

The population of root-knot nematode had been found to be positively correlated with *Phytophthora* wilt disease incidence in diseased betel vine gardens.

Combined application of *Pseudomonas* sp. (Pfbv 22) and *Bacillus* sp. (Bbv 57) gave significant reduction in nematode infestation (gall index, no. of egg laying females and soil population), wilt disease incidence and increase in leaf yield. The treatment also enhanced the biochemical markers responsible for induced systemic resistance such as peroxidase, polyphenol oxidase and phenylalanine ammonia lyase (Jonathan *et al.*, 2006) (Table 22.21).

**Table 22.21.** Effect of rhizobacterial formulations on leaf yield, nematode and wilt incidence in betel vine under glasshouse conditions

| Treatment/Dose ($2.5 \times 10^8$ cfu/g) | No. of leaves/ vine | Gall index (0-5 scale) | Wilt index (0-5 scale) |
|---|---|---|---|
| *Pseudomonas* spp. - Pfbv 22 | 195 | 3.0 | 2.4 |
| *Bacillus* spp. - Bbv 57 | 201 | 3.0 | 2.6 |
| *P. fluorescens* - Pf 1 | 190 | 3.5 | 2.4 |
| Pfbv 22 + Bbv 57 | 225 | 2.5 | 1.7 |
| Pfbv 22 + Pf 1 | 197 | 3.0 | 2.2 |
| Bbv 57 + Pf 1 | 190 | 3.5 | 2.4 |
| Pfbv 22 + Bbv 57 + Pf 1 | 191 | 3.0 | 2.4 |
| Metalaxyl (0.2%) + Carbofuran (2 g/vine) | 211 | 2.4 | 1.9 |
| Control | 163 | 5.0 | 3.3 |
| CD (P = 0.05) | 6.5 | 0.5 | 0.4 |

## 22.7. COCOA, *THEOBROMA CACAO*

### 22.7.1. High yielding varieties

**Table 22.22.** High yielding varieties of cocoa

| High yielding varieties | Yield (kg/ha) | Developed at |
|---|---|---|
| VTLCC-1 | 911 | CPCRI, Kasargod |

| | | |
|---|---|---|
| VTLCH-1 | 856 | CPCRI, Kasargod |
| CCRP-1 | 2.5 kg dry bean/tree | KAU, Vellanikkara |
| CCRP-2 | 2.4 kg dry bean/tree | KAU, Vellanikkara |
| CCRP-3 | 2.9 kg dry bean/tree | KAU, Vellanikkara |
| CCRP-4 | 3.9 kg dry bean/tree | KAU, Vellanikkara |
| CCRP-5 | 1.7 kg dry bean/tree | KAU, Vellanikkara |
| CCRP-6 | 3.1 kg dry bean/tree | KAU, Vellanikkara |
| CCRP-7 | 4.0 kg dry bean/tree | KAU, Vellanikkara |
| CCRP-8 | 3.5 kg dry bean/tree | KAU, Vellanikkara |
| CCRP-9 | 3.1 kg dry bean/tree | KAU, Vellanikkara |
| CCRP-10 | 3.5 kg dry bean/tree | KAU, Vellanikkara |

**Table 22.23.** Cocoa varieties resistant to diseases

| HYV | Yield (kg/ha) | Resistance to disease | Developed at |
|---|---|---|---|
| VTLCC-1 | 911 | Vascular streak dieback (T) | CPCRI, Kasargod |
| VTLCH-1 | 856 | Vascular streak dieback (T) | CPCRI, Kasargod |
| CCRP-1 | 2.5 kg dry bean/tree | Vascular streak dieback (T) | KAU, Vellanikkara |
| CCRP-2 | 2.4 kg dry bean/tree | Vascular streak dieback (T) | KAU, Vellanikkara |
| CCRP-3 | 2.9 kg dry bean/tree | Vascular streak dieback (T) | KAU, Vellanikkara |
| CCRP-4 | 3.9 kg dry bean/tree | Vascular streak dieback (T) | KAU, Vellanikkara |
| CCRP-5 | 1.7 kg dry bean/tree | Vascular streak dieback (T) | KAU, Vellanikkara |
| CCRP-6 | 3.1 kg dry bean/tree | Vascular streak dieback (T) | KAU, Vellanikkara |
| CCRP-7 | 4.0 kg dry bean/tree | Vascular streak dieback (T) | KAU, Vellanikkara |
| CCRP-8 | 3.5 kg dry bean/tree | Vascular streak dieback (T) | KAU, Vellanikkara |
| CCRP-9 | 3.1 kg dry bean/tree | Vascular streak dieback (T) | KAU, Vellanikkara |
| CCRP-10 | 3.5 kg dry bean/tree | Vascular streak dieback (T) | KAU, Vellanikkara |

# Chapter 23

# SPICE CROPS

From time immemorial, India is regarded by the rest of the world as the "magic land of spices". Spices are natural plant products used to improve flavour, aroma, taste and colour of food products, beverages, liquors, pharmaceutical, cosmetic and perfumery products. No other country in the world has such a diverse variety of spice crops as in India. Indian spices are also noted for their excellent aroma, flavour and pungency not so easily matched by any other country. Even in minute quantities, spices are a real delight to all our sensory organs, making the food dishes more palatable. Usage of spices in food processing industries is increasing quite fast in all the countries. Hence, the world demand for spices in recent years is continuously on the rise.

India is the largest producer, consumer and exporter of spices in the world. India is the largest producer in the world of ginger, turmeric and black pepper. The present annual production of spices in the country is 4.103 million MT (Bijay Kumar, 2009) (Tables 23.1 and 23.2) valued at US$ 1,500 million. India's share of the world spices trade is estimated at 45 to 50% by volume and 25% by value. The spices contributed 1.98% of total horticulture production in 2007-2008. About 8.5% of India's export earnings from agricultural and allied products come from spices. Indian spices flavour foods in over 134 countries. The important spice crops include black pepper, cardamom, ginger, garlic, turmeric and chilli.

**Table 23.1.** Crop-wise area and production of spice crops in India (2007-08) (Bijay Kumar, 2009)

| Spice crop | Area in '000 ha | Production in '000 MT | Productivity in MT/ha | % share of India production |
|---|---|---|---|---|
| Chilli | 765 | 1244 | 1.6 | 30 |
| Turmeric | 180 | 830 | 4.6 | 20 |
| Garlic | 168 | 810 | 4.8 | 20 |
| Ginger | 106 | 376 | 3.5 | 9 |

| | | | | |
|---|---|---|---|---|
| Coriander | 398 | 243 | 0.6 | 6 |
| Tamarind | 60 | 192 | 3.2 | 5 |
| Cumin | 429 | 172 | 0.4 | 4 |
| Black pepper | 242 | 68 | 0.3 | 2 |
| Fennel | 49 | 62 | 1.3 | 2 |
| Fenugreek | 55 | 55 | 1.0 | 1 |
| Others | 152 | 51 | 0.3 | 1 |
| Total | 2603 | 4103 | 1.6 | |

**Table 23.2.** State-wise area, production and productivity of major spices in India (2007-08) (Bijay Kumar, 2009)

| State | Area in '000 ha | Production in '000 MT | Productivity in MT/ha | % share of India production |
|---|---|---|---|---|
| Andhra Pradesh | 317.8 | 1235.2 | 3.9 | 30.1 |
| Rajasthan | 556.4 | 520.6 | 0.9 | 12.7 |
| Gujarat | 299.8 | 356.8 | 1.2 | 8.7 |
| Karnataka | 235.2 | 344.9 | 1.5 | 8.4 |
| Tamil Nadu | 126.8 | 279.2 | 2.2 | 6.8 |
| Madhya Pradesh | 194.6 | 249.8 | 1.3 | 6.1 |
| Orissa | 147.0 | 199.2 | 1.4 | 4.9 |
| Uttar Pradesh | 56.8 | 166.9 | 2.9 | 4.1 |
| Kerala | 312.3 | 159.3 | 0.5 | 3.9 |
| West Bengal | 79.0 | 114.4 | 1.4 | 2.8 |
| Maharashtra | 114.3 | 100.2 | 0.9 | 2.4 |
| Meghalaya | 18.4 | 80.9 | 4.4 | 2.0 |
| Others | 144.6 | 295.3 | 2.0 | 7.2 |
| Total | 2603.0 | 4102.7 | 1.6 | |

Since spices form a part of traditional/ethnic medicines, the demand for organically produced spices is growing. Export of organic spices from India has started in right earnest. The country at present exports around 50 MT of different varieties of organic spices. Export will get a significant boost in the coming years as more and more farmers switch to organic methods.

### Export of spice produce

India is the largest producer, consumer and exporter of spices and spice products in the world and produces more than 50 spices. Spice export figures for the year 1996-97 and 2006-07 shows the phenomenal growth of the Indian spice industry. The Indian spice export for the year 1996-97 was 225,000 MT which was worth $350 million; in the year 2006-07 the spice exports touched the figure of 375,000 MT which was worth US$ 800 million. Thus, these figures show the exponential growth of the Indian spice industry in a very short time. Spice exports from India have touched US $ one billion mark during 2007-08. The major destination of Indian spice exports is USA followed by European Union, Sri Lanka, Japan and Middle East countries.

## 23.1. BLACK PEPPER, *PIPER NIGRUM*

### 23.1.1. High yielding varieties

**Table 23.3.** High yielding varieties/hybrids of spice crops

| Variety/hybrid | Average yield (MT/ha) | Developed at |
|---|---|---|
| Panniyur-1 (hybrid) | 1.242 | Pepper Res. Stn., KAU, Panniyur |
| Panniyur-2 | 2.570 | Pepper Res. Stn., KAU, Panniyur |
| Panniyur-3 (hybrid) | 1.953 | Pepper Res. Stn., KAU, Panniyur |
| Panniyur-4 | 1.277 | Pepper Res. Stn., KAU, Panniyur |
| Panniyur-5 | 1.110 | Pepper Res. Stn., KAU, Panniyur |
| Panniyur-6 | 2.117 | Pepper Res. Stn., KAU, Panniyur |
| Panniyur-7 | 1.410 | Pepper Res. Stn., KAU, Panniyur |
| Subhakara | 2.352 | IISR, Calicut |
| Sreekara | 2.677 | IISR, Calicut |
| Panchami | 2.828 | IISR, Calicut |
| Pournami | 2.333 | IISR, Calicut |
| IISR Thevam | 1.787 | IISR, Calicut |
| IISR Malabar Excel | 1.065 | IISR, Calicut |
| IISR Girimunda | 2.112 | IISR, Calicut |
| IISR Shakthi | 5.755 | IISR, Calicut |
| Palode 2 | 2.475 | NRC Oil palm, Regional Stn., Palode |

**Table 23.4.** Resistant varieties/hybrids of spice crops

| Variety/hybrid | yield (MT/ha) | Resistance to disease | Developed at |
|---|---|---|---|
| Pournami | 2.333 | Root-knot nematode (T) | IISR, Calicut |
| IISR Thevam | 1.787 | Foot rot disease (T) | IISR, Calicut |
| IISR Shakthi | 5.755 | Foot rot disease (MR) | IISR, Calicut |

T – Tolerant, MR – Moderately Resistant

### 23.1.2. Integrated nutrient management

An experiment was conducted on eight year old pepper vines Panniyur 1 (P1) and Panniyur 2 (P2) on the INM aspects of pepper at TNAU, Coimbatore. The results revealed that application of *Azospirillum*, FYM and chemical fertilizers as N, P and K in addition to FYM resulted in the highest dry berry yield in pepper (Table 23.5).

**Table 23.5.** Organic and biofertilizers on yield of black pepper dry yield (kg/vine)

| Treatment | Dry yield (kg/vine) | |
|---|---|---|
| | Panniyur 1 | Panniyur 2 |
| 50% N as FYM + *Azospirillum* + 50% P as inorganic + P solubilizers + 100% K as inorganic | 1.001 | 2.077 |
| 50% N as FYM + *Azospirillum* + 50% N and P as inorganic + *Azospirillum* + P solubilizers + AMF + 100% K as inorganic | 1.355 | 1.427 |
| Recommended package of practices | 1.216 | 1.097 |

Studies on integrated plant nutrient management conducted at IISR, Calicut increased black pepper yield when coir compost and biofertilizers were included in the fertilizer schedule (Sadanandan, 2001) (Table 23.6).

**Table 23.6.** Effect of integrated nutrient management in pepper

| Treatment | Yield (MT/ha) |
|---|---|
| Check | 2.33 |
| NPK (100: 40: 140 kg/ha) | 3.62 |
| FYM (10 MT/ha) + Coir compost (2.5 MT/ha) + NPK (60: 50: 120 kg/ha) | 2.98 |
| FYM (10 MT/ha) + Coir compost (2.5 MT/ha)+ ½ NPK (50: 20: 70 kg/ha) + biofertilizers (20 kg/ha) | 4.03 |
| CD (P = 0.05) | 0.41 |

## 23.1.3. Integrated disease management

### (i) Quick wilt/foot rot, *Phytophthora capsici*

Application of AMF and Potassium phosphonate (Akomin) significantly reduced the vine mortality and increased the yield (Table 23.7).

**Table 23.7.** Effect of AMF and agrochemicals on foot rot and yield of black pepper

| Treatment | Mortality of vines (%) | | | Yield (kg/vine) | | |
|---|---|---|---|---|---|---|
| | AMF | Non AMF | Mean | AMF | Non AMF | Mean |
| Control | 60.9 | 44.4 | 52.7 | 1.226 | 1.323 | 1.275 |
| AMF | 10.8 | 27.7 | 19.7 | 6.313 | 1.828 | 4.070 |
| Copper oxychloride (Bordeaux mixture) | 33.8 | 27.6 | 30.4 | 2.973 | 8.920 | 2.460 |
| Metalaxyl 100 ppm (Ridomil mancozeb) | 38.8 | 21.9 | 30.4 | 2.821 | 0.456 | 1.639 |
| Potassium phosphonate (Akomin) | 10.8 | 21.9 | 16.4 | 4.745 | 3.633 | 4.181 |
| CD (P = 0.05) | NS | NS | 18.12 | 2.122 | --- | 1.294 |

### 23.1.4. Integrated nematode management

#### (i) Root-knot, *Meloidogyne incognita* and burrowing nematodes, *Radopholus similis*

The consortial formulation of *Pseudomonas fluorescens* (Pfbv 22) and *Bacillus subtilis* (Bbv 57) at 10 g/vine as soil application gave significant reduction in infestation of root-knot and burrowing nematodes and increase in yield (4.64 kg/vine) as compared to control (2.46 kg/vine) (Jonathan *et al.*, 2010).

## 23.2. CARDAMOM, *AMMOMUM SUBULATUM*

### 23.2.1. High yielding varieties

**Table 23.8.** High yielding varieties/hybrids of cardamom

| Variety/hybrid | Yield (kg/ha) | Developed at |
|---|---|---|
| PV-1 | 260 | Cardamom Res. Stn., KAU, Pampadumpara |
| PV-2 | 982 | Cardamom Res. Stn., KAU, Pampadumpara |
| Mudigere-1 | 275 | RRS, Mudigere (UHS, Bagalkot) |
| Mudigere-2 | 475 | RRS, Mudigere (UHS, Bagalkot) |
| ICRI-1 | 325 | ICRI (Spices Board), Myladumpara |
| ICRI-2 | 375 | ICRI (Spices Board), Myladumpara |
| ICRI-3 | 440 | ICRI (Spices Board), Myladumpara |
| ICRI-4 | 455 | ICRI (Spices Board), Myladumpara |
| IISR Avinash | 847 | IISR, CRS, Appangala |
| IISR Vijetha | 979 | IISR, CRS, Appangala |
| IISR Kodagu Suvasini | 745 | IISR, CRS, Appangala |
| CCS-1 | 745 | IISR, CRS, Appangala |
| RR-1 | 848 | IISR, CRS, Appangala |
| CPRI-1 | 375 | |
| CPRI-2 | 375 | |
| CPRI-3 | 439 | |

**Table 23.9.** Cardamom varieties resistant to pests/diseases

| Variety/hybrid | Yield (kg/ha) | Resistance to | Developed at |
|---|---|---|---|
| Mudigere-1 | 275 | Thrips (T), Hairy caterpillar (T), white grub (T) | RRS, Mudigere (UHS, Bagalkot) |
| PV-2 | 982 | Stem borer (T), Thrips (T) | Cardamom Res. Stn., KAU, Pampadumpara |
| ICRI-2 | 375 | Azukkal disease (T) | ICRI (Spices Board), Myladumpara |

| | | | |
|---|---|---|---|
| ICRI-3 | 440 | Rhizome rot (T) | ICRI (Spices Board), Myladumpara |
| ICRI-4 | 455 | Rhizome rot (T), Capsule borer (T) | ICRI (Spices Board), Myladumpara |
| IISR Avinash | 847 | Rhizome rot (T), Shoot and capsule borer (T) | IISR, CRS, Appangala |
| IISR Vijetha | 979 | Katte (Cardamom mosaic virus) | IISR, CRS, Appangala |

### 23.2.2. Fertigation

Significantly the highest yields were obtained with 75 and 100% water soluble fertilizer (WSF) (592 and 618 g of green capsules/clump) and were superior to rest of the treatments (320.5 to 521.6 g). Fertigation with 75% WSF holds promise (Table 23.10).

**Table 23.10.** Productivity of cardamom and levels of fertigation as influenced by sources and levels of fertilizers

| Treatment | Green capsules/clump (g) |
|---|---|
| Normal fertilizers 100% + Furrow irrigation | 321 |
| Normal fertilizers 75% + Drip irrigation | 369 |
| Normal fertilizers 75% Fertigation | 499 |
| Normal fertilizers 100% Fertigation | 522 |
| Water soluble fertilizers 75% Fertigation | 592 |
| Water soluble fertilizers 100% Fertigation | 619 |
| CD (P = 0.05) | 33.37 |

### 23.2.3. Integrated nutrient management

Application of boron in the form of disodium tetraborate at 20 kg/ha and molybdenum in the form of sodium molybdate at 0.25 kg/ha mixed with appropriate quantity of FYM at 1 kg/plant applied with onset of monsoon increased cardamom yield by 20%.

## 23.3. GINGER, *ZINZIBER OFFICINALE*

### 23.3.1. High yielding varieties

**Table 23.11.** High yielding varieties/hybrids of ginger

| Variety/hybrid | Yield (MT/ha) | Developed at |
|---|---|---|
| Suprabha | 16.6 | High Altitude Res. Stn., OUAT, Bhubaneshwar |
| Suruchi | 11.6 | High Altitude Res. Stn., OUAT, Bhubaneshwar |
| Surabhi | 17.5 | High Altitude Res. Stn., OUAT, Bhubaneshwar |
| IISR Varada | 22.6 | IISR, Calicut |

| | | |
|---|---|---|
| IISR Mahima | 23.2 | IISR, Calicut |
| IISR Rejatha | 22.4 | IISR, Calicut |
| Himagiri | 13.14 | Dept. of Veg. Crops, YSPUH&FD, Solan |

### 23.3.2. Integrated nutrient management

The maximum fresh rhizome yield of 18.17 MT/ha was recorded with application of inorganic N 100% RDF + *Azospirillum* at 5 kg/ha + FYM at 10 MT/ha with a net return of Rs.2.51 lakh/ha (Gupta *et al.*, 2006).

Application of micronutrients such as Zn, B and Mo at 5, 2 and 1 kg/ha also increased rhizome yield by 48 per cent (5.2 MT of dry rhizome/ha) and oleoresin content over control.

Studies on integrated plant nutrient management conducted at IISR, Calicut increased ginger yield when coir compost and biofertilizers were included in the fertilizer schedule (Sadanandan, 2001) (Table 23.12).

**Table 23.12.** Effect of integrated nutrient management in ginger

| Treatment | Yield (MT/ha) |
|---|---|
| Check | 1.76 |
| NPK (75: 50: 50 kg/ha) | 2.44 |
| FYM (10 MT/ha) + Coir compost (2.5 MT/ha) + NPK (60: 50: 129 kg/ha) | 3.20 |
| FYM (10 MT/ha) + Coir compost (2.5 MT/ha) + ½ NPK (37.5: 25: 25 kg/ha) + biofertilizers (20 kg/ha) | 3.44 |
| CD (P = 0.05) | 0.66 |

### 23.3.3. Integrated disease management

#### (i) Rhizome rot, *Pythium aphanidermatum, P. myriotylum*

Soil application of *Aspergillus niger, A. terreus, Pencillium* sp. and *Absidia cylindrospora* showed reduction in the disease incidence ranging from 7-100% (Table 23.13) (Balakrishnan *et al.*, 1997).

**Table 23.13.** Effect of antagonists on rhizome rot of ginger

| Antagonist | Germination (%) | Disease incidence (%) | Disease severity (%) | Yield (MT/ha) |
|---|---|---|---|---|
| *Aspergillus niger* | 100.0 | 0.0 | 0.00 | 33.10 |
| *A. terreus* | 97.5 | 20.0 | 12.00 | 23.30 |
| *Pencillium* sp. | 100.0 | 25.0 | 19.25 | 19.27 |
| *Absidia cylindrospora* | 80.0 | 30.0 | 30.00 | 14.03 |
| Control | 75.0 | 72.5 | 64.83 | 4.35 |

Soil solarization prior to planting followed by application of biocontrol agents showed synergistic effect on disease reduction and increased yield (Sarma *et al.*, 1996) (Table 23.14).

**Table 23.14.** Effect of the biocontrol agents on germination, rhizome rot incidence and yield of ginger

| Biocontrol agent | Solarized soil | | Non- solarized soil | |
|---|---|---|---|---|
| | Disease index (%) | Yield (kg/plot) | Disease index (%) | Yield (kg/plot) |
| *Trichoderma viride* | 13.70 | 2.846 | 35.25 | 1.225 |
| *T. harzianum* 1 | 15.80 | 3.552 | 32.07 | 2.673 |
| *T. harzianum* 2 | 19.38 | 2.910 | 43.61 | 1.439 |
| *T. hamatum* | 16.16 | 2.744 | 39.26 | 1.602 |
| *Gliocladium virens* | 18.25 | 2.641 | 36.63 | 1.705 |
| Mancozeb | 36.22 | 2.260 | 46.70 | 0.818 |
| Control | 57.58 | 1.692 | 53.80 | 0.992 |
| Mean | 25.30 | 2.278 | 41.08 | 1.485 |
| CD (P = 0.05) | 6.26 | 0.430 | 6.26 | 0.430 |

## 23.4. TURMERIC, *CURCUMA LONGA*

### 23.4.1. High yielding varieties

**Table 23.15.** High yielding varieties/hybrids of turmeric

| Variety/hybrid | Yield (MT/ha) | Developed at |
|---|---|---|
| CO-1 | 35 | TNAU, Coimbatore |
| BSR-1 | 31 | ARS, TNAU, Bhavanisagar |
| BSR-2 | 32 | ARS, TNAU, Bhavanisagar |
| Krishna | 15.0 | Maharashtra Agri. Univ., Kasaba, Digraj |
| Sugandham | 17.4 | Spices Res. Stn., GAU, Jagudan |
| Suvarna | 17.4 | IISR, Calicut |
| Suguna | 29.3 | IISR, Calicut |
| Sudarsana | 28.8 | IISR, Calicut |
| IISR Prabha | 37.5 | IISR, Calicut |
| IISR Prathiba | 39.1 | IISR, Calicut |
| IISR Alleppey Supreme | 5.58 kg/plant | IISR, Calicut |
| IISR Kedaram | 5.28 kg/plant | IISR, Calicut |
| Roma | 20.7 | High Altitude Res., OUAT, Pottangi |
| Suroma | 20.0 | High Altitude Res., OUAT, Pottangi |
| Rasmi | 31.3 | High Altitude Res., OUAT, Pottangi |
| Ranga | 29.0 | High Altitude Res., OUAT, Pottangi |
| Rajendra Sonia | 4.8 kg/plant | Tirhut Agri. College, RAU, Dholi |
| Kanthi | 7.9 kg/plant | Hort. College, KAU, Trichur |
| Sobha | 5.74 kg/plant | Hort. College, KAU, Trichur |

**Table 23.16.** Turmeric varieties resistant to pests/diseases

| Variety/hybrid | Yield (MT/ha) | Resistance to | Developed at |
|---|---|---|---|
| BSR-2 | 9.2 | Scale insects | ARS, TNAU, Bhavanisagar |
| Suranjana | | Rhizome scales, Shoot borer (MR), Leaf blotch (T), Rhizome rot (T) | UBKVV, Pundibari |
| Suguna | 29.3 | Rhizome rot (T) | IISR, Calicut |
| IISR Alleppey Supreme | 5.58 kg/plant | Leaf blotch (T) | IISR, Calicut |
| IISR Kedaram | 5.28 | Leaf blotch (T) | IISR, Calicut |
| Sona | | Leaf blotch (T) | Hort. College, KAU, Trichur |
| Varna | | Leaf blotch (T) | Hort. College, KAU, Trichur |

### 23.4.2. Integrated nutrient management

Application of FYM at 25 MT/ha and vermicompost at 2.5 MT/ha gave higher yield of turmeric compared to control (Table 23.17).

**Table 23.17.** Effect of organic manures on yield of turmeric.

| Treatment | Cured rhizome yield (MT/ha) |
|---|---|
| FYM at 25 MT/ha | 3.34 |
| Vermicompost at 2.5 MT/ha | 4.79 |
| Control | 3.01 |
| CD (P = 0.05) | 0.97 |

Turmeric plants raised with NPK at 75% RDF + *Azospirillum* + AMF gave maximum yield (32.92 kg/2.25 m$^2$) followed by NPK at 100% RDF + *Azotobacter* + AMF (31.53 kg/2.25 m$^2$) and NPK at 100% RDF + *Azospirillum* + AMF (29.37 kg/2.25 m$^2$) (Roy and Hore, 2009).

Studies taken up at TNAU, Coimbatore, on effect of nitrogen on turmeric revealed that the response was higher at 100 to 140 kg/ha (Ahmed Shah and Muthuswamy, 1981; Balashanmugam and Chezhiyan, 1986) (Table 23.18).

**Table 23.18.** Yield of turmeric as influenced by Nitrogen levels

| Treatment | Yield of fresh rhizomes (MT/ha) | % increase over control |
|---|---|---|
| No Nitrogen ( control) | 14.6 | --- |
| N at 100 kg/ha | 22.32 | 52.88 |
| N at 120 kg/ha | 21.98 | 50.55 |
| N at 140 kg /ha | 22.91 | 56.92 |
| N at 160 kg/ha | 21.92 | 50.14 |

*Note:* FYM at 10 MT + 60 kg each of $P_2O_5$ and $K_2O$/ha was common for all the treatments (Ahmed Shah and Muthuswami, 1981)

Field experiments were conducted during 1987-88 at Agricultural Research Station, TNAU, Bhavanisagar to study the effect of potash fertilization on growth and yield attributes of turmeric cv. BSR 1. The experiment involved three dosages of potash viz., 30, 60 and 90 kg/ha applied in either full or in three or four or five splits. A common dosage of 120 kg N/ha was applied in five splits at monthly intervals up to 120 days and 60 kg $P_2O_5$/ha applied basally. It was found that the application of 90 kg $K_2O$/ha in four splits resulted in higher uptake and yield of 29.7 MT/ha (Balashanmugam and Subramanian, 1991) (Table 23.19).

**Table 23.19.** Influence of K levels on yield of BSR 1 turmeric and K uptake

| K levels | Yield (MT/ha ) |
|---|---|
| 30 kg $K_2O$/ha | 23.5 |
| 60 kg $K_2O$/ha | 24.9 |
| 90 kg $K_2O$/ha | 29.7 |

Application of FYM, FYM + Zn solubilizer and soil and foliar application of 0.5% Zn $SO_4$ and 1% $FeSO_4$ enhanced the uptake of all the major nutrients right from the early phase of crop growth to harvest (Senthil Kumar *et al.*, 2004) (Table 23.20).

**Table 23.20.** Effect of FYM and Zinc on turmeric

| Treatment | Rhizome yield (MT/ha) | Curcumin content ( %) |
|---|---|---|
| Control | 23.5 | 4.04 |
| NPK + FYM | 25.6 | 4.38 |
| NPK + FYM + Zn solubilizer | 28.6 | 4.53 |

Studies on integrated plant nutrient management conducted at IISR, Calicut increased turmeric yield when coir compost and biofertilizers were included in the fertilizer schedule (Sadanandan, 2001) (Table 23.21).

**Table 23.21.** Effect of integrated nutrient management in turmeric

| Treatment | Yield (MT/ha) |
|---|---|
| Check | 2.60 |
| NPK (60: 50: 120 kg/ha) | 3.85 |
| FYM (10 MT/ha) + Coir compost (2.5 MT/ha) + NPK (60: 50: 120 kg/ha) | 4.59 |
| FYM (10 MT/ha) + Coir compost (2.5 MT/ha) + NPK (30: 25: 60 kg/ha) + biofertilizers (20 kg/ha) | 3.78 |
| CD (P = 0.05) | 0.90 |

### 23.4.3. Fertigation

Fertigation studies taken up at ARS, TNAU, Bhavanisagar revealed the significance of fertigation. The required N and K fertilizers were divided into 20 equal splits and applied at weekly intervals through fertigation commencing from 30 days after sowing. The studies revealed that even at 50% of the recommended dosage of N and K when supplied through drip resulted in higher yields as compared to application of 100% recommended NPK dose through soil application of fertilizers and surface irrigation. This study clearly reveals that it is possible to enhance nutrient and water use efficiency and improve yield by adopting fertigation in turmeric (Table 23.22).

**Table 23.22.** Influence of irrigation and fertigation levels on turmeric yield

| Irrigation levels | Turmeric yield (MT/ha) at different fertigation levels | | | |
|---|---|---|---|---|
| | 100 % N&K | 75 % N&K | 50 % N&K | Mean |
| Irrigation at 80 % PE | 32.91 | 34.76 | 33.62 | 33.76 |
| Irrigation at 60 % PE | 33.26 | 33.69 | 34.12 | 33.69 |
| Irrigation at 40 % PE | 29.70 | 29.98 | 29.49 | 29.72 |
| Mean | 31.96 | 32.81 | 32.41 | |

## 23.5. CORIANDER, *CORIANDRUM SATIVUM*

### 23.5.1. High yielding varieties

**Table 23.23.** High yielding varieties of coriander

| High yielding varieties | Yield (MT/ha) | Developed at |
|---|---|---|
| CO 1 | 0.500 | TNAU, Coimbatore |
| CO 2 | 0.600-0.700 | TNAU, Coimbatore |
| CO 3 | 0.650 | TNAU, Coimbatore |
| CO 4 | 0.540-0.590 | TNAU, Coimbatore |
| Gujarat Coriander-1 | 1.100 | Main Spices Res. Stn., GAU, Jagudan |
| Gujarat Coriander-2 | 1.500 | Main Spices Res. Stn., GAU, Jagudan |
| Rajendra Swati | 1.200-1.400 | Tirhut Hort. College, RAU, Dholi |
| RCr-41 | 1.200 | SKN Agri. College, RAU, Jobner |
| Swathi | 0.885 | RARS, ANGRAU, Lam, Guntur |
| Sadhana | 1.000 | RARS, ANGRAU, Lam, Guntur |
| NRCSS ACr-1 | 1.250 | NRC Seed Spices, Ajmer |

**Table 23.24.** Coriander varieties resistant to pests/diseases

| Variety/hybrid | Yield (MT/ha) | Resistance to pest/disease | Developed at |
|---|---|---|---|
| Gujarat Coriander-1 | 1.100 | Fusarium wilt (T), Powdery mildew (T) | Main Spices Res. Stn., GAU, Jagudan |

| | | | |
|---|---|---|---|
| Gujarat Coriander-2 | 1.500 | Powdery mildew (T) | Main Spices Res. Stn., GAU, Jagudan |
| CO-3 | | Fusarium wilt (T), Powdery mildew (T) | TNAU, Coimbatore |
| RCr-41 | 1.200 | Stem gall, Fusarium wilt, Powdery mildew (MR) | SKN Agri. College, RAU, Jobner |
| RCr-20 | | Stem gall (MR) | SKN Agri. College, RAU, Jobner |
| RCr-435 | | Root-knot nematode (MR), Powdery mildew (MR) | SKN Agri. College, RAU, Jobner |
| RCr-436 | | Root rot, Root-knot nematode | SKN Agri. College, RAU, Jobner |
| Sadhana | 1.000 | White fly (T), Mites (T), Aphids (T) | RARS, ANGRAU, Lam, Guntur |
| Sindhu | | Fusarium wilt (T), Powdery mildew (T) | RARS, ANGRAU, Lam, Guntur |
| Hisar Sugandh | | Stem gall | HAU, Hisar |
| Azad Dhania-1 | | Powdery mildew (T), Aphids (T) | CSAUAT, Kanpur |
| Pant Haritima | | Stem gall | GBPUAT, Pantnagar |
| DWA 3 | | Powdery mildew (T) | UAS, Dharwad |
| NRCSS ACr-1 | 1.250 | Powdery mildew (T) | NRC Seed Spices, Ajmer |

MR – Moderately Resistant, T - Tolerant

### 23.5.2. Integrated nutrient management

The nutrient combination of vermicompost at 2.5 MT/ha and NPK at 25: 35: 15 kg/ha gave maximum seed yield of 1.025 MT/ha. Essential oil content was maximum (0.62%) in combination treatment with vermicompost at 1.5 MT/ha and NPK at 25: 35: 15 kg/ha (Pariari and Khan, 2009).

The maximum seed yield (1.779 MT/ha) was recorded with the application of Rhizobacteria FL-18 (seed treatment + soil application), followed by *Trichoderma* MTCC-5179 (1.611 MT/ha) (Malik *et al.*, 2009).

### 23.5.3. Integrated disease management

#### (i) Wilt, *Fusarium* sp.

Of late, coriander wilt is becoming a serious disease which causes significant yield loss.

Seed treatment with *P. fluorescens* at 10 g/kg gave the lowest Fusarial wilt incidence of 14.4% (36.6% in control) with an yield of 603 kg/ha (423 kg/ha in control).

*T. viride* as seed treatment (4 g/kg) and soil application (5 kg/ha) reduced the Fusarial wilt incidence by 58% with a yield of 575 kg/ha compared to 433 kg/ha in control (Anon, 2005).

Seed treatment and soil application of *Trichoderma viride* was effective for the management of wilt disease. *T. harzianum* was also found effective in reducing coriander wilt and increasing the yield (Table 23.25).

**Table 23.25.** Effect of seed treatment with biocontrol agents on wilt incidence in coriander

| Treatment | Wilt incidence (%) | Yield (MT/ha) |
|---|---|---|
| Seed treatment + Soil drenching with carbendazim | 11.4 | 0.307 |
| Seed treatment with *Trichoderma viride* | 6.3 | 0.356 |
| Seed treatment with *T. harzianum* | 12.3 | 0.300 |
| Seed treatment with *Bacillus subtilis* | 14.8 | 0.290 |
| Seed treatment with *Pseudomonas fluorescens* | 29.5 | 0.172 |
| Seed treatment with carbendazim | 14.2 | 0.285 |
| Control | 28.9 | 0.168 |
| CD (P = 0.05) | 5.8 | 26.4 |

## 23.6. FENUGREEK, *TRIGONELLA FOENUM-GRAECUM*

### 23.6.1. High yielding varieties

**Table 23.26.** High yielding varieties of fenugreek

| High yielding varieties | Grain yield (MT/ha) | Developed at |
|---|---|---|
| CO-1 | 0.685 | TNAU, Coimbatore |
| CO-2 | 0.500 | TNAU, Coimbatore |
| Rajendra Kranti | 1.200-1.400 | Tirhut Agri. College, RAU, Dholi |
| RMt-1 | 1.500 | SKN Agri. College, RAJAU, Jobner |
| Lam Sel-1 | 1.200 | RARS, ANGRAU, Guntur |
| NRCSS AM-1 | 2.720 | NRC Seed Spices, Ajmer |
| NRCSS AM-2 | 1.810 | NRC Seed Spices, Ajmer |

**Table 23.27.** Fenugreek varieties resistant to pests/diseases

| Variety/hybrid | Yield (MT/ha) | Resistance to pest/disease | Developed at |
|---|---|---|---|
| CO-1 | 0.685 | Root rot (T) | TNAU, Coimbatore |
| CO-2 | 0.500 | Rhizoctonia root rot (T) | TNAU, Coimbatore |
| Rajendra Kranti | 1.200-1.400 | Cercospora leaf spot (T), Powdery mildew (T), Aphids (T) | Tirhut Agri. College, RAU, Dholi |
| RMt-1 | 1.500 | Root-knot nematode (MR), Powdery mildew (MR), Aphids (MR) | SKN Agri. College, RAJAU, Jobner |

| | | |
|---|---|---|
| RMt-143 | Powdery mildew (MR) | SKN Agri. College, RAJAU, Jobner |
| RMt-305 | Root-knot nematode, Powdery mildew | SKN Agri. College, RAJAU, Jobner |
| Hisar Sonali | *Rhizoctonia* root rot (MR), Aphids (MR) | Dept. of Veg. Crops, CCSHAU, Hisar |
| Hisar Suvarna | Cercospora leaf spot (MR), Powdery mildew (MR) | Dept. of Veg. Crops, CCSHAU, Hisar |
| Hisar Madhavi | Powdery mildew (MR), Downy mildew (MR) | Dept. of Veg. Crops, CCSHAU, Hisar |
| Hisar Mukta | Powdery mildew (MR), Downy mildew (MR) | Dept. of Veg. Crops, CCSHAU, Hisar |
| Pant Ragini | Downy mildew, *Rhizoctonia* root rot | GBPUAT, Pantnagar |

MR – Moderately Resistant, T - Tolerant

### 23.6.2. Integrated disease management

#### i) Root Rot, *Rhizoctonia solani*

The efficacy of seed pelleting with *Trichoderma* spp. and soil application of *Trichoderma* spp. along with neem cake at 150 kg/ha showed greater protection against root rot in fenugreek and increased the yield (Table 23.28).

**Table 23.28.** Biocontrol of root rot disease of fenugreek

| Treatment | Root rot incidence (%) | | Yield (MT/ha) | |
|---|---|---|---|---|
| | Kharif 1992 | Rabi 1992 | Kharif 1992 | Rabi 1992 |
| T 1. Seed treatment + soil drenching with carbendazim | 4.8 | 10.8 | 0.422 | 0.315 |
| T 2. Seed treatment with *Trichoderma viride* | 4.8 | 4.4 | 0.384 | 0.365 |
| T 3. *T. viride* 20 days before sowing | 26.3 | 20.4 | 0.288 | 0.285 |
| T 4. Neem cake at 150 kg/ha | 3.9 | 3.2 | 0.427 | 0.385 |
| T 5. T4 + T2 | 3.2 | 3.4 | 0.424 | 0.360 |
| T 6. T4 + T3 | 5.4 | 12.4 | 0.288 | 0.325 |
| T 7. T4 + T1 | 12.8 | 14.3 | 0.345 | 0.340 |
| T 8. Seed treatment with carbendazim | 27.9 | 24.5 | 0.294 | 0.265 |
| T 9. Control | 36.2 | 32.8 | 0.163 | 0.184 |
| CD (P = 0.05) | 5.7 | 4.3 | 52.4 | 27.0 |

## 23.7. FENNEL, *FOENICULUM VULGARE*

### 23.7.1. High yielding varieties

**Table 23.29.** High yielding varieties of fennel

| High yielding varieties | Grain Yield (MT/ha) | Developed at |
|---|---|---|
| NRCSS AF-1 | 1.950-2.510 | NRC Seed Spices, Ajmer |
| CO 1 | 0.567 | TNAU, Coimbatore |

**Table 23.30.** Fennel varieties resistant to diseases

| Variety/hybrid | Yield (MT/ha) | Resistance to disease | Developed at |
|---|---|---|---|
| NRCSS AF-1 | 1.95-2.51 | *Ramularia* blight (T), *Alternaria* blight (T) | NRC Seed Spices, Ajmer |
| S-7-9 | | *Alternaria* blight (T) | Spices Res. Stn.,GAU, Jagudan |
| FF 35 | | Leaf spot (T), *Alternaria* leaf blight (T), Sugary disease (T) | Spices Res. Stn.,GAU, Jagudan |
| Gujarat Fennel 1 | | Sugary disease (T) | Spices Res. Stn.,GAU, Jagudan |
| RF 125 | | Sugary disease (T) | SKN Agri. College, RAJAU, Jobner |

T - Tolerant

### 23.7.2. Integrated nutrient management

The highest seed yield (924 kg/ha) was obtained with treatment irrigation at 6 cm pan evaporation combined with application of N at 150 kg/ha, while essence yield was maximum (4.9 litres/ha) with treatment irrigation at 12 cm pan evaporation combined with application of N at 150 kg/ha (Mirshekari *et al.*, 2009b).

### 23.7.3. Bioregulators

The highest seed yield in fennel (958 kg/ha) was obtained in the treatment of seed inoculation with *Azospirillum* strain 21 as compared to 729 kg/ha in control (uninoculated treatment) (Mirshekari *et al.*, 2009a).

## 23.8. CINNAMON, *CINNAMOMUM CASSIA*

### 23.8.1. High yielding varieties

**Table 23.31.** High yielding varieties of cinnamon

| HYV | Bark yield (kg/ha) | Developed at |
|---|---|---|
| IISR Navashree | 250 | IISR, Calicut |
| IISR Nithyashree | 250 | IISR, Calicut |

| | | |
|---|---|---|
| Konkan Tej | 85 | Regional Coconut Res. Stn., Dr. BSKVV, Vengurla |
| YCD-1 | 360 | HRS, TNAU, Yercaud |
| PPI(Ci) 1 | 980 | TNAU, Coimbatore |

## 23.9. CUMIN, *CUMINUM CYMINUM*

### 23.9.1. High yielding varieties

**Table 23.32.** High yielding varieties of cumin

| High yielding varieties | Resistance to | Developed at |
|---|---|---|
| S-404 | Powdery mildew (T) | Spices Res. Stn., GAU, Jagudan |
| MC 43 | Fusarium wilt (T), Alternaria blight (T), Powdery mildew (T) | Spices Res. Stn., GAU, Jagudan |
| Gujarat Cumin 1 | Fusarium wilt (T), Alternaria blight (T), Powdery mildew (T) | Spices Res. Stn., GAU, Jagudan |
| Gujarat Cumin 2 | Fusarium wilt (T), Alternaria blight (T), Powdery mildew (T) | Spices Res. Stn., GAU, Jagudan |
| Gujarat Cumin 3 | Fusarium wilt | Spices Res. Stn., GAU, Jagudan |
| Gujarat Cumin 4 | Fusarium wilt | Spices Res. Stn., GAU, Jagudan |
| RZ 19 | Fusarium wilt (T), Alternaria blight (T) | SKN Agri. College, RAJAU, Jobner |
| RZ 209 | Alternaria blight | SKN Agri. College, RAJAU, Jobner |
| RZ 223 | Fusarium wilt | SKN Agri. College, RAJAU, Jobner |

### 23.9.2. Integrated weed management

Pre-emergence application of oxadiargyl at 75 g/ha + one hand weeding at 45 DAS was found as the best economically feasible practice to keep weed infestation at minimum level and to ensure higher economic yield (591 kg/ha) in cumin (Meena *et al.*, 2009).

## 23.10. AJOWAN

### 23.10.1. High yielding varieties

**Table 23.33.** High yielding varieties of ajowan

| HYV | Yield (kg/ha) | Developed at |
|---|---|---|
| NRCSS-AA 1 | 1426 (irrigated), 580 (rainfed) | NRC Seed Spices, Ajmer |
| NRCSS-AA 2 | 1283 (irrigated), 520 (rainfed) | NRC Seed Spices, Ajmer |

**Table 23.34.** Ajowan varieties resistant to diseases

| HYV | Yield (kg/ha) | Resistance to | Developed at |
|---|---|---|---|
| NRCSS-AA 2 | 1283 (irrigated)<br>520 (rainfed) | Powdery mildew | NRC Seed Spices, Ajmer |

## 23.11. DILL

### 23.11.1. High yielding varieties

**Table 23.35.** High yielding varieties/hybrids of dill

| HYV | Yield (kg/ha) | Developed at |
|---|---|---|
| NRCSS-AD 1 | 1470 (irrigated) | NRC Seed Spices, Ajmer |
| NRCSS-AD 2 | 1460 (irrigated), 580 (rain fed) | NRC Seed Spices, Ajmer |

## 23.12. OTHER SPICE CROPS

### 23.12.1. High yielding varieties

**Table 23.36.** High yielding varieties/hybrids of spice crops

| Spice crop | HYV | Seed yield (MT/ha) | Developed at |
|---|---|---|---|
| Celery | NRCSS-ACel 1 | 0.8-1.2 | NRC Seed Spices, Ajmer |
| Nigella | NRCSS-AN 1 | 0.8-1.2 | NRC Seed Spices, Ajmer |
| Anise | NRCSS ANi 1 | 0.73-1.15 | NRC Seed Spices, Ajmer |

## Chapter 24

# A ROADMAP AHEAD FOR GOLDEN REVOLUTION

### 24.1. Enhancing productivity

The average productivity of most horticultural crops in India is low. There is a wide gap between yields obtained and potential yields with improved varieties/hybrids and good agricultural practices (Table 24.1). Therefore, programmes need to be taken up to reduce the yield gap by improving productivity. The main emphasis should be on enhancing the productivity by use of disease free, quality planting material of only released and recommended varieties/hybrids, both in the public and private sector. High density planting by reduction in planting distance or by use of plant growth inhibitors and dwarfing rootstocks as recommended for crops like mango, citrus, apple, banana, pineapple and some temperate fruits will go a long way in increasing the productivity. Use of protected cultivation under controlled conditions for growing fruits like strawberry, vegetables such as cucumber, cabbage, capsicum, tomato and temperate vegetables in plain areas and high value cut flowers for domestic use and export need to be promoted. The cultivation of crops, which produce higher biomass/unit area/unit time like, banana, pineapple, papaya, potato, sweet potato, tapioca, elephant foot yam should be taken up in areas requiring poverty alleviation and nutritional security.

**Table 24.1.** Scope to enhance yield of horticultural crops

| Horticultural crop | World average yield (MT/ha) | Indian average yield (MT/ha) | Additional scope to enhance yield (MT/ha) |
|---|---|---|---|
| Apple | 20.62 | 8 | 12 |
| Mango | 8.09 | 6 | 2 |
| Pineapple | 26.45 | 15 | 11 |
| Fruit crops | 13.09 | 11 | 2 |
| Brinjal | 18.45 | 17.00 | 1.5 |
| Cabbage | 30.72 | 22.20 | 8.5 |

| | | | |
|---|---|---|---|
| Onion | 27.54 | 15.10 | 12.5 |
| Potato | 25.90 | 19.3 | 6.5 |
| Tomato | 39.00 | 17.9 | 21.0 |
| Vegetable crops | 23.45 | 16.00 | 7.5 |

## 24.2. Reducing cost of production

In view of the global competition, as a result of implementation of WTO w.e.f. April 1, 2001; reduction in unit cost of production of horticultural commodities which are exported, has become inevitable. There is also a danger of large-scale imports of horticulture commodities from abroad if our local production costs are not lower/ comparable. Appropriate programmes are therefore, required to reduce the cost of production by reducing the cost of planting material by mass multiplication of horticulture crops using micro-propagation, wherever feasible. For instance in banana, cardamom, vanilla, ornamental and medicinal plants, reduction in cost of fertilizer by determination of plant needs through leaf nutrient standards and applying only required quantities, efficient utilization and conservation of water and applied nutrients through drip/micro irrigation and fertigation, weed control, moisture conservation and solarization techniques. Integrated nutrient management should be followed by cover cropping with leguminous crops in perennial plantations and incorporating them to improve soil fertility, thus, supplementing the fertilizer needs; and thereby reducing the cost of production. Promotion of integrated pest and disease management, thus, reducing costs of chemical pesticides and fungicides should be taken up. There is a need to reduce post harvest losses by proper pre- and post-harvest handling, proper packaging and creating suitable infrastructure like low cost cold storage for fruits and vegetables, storage for potato and onion cold chain transport and other technologies.

## 24.3. Area value share of horticultural crops in crop sector

It is relevant to analyze the area and value share of horticultural crops in crop sector (Birthal *et al*., 2008) (Table 24.2).

**Table 24.2.** Area share, value share and share of horticultural crops in crop sector

| Crop groups | Area share* (%) | Value share** (%) | Annual growth rate (%) | | Share in overall growth (%) | |
|---|---|---|---|---|---|---|
| | | | 1981-82 to 1995-96 | 1995-96 to 2004-05 | 1981-82 to 1995-96 | 1995-96 to 2004-05 |
| Condiments | 1.2 | 3.7 | 9.6 | 2.6 | 7.7 | 5.5 |
| Fruits & vegetables | 5.8 | 25.6 | 2.9 | 5.6 | 20.7 | 63.7 |
| Total | 7.0 | 29.3 | 12.5 | 6.2 | 28.4 | 69.2 |

* Area share in gross cropped area

** Value share in value of crop sector output computed at 1993-94 prices

Three points are relevant in the context of the issue whether growth of crop sector was mainly horticulture-led growth or not.

- With an area share of about 7%, horticulture contributed about 29% of value of crop output. Obviously, the contribution of fruits and vegetables was quite dominant.
- The annual compound growth rates of condiments had declined considerably during second period (1995-96 to 2004-05), whereas the growth rate of fruits and vegetables had almost doubled during the same period.
- The share of horticulture in the total value of crop output had increased from about 28% in the first period (1981-82 to1995-96) to about 69% in the second period (1995-96 to 2004-05). The robust growth in horticulture sector was mainly demand-led due to change in the composition of consumer basket of goods within the country, and growing demand in the international market.

## 24.4. Growth rate analysis of crop groups in agriculture and allied sectors

The annual growth rates in case of fruits and vegetables increased from 2.84% in the first period (1981-82 to 1990-91) to 6.07% during the second period (1991-92 to1995-96), and declined to 3.54% during the third period (1996-97 to 2006-07) (Table 24.3). This is a disturbing trend, which needs to be arrested. Details of the needed interventions shall be discussed later.

**Table 24.3.** Annual growth rate analysis of crop groups in agriculture and allied sectors (Singh and Mathur, 2008)

| Crop groups | Annual growth rate (%) | | |
|---|---|---|---|
| | 1981-82 to 1990-91 | 1991-92 to 1995-96 | 1996-97 to 2006-07 |
| Cereals | 3.52 | 2.36 | 1.39 |
| Pulses & oil seeds | 5.41 | 2.92 | 1.43 |
| Fruits & vegetables | 2.84 | 6.07 | 3.54 |
| All crops | 2.97 | 3.09 | 2.36 |
| Live stock | 4.78 | 4.00 | 3.61 |
| Fishery | 5.74 | 7.05 | 2.93 |

## 24.5. Disaggregate analysis of growth performance of horticultural crops

The insights into the production and productivity performance of horticulture crop groups for two periods of post-reform regime is presented in Table 24.4.

**Table 24.4.** Growth rates of output and productivity of major horticulture crop groups during two periods (Singh and Mathur, 2008)

| Horticulture crop groups | Production (%) | | Productivity (%) | |
|---|---|---|---|---|
| | 1991-93 to 1996-97 | 1997-98 to 2005-06 | 1991-93 to 1996-97 | 1997-98 to 2005-06 |
| Fruit crops | 7.27 | 2.89 | 3.49 | -1.96 |
| Vegetable crops | 4.71 | 3.40 | 4.34 | 0.89 |
| Plantation crops | 5.36 | 1.81 | 0.83 | 0.01 |
| Spice crops | 6.14 | 7.70 | 4.33 | 2.69 |
| Average | 5.66 | 3.27 | 3.34 | -0.45 |

In aggregate, the growth rate of horticulture crops had declined from 5.66% during 1991-93 to 1996-97 to 3.27% during 1997-98 to 2005-06. This decelerated growth is a disturbing trend, as discussed earlier. But with disaggregate analysis, the output growth rates decelerated more in case of fruits and plantation crops than in the case of vegetables.

With respect to negative productivity of horticulture crops in general, the trend is observed to be much more in case of fruits. In case of vegetables and spices, productivity even though positive, had decelerated considerably. What intervention strategies needed for reversing this trend towards negative growth rates and decelerated growth rates has to be worked out.

## 24.6. Technology interventions needed in horticultural sector

There are potentials in sunrise sectors especially horticulture sector to promote rural growth and to enhance the scope for poor people to experience livelihood security. Even though growth performance of horticulture has been responsible for preventing high rate of productivity and production growth deceleration in crop production sector, production and productivity of horticulture sector have started decelerating since 1997-98. This is a concern that needs to be addressed.

- Growth-mediated orientation to nutritional security would need improvement in productivity of horticultural crops. One of the disquieting features of Indian horticulture sector is the low and declining productivity of fruits and vegetables. The average productivity is much lower in India than in other fruit producing countries of the world. As regards vegetables, average productivity is the lowest in the world. With this low and decelerated productivity in horticulture sector, the Indian horticulture sector cannot perform well in international market.
- Due to recent decline in growth rates of fruits and vegetables, the ability of horticulture sector as a remedy for poverty and hunger is reduced (Chand *et al.*, 2008).

- Post-harvest losses of fruits and vegetables are estimated to be 17-35%. In fact, the reduction in post-harvest losses is a complementary means for increasing productivity.
- Depletion and degradation of natural resource base in crop production sector would very much require technologies to maintain long run productivity of natural resource base.
- There are several unsolved chronic disorders like irregular bearing, spongy tissue and malformation in mango; wilt in guava; bacterial blight in pomegranate; decline in citrus; virus diseases in vegetable crops; and so on which needs to be tackled on war footing.

### 24.6.1. Genetic improvement

**Goal:** To create and disseminate new varieties that meet environmental, processing and market requirements.

**R&D objectives:**

- Develop improved lines for enhanced productivity, nutrition, product quality and safety.
- Determine gene products, metabolic processes, regulation and phenotypic expression through genomics and other modern applications of biotechnology.
- Incorporate new genes from distant relatives through tools of modern biotechnology.
- Foster the development of small and medium-scale private seed companies and nurseries through the development and dissemination of improved lines and through training.

**R&D activities**

- Develop molecular markers for key tomato traits, including resistance to bacterial wilt, tomato leaf curl virus, heat and drought.
- Develop tomato lines with increased content of lycopene, beta-carotene and other anti-oxidants.
- Develop chilli lines with multiple resistance to chilli veinal mosaic virus, cucumber mosaic virus, bacterial wilt, anthracnose, and *Phytophthora* blight.
- Develop improved cucurbit lines with resistance to multiple viruses (for all cucurbits) and higher beta-carotene content (for squash).
- Introduce *Bt*-crystal protein genes to control bean pod borer in beans.
- Identify molecular markers for the development of bean lines resistant to moonbeam yellow mosaic virus, angular leaf spot, common bacterial blight, anthracnose, root rot, thrips, and pod weevils.

- Adapt vegetable soybean lines for enhanced environmental range and utilization.
- Biofortify micronutrient-rich indigenous vegetables to support HIV/AIDS-afflicted communities.
- Develop efficient gene variety enhancement protocols to improve post-harvest quality of papaya and mango.
- Transform selected fruit species with genes known to control post-harvest deterioration.
- Develop high quality papaya lines with resistance to ring spot virus and high quality guava lines with resistance to wilt disease.

### 24.6.2. Sustainable production technologies

**Goal:** To develop production systems that stabilize yields, sustain the environment, and satisfy market demands.

### R & D objectives

- Identify crops and develop technologies for improved water use efficiency.
- Reduce pesticide abuse and environmental degradation in crop production.
- Develop production systems that reduce seasonality of production and market supplies.
- Encourage public/private sector cooperation to increase availability of improved varieties and other technologies to poor farmers.

### R & D activities

- Adapt low-cost micro-irrigation/fertigation technologies for horticulture crop production in marginal areas.
- Establish water management practices that enhance quality of citrus and other tree fruits.
- Develop biopesticide, pheromone and other bio-intensive IPM technologies for fruit and vegetable crop production.
- Develop IPM and organic technologies to reduce pesticide abuse in selected peri-urban communities.
- Develop organic production systems for peri-urban and urban areas that safely utilize urban wastes.
- Integrate beans into the cereal-based cropping systems.
- Develop IPM practices for ornamental cut flowers that promote worker safety and reduce pesticide abuse.
- Develop micro-irrigated and reduced labour home garden models to improve livelihoods of HIV/AIDS-afflicted households.

- Design low-cost protective net-tunnels and shelters to reduce the need for insecticides and allow for year-round cultivation.
- Develop production technologies to support commercial production of promising medicinal herbs in the tropics and subtropics.
- Develop participatory methodologies to select and develop superior clones of selected fruit species.
- Develop rapid clonal propagation techniques for underutilized tropical fruit crops with market potential.
- Develop production practices that induce flowering and higher productivity of carambola, guava, lime and mango.
- Develop fruit varieties in which flowering can be controlled at will, using transformed mango as a model.
- Develop production practices for underutilized indigenous fruits.

### 24.6.3. Post-harvest storage, processing and marketing

**Goal:** To link small-scale farmers to profitable food chains through improvements in post-harvest handling and marketing.

#### R & D objectives

- Improve the movement of horticultural crops from farms to markets through better handling, cold chain management, and packaging.
- Support the establishment and implementation of international sanitary and phytosanitary (SPS), quality and trade regulations and systems.
- Link small-scale farms to high-value urban and export markets, processors and supermarkets.

#### R & D activities

- Develop low-cost processing, packaging, storing and shipping technologies.
- Assist farmers in developing countries to become certified for selling their produce to European and American markets.
- Improve post-harvest handling, processing and marketing of indigenous fruit and vegetable crops for domestic and international markets.
- Determine the critical features of successful farmers' cooperatives as potential models for establishment of farmers' cooperatives.
- Promote farmers' cooperatives for more efficient production and marketing of horticultural crops as well as improved farmer access to micro-credit, extension and technical information.

- Promote the grading of produce to reduce excessive handling and post-harvest losses.
- Develop models of modern retail/wholesale fruit and vegetable markets for developing countries.
- Develop equitable contract farming schemes that equate supply and demand for perishable horticultural crops.
- Provide training in SPS procedures and regulations.

Hence, research and development efforts should be priority items of agenda in promoting and sustaining the growth and productivity of horticulture sector. Horticulture research and development has the potential to boost productivity of horticulture and allied sectors, promote sustainable management of natural resources, and support nutritional security of poor people through the development of new technologies (Swaminathan, 2008). While enhancing allocation of budget to research and development, a critical review of technology subsystems viz., technology generation (research), technology transfer (extension) and technology utilization (farm level) would be indispensable to set the research and development priorities, streamline the extension, and to overcome the constraints facing farmers in technology adoption.

After the success of 'Green Revolution' in agriculture, India is now moving towards the **'Golden Revolution'** comprising fruits, vegetables and flowers. Commercialization of horticulture along with post harvest management, processing and marketing assume great significance and add value not only in the context of farmers' income, but also its contribution to the agriculture GDP. Commercialization of horticulture is largely dependent upon use of appropriate technology and market oriented production.

# REFERENCES


Agarwal, K.N., Singh, R.K. and Satpathy, K.K. 2003. Surface covered cultivation for higher production in NEH Region. *Workshop on Green House Technology for Cold Desert Climate*, CIPHET, Ludhiana.

Ahluwalia, M.S., Singh, B. and Gill, B.S. 1993. Drip irrigation system: Its hydraulic performance and influence on tomato and cauliflower. *J. Water Mangmt.* **1**: 6-9.

Ahmed Shah, H. and Muthuswamy, S. 1981. Studies on the influence of N on the yield and yield components of turmeric. *Indian Cocoa Arecanut Spices J.* **5**: 9-10.

Amerika Singh, Trivedi, T.P., Sardana, H.R., Sabir, N., Krishnamoorthy, P.N., Pandey, K.K., Sengupta, A., Ladu, L.N. and Singh, D.K. 2004. Integrated pest management in horticultural crops – A wide area approach. In *Crop Improvement and Production Technology of Horticultural Crops* (K.L. Chadha, B.S. Ahluwalia, K.V. Prasad and S.K. Singh, eds.), pp. 621-636.

Ananthi, S. 2002. Comparative Efficacy of Sulphate of Potash and Muriate of Potash on Yield and Quality of Chilli (*Capsicum annuum* L.). M.Sc. (Hort.) thesis, TNAU, Coimbatore.

Anita, B. and Selvaraj, N. 2010. Management of lettuce root-knot disease by biofumigation and biocontrol agents. *National Conf. on Innovations in Nematological Research for Agricultural Sustainability – Challenges and A Roadmap Ahead.* Tamil Nadu Agri. Univ., Coimbatore, pp. 99-115.

Anita, B. and Subramanian, S. 1998. Management of the reniform nematode *Rotylenchulus reniformis* in tomato. In *Nematology - Challenges & Opportunities in 21st Century* (Usha K. Mehta, ed.), pp. 249-250. Sugarcane Breeding Inst., Coimbatore.

Anita, B. and Vadivelu, S. 1997. Management of root-knot nematode, *Meloidogyne hapla* on scented geranium, *Pelargonium graveolans*. *Indian J. Nematol.* **27**:123-125.

Annapurna, K., Bojappa, K.M. and Bhargava, B.S. 1988. Leaf sampling guide for sapota (*Manilkara achras* M. Foseberg) cv. Cricket Ball. *Crop Res.* **1**: 69-75.

Anon. 1985. Research Report. All India Co-ordinated Research Project. Dept. of Hort., Assam Agri. Univ., Jorhat, 29 pp.

Anon. 1987. Research papers and reports. *IVth National Workshop on Arid Zone Fruit Res.*, Agri. College & Res. Inst., Tamil Nadu Agri. Univ., Madurai. Tech. Doc. No. 20.

Anon. 1989. Biennial Report. *Vth National Workshop on Arid Zone Fruit Res.*, Guj. Agri. Univ., Sardarkrishinagar.

Anon. 1990. Review of results of winter maize. *Group Meeting*. Haryana Agri. Univ., Hisar, pp. 3-4.

Anon. 1996. Annual Report, AICRP on Pulses, Gulburga Centre, Karnataka.

Anon. 1998. Studies on fertigation in cardamom in Smiley Greens Hoskur. In *Fertigation Studies with Water Soluble NPK Fertilizers in Crop Production* (K. Shivashankar, M.M. Khan and A.A. Farooki, eds.), p. 12. Kremia Oy. Finland and Micro-Irrigation Soc. of India.

Anon. 2005. Annual Report for 2003-04 of All India Coordinated Research Project on Spices. Indian Inst. of Spices Res., Calicut, Kerala, India.

Anon. 2007. Efficacy of sulphate of potash on growth, yield, quality and shelf life of tomato. Final report submitted to K+S Kali GmbH, Kassel, Germany, Department of Vegetable Crops, Hort. College & Res. Inst., Tamil Nadu Agri. Univ., Coimbatore.

Antogaozzi, E., Luchetti, A. and Romani, F. 1988. Possibilities of high density apple planting in Umbria. *Annali-della-di-Agraria-Universilu-degli-Studi-di Perugia* **42**: 229-242.

Anwar, M.A. and Khan, M.R. 2010. Biological control of root-knot of tomato with the application of *Aspergillus niger* isolates. National Conf. on Innovations in Nematological Research for Agricultural Sustainability – Challenges and A Roadmap Ahead. Tamil Nadu Agri. Univ., Coimbatore, p. 100.

Aphsara, S.E. and Sathiamoorthy, S. 2003. Effect of planting more than one sucker per pit on growth and yield of banana cv. Nendran (AAB). *Indian J. Hort.* **60**: 335-338.

Arora, R.K. 2000. Combined effect of soil solarization and seed treatment on control of potato black scurf. In *Potato – Global Research and Development* (S.M. Paul Khurana *et al.*, eds.), pp. 390-392. Indian Potato Association, Central Potato Res. Inst., Shimla, Himachal Pradesh, India.

Arora, R.K. and Somani, A.K. 2001. A bioformulation for control of black scurf disease of potato. *J. Indian Potato Assoc.* **28**: 88-89.

Arora, Y.K. and Mohan, S.C. 1985. Water harvesting and water management for fruit crops in wastelands. *National Workshop on Arid Zone Fruit Res.*, Mahatma Phule Agri. Univ., Rahuri. Tech. Doc. No. **17**: 104-112.

Ashok, A.D., Arun, D.S. and Rengasamy, P. 1999. Influence of graded levels and sources of N fertigation on flowering of cut rose cv. First Red under protected conditions. *S. Indian Hort.* **47**: 115-118.

Athani, S.I., Hulmani, N.C., Shirol, A.M., Gasti, V.D., Kanamadi, V.C. and Kuradikeri, M.B. 2006a Effect of organic and inorganic fertilizers on growth, crop duration, quality and yield attributes in banana cv. Rajapuri (*Musa* AAB). *National Symp. on Improving Input Use Efficiency in Hort.*, Indian Inst. of Hort. Res., Bangalore, p. 156.

Athani, S.I., Ustad, A.I., Prabhuraj, H.S., Swamy, G.S.K., Sabarad, A.I., Kotikal, Y.K., Patil, P.B. and Dasaraddi, S.V. 2006b. Variation in quality and yield as influenced by vermicompost in guava fruits. *National Symp. on Improving Input Use Efficiency in Hort.*, Indian Inst. of Hort. Res., Bangalore, p. 156.

Awasthi, R.P. and Mehta, K. 2000. Strategies for developing high density planting in horticultural crops. In *Souvenir of National Seminar on High-tech Horticulture*, Bangalore, pp. 29-33.

Badiyala, S.D. and Aggarwal, G.C. 1981. Note on the effect of mulches on strawberry production. *Indian J. Agric. Sci.* **51**: 832-834.

Bafna, A.M., Dafatdar, S.Y., Khade, K.K. and Dhotre, R.S. 1993. Utilization of nitrogen and water by tomato under drip irrigation. *J. Water Mangmt.* **1**: 1-5.

Bagle, B.G., Joshi, H.K., Hiwale, S.S. and Appa Rao, V.V. 2006. Management of fruit fly, fruit borer and powdery mildew in ber cv. Umran. *National Symp. on Improving Input Use Efficiency in Hort.*, Indian Inst. of Hort. Res., Bangalore, p. 182.

Bahadur, A., Singh, J., Singh, K.P., Nirmal De and Mathura Rai. 2006. Influence of organic inputs on physiological parameters, yield and quality traits of lettuce (*Lactuca sativa* L.). *National Symp. on Improving Input Use Efficiency in Hort.*, Indian Inst. of Hort. Res., Bangalore, pp. 168-169.

Bahadur, V., Daheria, R., Singh, D. and Singh, D.B. 2009. Effect of nitrogen and biofertilizers on growth and yield of garlic (*Allium sativum*) cv. G-282 (Abstr.). *Int. Conf. on Hort.*, Bangalore, p.144.

Balakrishna, Arora, R.L. and Srivastava, R. 1998. Planting system cum high density studies in plum cv. Titron. *Prog. Hort.* **31**: 31-35.

Balakrishnan, P., Usman, N.M. and Sarma, Y.R. 1997. Management of rhizome rot disease of ginger by fungal antagonists. In *Biotechnology of Spices, Medicinal and Aromatic Plants* (S. Edison, K.V. Ramana, B. Sasikumar, K. Nirmal Babu and S.J. Eapen, eds.), pp. 146-149. Indian Soc. for Spices, Indian Inst. of Spices Res., Calicut, India.

Balashanmugam, P.V. and Chezhiyan, N. 1986. Effect of differential application of nitrogen on growth and yield of turmeric (*Curcuma longa* L.). *Madras Agric. J.* **73:** 439-442.

Balashanmugam, P.V. and Subramanian, K. S. 1991. Effect of split application of potassium in turmeric. *S. Indian Hort.* **39**: 139-142.

Balasubramanian, P. and Vadivelu, E. 2009. Effect of different systems of irrigation and fertigation on growth and yield of tomato (Abstr.). *Int. Conf. on Hort.*, Bangalore, p. 115.

Bansod, R.D. and Shukla, K.N. 2006a. Effect of micro sprinkler and surface irrigation on biometric development and yield parameters of cauliflower. *National Symp. on Improving Input Use Efficiency in Hort.*, Indian Inst. of Hort. Res., Bangalore, pp. 107-108.

Bansod, R.D. and Shukla, K.N. 2006b. Response of litchi to different methods of irrigation. *National Symp. on Improving Input Use Efficiency in Hort.*, Indian Inst. of Hort. Res., Bangalore, p. 111.

Baswana, K.S. and Rana, M.K. 2009. Effect of organic sources and biofertilizers on growth and yield of garden pea (Abstr.). *Int. Conf. on Hort.*, Bangalore, p. 352.

Bevington, K.B. 1994. Initial growth and yield of high density plantings of Valencia orange as influenced by micro-sprinkler and drip irrigation. *VIIth Int. Citrus Cong.*, Acireale, Italy, pp. 709-711.

Bhalla, H.S. 1988. Tomato response to trickle irrigation and black polythene mulch. *J. Amer. Soc. Hort. Sci.* **113**: 543-546.

Bhambal, S.B. 1987. Effect of Foliar Application of Micronutrients on Growth, Yield, Fruit Quality and Leaf Nutrient Status of Pomegranate (*Punica granatum* L.) cv. Ganesh. M. Sc. (Ag.) thesis, Mahtma Phule Agri. Univ., Rahuri, Maharashtra.

Bharadwaj, L.N. and Sharma, S.K. 2000. Y.S. Parmar University of Horticulture and Forestry, Solan, Himachal Pradesh. In Progress Report for the Period Sept. 1997-Aug. 2000 of National Network Project on Phytophthora Diseases of Horticultural Crops (S.R. Sarma, ed.), pp. 26-40. Indian Inst. of Spices Research, Calicut, Kerala.

Bhargava, B.S. and Chadha, K.L. 1993. Leaf nutrient guide for fruit crops. In *Advances in Horticulture*, Vol. 2 (K.L. Chadha and O.P. Pareek, eds.), pp. 973-1029. Malhotra Publishing House, New Delhi.

Bhargava, B.S., Raturi, G.B. and Hiwale, S.S. 1990. *Singapore J. Prim. Ind.* **18**: 85.

Bhaskaran, R. 1993. Integrated management of basal stem rot disease of coconut. *Indian Coconut J.* **24**: 5-8.

Bhasker, S., Arun, M.N. and Shivananda, T.N. 2009. Growth and yield of patchouli (*Pogostemon patchouli* Pellet) as influenced by irrigation and nutrition (Abstr.). *Int. Conf. on Hort.*, Bangalore, p. 197.

Bhat, K. 1989. M. Sc. (Agri.) thesis, Univ. of Agri. Sci., Dharwad, Karnataka.

Bhat, M.G. and Venkata Kumar, R. 2005. Present status and future prospects of cashew research in India. In *Souvenir of Rashtriya Bagwani Kisan Mela* (R.R. Suresh, D. Sreenivasa Murthy, Leela Sahijram and V. Keshava Rao, eds.), pp. 92-94. Indian Inst. of Hort. Res., Bangalore.

Bhatia, A.K., Arora, S.K. and Malik, T.P. 2009. Screening cucumber hybrids for greenhouse production (Abstr.). *Int. Conf. on Hort.*, Bangalore, p.p. 148-149.

Bhattacharya, A., Prasad, V.M. and Chattopadhyay, N. 2009. Effect of *Azotobacter chroococcum* through basal and foliar application on growth and yield of leek (*Allium porrum* L.) cv. Lincoln (Abstr.). *Int. Conf. on Hort.*, Bangalore, p.145.

Bhattacharya, P. 2006. Influence of *Azospirillum brasiliense*, chemical nitrogen and their combination on growth and yield of brinjal (*Solanum melongena* L.) *in vivo*. *National Symp. on Improving Input Use Efficiency in Hort.*, Indian Inst. of Hort. Res., Bangalore, pp. 134-135.

Bhattacharya, R.K. and Rao, V.N.M. 1985. Soil cover increases banana production. *Indian J. Hort.* **30**: 5-8.

Bhella, H.S. and Wilcox, G.E. 1986. Yield and composition of muskmelon as influenced by pre-plant and trickle applied nitrogen. *Hort. Sci.* **21**: 86-88.

Bhriguvanshi, S.R. and Singh, G. 2006. Studies on yield, quality and water use of mango as influenced by drip and basin method of irrigation. *National Symp. on Improving Input Use Efficiency in Hort.*, Indian Inst. of Hort. Res., Bangalore, p. 114.

Birhade, R.M., Badgujar, C.D., Deshmukh, S.S., Palwe, C.R. and Patil, N.M. 1999. Standardization of banana sucker size and age of tissue culture seedlings as planting material. *Seminar on Technological Advancement in Banana Production, Handling and Processing Management*, Jalgaon, Maharashtra, pp. 27-28.

Bijay Kumar. 2009. *Indian Horticulture Database 2007-08*. National Hort. Board, Gurgaon, Haryana, India, 324 pp.

Birthal, P.S., Joshi, P.K., Chauhan, S. and Singh, H. 2008. Can horticulture revitalise agricultural growth? *Indian J. Agri. Econ.* **63**: 310-321.

Biswas, B., Sen, S.K. and Maiti, S.C. 1989. Effect of plant density on growth, yield and chemical composition on papaya fruit varieties. *Prog. Hort.* **21**: 280-284.

Blomme, R. and Dambre, P. 1984. The culture of gerberas. Acidity and salt concentration as a function of manurial regime. *Verbondsnieaws Voor de Belgische Sierteult.* **28**: 684-687.

Borah, S.C., Barbora, A.C. and Bhattacharyya, D. 2001. Effect of organic and inorganic manuring on growth, yield and quality of Khasi mandarin (*Citrus reticulata*). *S. Indian Hort.* **49:** 115-118.

Bunea, A., Inglese, P. Marra, F.P. and Sottile, F. 1999. Investigations on optimum tree density in apricot orchard. *Acta Hort.* **121**: 323-326.

Caruso, T., Inglese, P., Marra, F.P. and Sottile, F. 1999. Effect of planting system on productivity, dry matter partitioning and carbohydrate content in above-ground components of 'Florida Prince' peach trees. *J. Amer. Soc. Hort. Sci.* **124**: 39-45.

Castillo, M.B. 1985. Some studies on the use of organic soil amendments for nematode control. *Philippine Agriculturist* **68**: 76-93.

Ceausescu, I. and Isac, I. 1986. Results obtained in very high density apple orchards in Romania. *Acta Hort.* **16**: 269-277.

Chacko, E.K. and Reddy, A. 1981a. Effect of planting distance and intercropping with cowpea on growth in banana. In *Proc. of the National Symp. on Banana Production Technol.* (C.R. Muthukrishnan and J.B.M. Abdul Kahader, eds.), pp. 85-88. Tamil Nadu Agri. Univ., Coimbatore.

Chacko, E.K. and Reddy, A. 1981b. Effect of planting distance and intercropping with cowpea on weed growth in banana. *Proc. VIII Asian Pacific Weed Sci. Conf.*, Bangalore, pp. 137-141.

Chadha, K.L. 1977. *A Feasibility Report on Production, Processing and Export of Pineapple from India.* Dept. of Export Promotion, Ministry of Commerce, Govt. of India, New Delhi.

Chadha, K.L., Melanta, K.R. and Shikamany, S.D. 1974. Effect of type and size of planting material on the vigour of the subsequent plant's yield and quality in Kew pineapple (*Annanas comosus* L.). *Indian J. Hort.* **31**: 9-15.

Chadha, S., Rana, S.S. and Parmar, D.K. 2006. Influence of FYM and micronutrients on the productivity of seed potato (*Solanum tuberosum*) under cold desert conditions of Himachal Pradesh. *National Symp. on Improving Input Use Efficiency in Hort.*, Indian Inst. of Hort. Res., Bangalore, p. 120.

Chand, R., Raju, S.S. and Pandey, L.M. 2008. Progress and potential of horticulture in India. *Indian J. Agri. Econ.* **63**: 299-309.

Chandra, R. and Govind, S. 1995. Influence of time and intensity of pruning on growth, yield and fruit quality of guava under high-density planting. *Trop. Agri.* **72**: 110-113.

Chattopadhyay, P.K., Bhowmik, D.J., Maiti, S.C. and Bose, T.K. 1985. Optimum planting density for plant and ratoon crops of Giant Governer Cavendish banana in West Bengal. *Indian J. Agric. Sci.* **55**: 17-21.

Chaudhri, S.M., Shinde, S.H., Dahiwalkar, S.D., Danawale, N.J., Shrisanth, H.K. and Berad, S.M. 2001. Effect of fertigation through drip on productivity of papaya. *J. Maharashtra Agri. Univ.* **26**: 18-20.

Chaudhri, S.M. and More, T.A. 2002. Fertigation, fertilizer and spacing requirement of tropical gynoecious cucumber hybrids. *Acta Hort.* **588**: 233-240.

Chaurasia, S.N.S., Singh, K.P. and Mathura Rai. 2006. Response of cauliflower (*Brassica oleracea* L. var. *botrytis*) to biofertilizers and nitrogen. *National Symp. on Improving Input Use Efficiency in Hort.*, Indian Inst. of Hort. Res., Bangalore, pp. 170-171.

Chavan, S.K., Ali, M.I., Masalkar, S.D. and Pachankar, P.B. 2009. Effect of organic production on growth and yield of tomato (*Lycopersicon esculentum* Mill.) (Abstr.). *Int. Conf. on Hort.*, Bangalore, p. 349.

Chawla, J.K. and Narda, N.K. 2000. Growth parameters of trickle irrigated potato (*Solanum tuberosum*). *Indian J. Agric. Sci.* **70**: 747-752.

Chaya, M.K., Rao, M.S. and Ramachandran, N. 2010. Biomanagement of disease complex in okra (*Abelmoschus esculentus* L. Moench) using microbial antagonists. *National Conf. on Innovations in Nematological Research for Agricultural Sustainability – Challenges and A Roadmap Ahead*, Tamil Nadu Agri. Univ., Coimbatore, pp. 96-97.

Chhabra, H.K., Bindra, O.S. and Singh, I. 1977. Further studies on the control of the citrus nematode, *Tylenchulus semipenetrans* Cobb, 1913 infecting *Citrus maxima* Merrill (Abstr.). *Proc. Int. Symp. Citriculture,* Bangalore, P. 38.

Chundawat, B.S. 1997. *Nutrient Management in Fruit Crops.* Agrotech Publishing Academy, Udaipur.

Chundawat, B.S., Dave, S.K. and Patel, N.L. 1982. Effect of close planting on the yield and quality of Lacatan banana. *Indian J. Agric. Sci.* **53**: 470-472.

Chundawat, B.S., Kirani, K.P., Verma, L.R. and Jadav, R.G. 1992. Studies on hedgerow plantation in guava cv. Allahabad Safeda. *Indian J. Hort.* **49**: 134-137.

Dabke, D.J., Talashilkar, S.C., Dhane, S.S. and Mehta, V.B. 2006. Effect of sources and rates of potassium on increasing the yield and quality of Alphanso mango. *National Symp. on Improving Input Use Efficiency in Hort.*, Indian Inst. of Hort. Res., Bangalore, p. 154.

Dahiya, M.S., Phor, S.K, Bhatia, A.K. and Nandal, J.K. 2009. Response of nitrogen and phosphorus levels and spacing on yield of garlic (Abstr.). *Int. Conf. on Hort.*, Bangalore, p. 143.

Das, B., Mitra, S.K. and Bose, T.K. 1987. Response of Kew pineapple to plant densities and Calcium carbide. *Sci. Cult.* **53**: 120-121.

Dereuddre, J., Blandin, S. and Hassen, N. 1991. Resistance of alginate-coated somatic embryos of carrot (*Daucus carota* L.) to desiccation and freezing in liquid nitrogen: 1. Effects of preculture. *Cryo-Letters* **12**: 125-134.

Devarajan, K., Subramanian, S. and Sivakumar, M. 2010. Management of root-knot nematode, *Meloidogyne incognita* in snake gourd through organic amendments. *National Conf. on Innovations in Nematological Research for Agricultural Sustainability – Challenges and A Roadmap Ahead.* Tamil Nadu Agri. Univ., Coimbatore, p. 92.

De Verna, J.W. and Paterson, A.H., 1990. Genetics of *Lycopersicon*. In *Genetic Improvement of Tomato. Monographs on Theoretical and Applied Genetics*. Vol. 14. (G., Kalloo, ed.), p. 21. Springer Verlag, Berlin.

Devyatov, A.S. 1991. Fruiting potentials of apple orchards planted at different densities. *Fruit Sci. Reports* **18**: 111-117.

Dhanpal, R., Mathew, A.C., Palaniswami, C. and Jerard, B.A. 2006. Soil and moisture conservation measures in coconut plantations. *National Symp. on Improving Input Use Efficiency in Hort.*, Indian Inst. of Hort. Res., Bangalore, p. 112.

Dutta, A., Sen, H., Chattopadhyay, A. and Tarfdar, S. 2002. Studies on the effect of bio-fertilizer for increasing nitrogen use efficiency in sweet potato under alluvial soils of West Bengal. In *Impact of Vegetable Research in India* (Sant Kumar, P.K. Joshi and Suresh Pal, eds.), p. 218. National Centre for Agricultural Economics and Policy Research, New Delhi.

Dutta, P., Kundu, S., Biswas, S. and Pramanik, B. 2009. Effect of biofertilizers on production of organic litchi for safe nutrition to human health (Abstr.). *Int. Conf. on Hort.*, Bangalore, pp. 346-347.

Edison, S. 2005. Tropical tubers – the omnipotent crops – production, processing and commercial utilities. In *Souvenir of Rashtriya Bagwani Kisan Mela* (R.R. Suresh, D. Sreenivasa Murthy, Leela Sahijram and V. Keshava Rao, eds.), pp. 80-84. Indian Inst. of Hort. Res., Bangalore.

Edward Raja, M., Chandre Gowda, M.J., Krishna Moorthy, P.N. and Ramachander, P.R. 1999. Foliar nutrition as a tool to increase nutrient use efficiency and to reduce input cost in banana. *Proc. of National Seminar on Participatory Approaches for Horticulture Development*, Thiruvanathapuram. Kerala Hort. Devp. Programme, Kerala, pp. 224-226.

Elangovan, M., Suthanthirapandian, I.P. and Sayeed, S. 1982. Intercropping of onion in chilli. *S. Indian Hort.* **30**: 48-50.

Elizabeth Syriac, K., Raghavan Pillai, G. and Varughese, K. 1999. Nutrient-growth regulator interaction in snake gourd (*Trichosanthes anguina* L.) under drip irrigation system. *Proc. of National Seminar on Participatory Approaches for Horticulture Development*, Thiruvanathapuram. Kerala Hort. Devp. Programme, Kerala, pp. 227-233.

Elkner, K. and Kanizseski, S. 1995. Effect of drip irrigation and mulching on quality of tomato. *Acta Hort.* **379**: 175-180.

Farooki, A.A., Nalawadi, U.G. and Sulladmath, U.V. 1971. Effect of growth regulators on germination of sapota seeds. *Mysore J. Agric. Sci.* **5**: 3-4.

Firake, N.N. and Deolankar, K.P. 2000. Response of pomegranate to soluble fertilizers through drip. *J. Maharashtra Agri. Univ.* **25**: 196-197.

Firake, N.N. and Kumbhar, D.B. 2002. Effect of different levels of N, P and K fertigation on yield and quality of pomegranate. *J. Maharashtra Agri. Univ.* **27**: 146-148.

Furukawa, Y., Kataoka, T., Ogata, T. and Shimomura, M. 1992. Productivity of high density peach orchard using a rootstock (*Prunus persica* Thunb.). *Acta Hort.* **315**: 97-102.

Ganeshmurthy, A.N. and Hebbar, S.S. 2006. Nutrient use efficiency and yield response in hybrid brinjal across fertility gradient on alfisols. *National Symp. on Improving Input Use Efficiency in Hort.*, Indian Inst. of Hort. Res., Bangalore, pp. 147-148.

Gaosh, D. 1986. Japanese plum (*Prunus salicina* L.) in high density plantings. *Acta Hort.* **160**: 315-319.

Garg, N., Singh, G., Yadav, P. and Goel, N. 2006. Efficacy of mulching on improving the microbial population in high density guava orchard soil. *National Symp. on Improving Input Use Efficiency in Hort.*, Indian Inst. of Hort. Res., Bangalore, pp. 166.

Geetha, K. 2006. Integrated nutrient management of banana for high returns. *National Symp. on Improving Input Use Efficiency in Hort.*, Indian Inst. of Hort. Res., Bangalore, p. 123.

Ghar, Ghi Suta, A.K., Millium, A., Burzo, I. and Panit, E. 1994. Studies on quality and the storability of apple from a very intensive orchard in relation to crown shape and tree density. *Lucrane Stiintifice* **15**: 103-108.

Gheisen, B. 1990. First known data on sweet cherries as hedges. *Fruitteett-Den-Haag* **80**: 11.

Ghosh, B., Bose, T.K. and Mitra, S.K. 1986. Chemical induction of flowering and control of fruit drop in litchi (*Litchi chinensis* Sonn.). *Proc. 22nd Int. Hort. Cong.*, California, Abstr. 1189.

Ghosh, C., Mandal, J. and Chattopadhyay, G.N. 2009. Effect of vermicompost on the growth and yield of cabbage (*Brassica oleracea* L. var. *capitata*) (Abstr.). *Int. Conf. on Hort.*, Bangalore, p. 353.

Gopalakrishnan, C. and Ajith Kumar, K. 2006. Increasing the field efficacy of *Pseudomonas fluorescens* in the management of bacterial wilt in tomato, *Lycopersicon esculentum* Mill. *National Symp. on Improving Input Use Efficiency in Hort.*, Indian Inst. of Hort. Res., Bangalore, pp. 197-198.

Gopinath, G., Chandrashekar, S.Y. and Hemla Naik, B. 2009. Yield of carnation as influenced by levels of fertigation and sources of nutrients and economics of growing standard carnation var. 'Trendy' under cost effective greenhouse (Abstr.). *Int. Conf. on Hort.*, Bangalore, p. 188.

Grosser, J.W., 1990. Citrus rootstock improvement by cell fusion. *Citrus and Vegetable Magazine* **9**: 28-32.

Guerriero, R., Loreti, K., Morini, S. and Natali, S. 1981. Eight years observations on peach double row orchard. *Acta Hort.* **114**: 362-382.

Guerriero, R. and Scalabrelli, G. 1981. Six years observations on apricot high density plantings. *Acta Hort.* **243**: 365-372.

Gulati, J.M.L., Mishra, M.M., Paul, J.C. and Sahu, G.S. 1995. Production potential of chilli under levels of irrigation and nitrogen. *Indian J. Agron.* **40**: 145-146.

Gupta, C.R., Singh, A.K. and Banwasi, R. 2006. Efficacy of using *Azospirillum* in ginger. *National Symp. on Improving Input Use Efficiency in Hort.*, Indian Inst. of Hort. Res., Bangalore, p. 140.

Gupta, R.C., Pandey, S. and Srivastava, K.J. 2009. Integrated crop management for the production of export quality onion bulb (*Allium cepa* L.) (Abstr.). *Int. Conf. on Hort.*, Bangalore, p. 138.

Gutal, B.B., Bhilare, R.M. and Takte, R.L. 1992. Mulching effect on yield of tomato crop. In *Int. Agri. Engg. Conf.*, Bangkok, Vol. **3**: 883-887.

Hanumashetti, S.I., Rao, M.M. and Bankapur, V.M. 1981. Preliminary studies on the effect of growth regulators and chemicals on the improvement of colouration in Gulabi grapes. *Curr. Res.* **10**: 45-46.

Hasan, M.A. and Chattopadhyay, P.K. 1990. Effect of different soil moisture regimes on growth and yield of litchi cv. Bombai. *Indian J. Hort.* **47**: 397-400.

Haseeb, A. and Shukla, P.K. 2005. Wilt disease complex of pigeon pea and its management. In *Plant Diseases: Biocontrol Management* (S. Nehra, ed.), pp. 84-96. Avishkar Publishers, Distributors, Jaipur.

Hazan, A., Zacks, M., Mezuman, A., Berg, Z. and Alkayim, A. 1994. Fertilizing roses grown on rock wool. *Int. Water and Irrigation Rev.* **14**: 5-6.

Hemla Naik, B., Patil, A.A., Patil, V.S., Basavaraj, M. and Hiremath, S.M. 2004. Effect of pinching and chemicals on xanthophylls yield in African marigold (*Tagetes erecta* L.). *J. Orn. Hort.* **7**: 182-190.

Hemla Naik, B., Patil, A.A., Pradeep Kumar, B.A., Poornima, G. and Thippesh, D. 2009. Effect of nutrition on marigold xanthophylls (Abstr.). *Int. Conf. on Hort.*, Bangalore, p. 185.

Hilli, J.S., Sarojini, K. and Mamledesai, N.R. 2006. Stabilization of flower yield and quality in chrysanthemum flower crop with incorporation of vermicompost. *National Symp. on Improving Input Use Efficiency in Hort.*, Indian Inst. of Hort. Res., Bangalore, pp. 143-144.

Hirano, S., Morioka, S. and Tachibana, S. 1982. Tree density and age as related to fruit yield or field area in Satsuma mandarin – Estimation for an optimum leaf area index on field basis. *Proc. of the Int. Soc. of Citriculture* **1**: 184-186.

Hiremath, S.M., Nagaraju, H.R., Kusagur, P.V. and Gowda, D.S.M. 2006. Copper mine ore waste – rich micronutrient source for cole crops. *National Symp. on Improving Input Use Efficiency in Hort.*, Indian Inst. of Hort. Res., Bangalore, pp. 152-153.

Hiwale, S.S., Apparao, V.V., Dhandar, D.G. and Bagle, B.G. 2006. Effect of nutrient replenishment through organic fertilizers in sapota cv. Kalipatti. *National Symp. on Improving Input Use Efficiency in Hort.*, Indian Inst. of Hort. Res., Bangalore, p. 124.

Hokkanen, H.M.T. 1998. Ecological impact of transgenic, insect resistant crops. In *Génie génétique: perspectives, inconnues et risques. Les applications en agriculture et dans l'alimentation.* Colloque international, Groupe des Verts au Parlement Européen Bruxelles, p. 8.

Holm, L.G., Plucknett, D.L., Pancho, J.V. and Herberger, J.P. 1977. *The World's Worst Weeds. Distribution and Biology.* Published for the East-West Centre by the University Press of Hawaii, Honolulu.

Hooda, R., Singh, S.J., Malik, Y.S. and Batra, V.K. 1999. Influence of direct seeding, transplanting time and mulching on tomato yield. *Veg. Sci.* **26**: 140-142.

Hooda, R.S., Sidhu, A.S. and Pandita, M.L. 1985. Effect of seedling treatment and foliar application of cytozyme on growth and yield of tomato, brinjal and chillies. *J. Res. Punjab Agri. Univ.* **15**: 329-331.

Hussain, Z. and Bora, B.C. 2008. Integrated management of *Meloidogyne incognita* and *Ralstonia solanacearum* complex in brinjal. *Indian J. Nematol.* **38**: 159-164.

Hutton, R.J. and Cullis, B.R. 1982. Tree spacing effect on productivity of high density dwarf orange trees. *Proc. of the Int. Soc. of Citriculture* **1**: 186-190.

Hutton, R.J., McFadyen, L.M. and Lill, W.J. 1987. Relative productivity and yield efficiency of canning peach trees in tree intensive growing system. *Hort. Sci.* **22**: 552-560.

Irulappan, I., Abdul Khader, J.B.M. Md. and Muthukrishnana, S. 1984. Papaya research in Tamil Nadu. *Proc. National Seminar on Papaya and Papain Production.* Faculty of Horticulture, TNAU, Coimbatore, p.15-20.

Jacob, M and Vonvexkull. 1960. *Fertilizer Use, Nutrition and Manuring of Tropical Crops.* Second edition, Mannover.

Jain, R.K. and Bhatti, D.S. 1987. Population development of root-knot nematode (*Meloidogyne javanica*) and tomato yield as influenced by summer ploughing. *Trop. Pest Mangmt.* **33**: 122-125.

Jain, R.K. and Gupta, D.C. 1991. Integrated management of root-knot nematode, *Meloidogyne javanica* infecting *Solanum melongena. Afro-Asian J. Nematol.* **1**: 156-160.

Jatala, P. 1985. Biological control of nematodes. In *Advanced Treatise on* Meloidogyne. *Biology and Control, Vol. 1* (J.N. Sasser and C.C. Carter, eds.), pp. 303-308. North Carolina State Univ. Graphics, Raleigh.

Jatala, P. 1986. Biological control of nematodes. *Ann. Rev. Phytopath.* **24**: 453-489.

Jain, S.M., Shahin, E.A. and Sam, S. 1988. Interspecific protoplast fusion for the transfer of atrazine resistance from *Solanum nigrum* to tomato (*Lycopersicon esculentum* L.). *Plant Cell Tissue Organ Culture* **12**: 189-192.

Jeeva, S., Kulasekaran, M., Shanmugavelu, G.K. and Obilisami, G. 1988. Effect of *Azospirillum* on growth and development of banana cv. Poovan (AAB). *S. Indian Hort.* **36**: 1-4.

Jeyabaskaran, K.J., Pandey, S.D. and Murugan, V. 2006. Integration of organic manures, bio-fertilizers and inorganic fertilizers for high nutrient use efficiency in banana cultivation. *National Symp. on Improving Input Use Efficiency in Hort.*, Indian Inst. of Hort. Res., Bangalore, p. 128.

Jeyakumar, P and Kumar, N. 2002. Bioregulators for increased fertilizer use efficiency in banana cv. Dwarf Cavendish. In *Proceedings of the 13th International Symposium of the CIEC*, Gazi Osmanpasa University, Tokat, Turkey, pp.70-82.

Jeyakumar, P., Kumar, N and Kavino, M. 2003. Physiological response of banana cv. Robusta (AAA) to foliar applied growth regulators on productivity. *Madras Agric. J.* **90**: 702-706.

Jeyakumar, P., Kumar, N. and Soorianathasundaram, K. 2001. Fertigation studies in papaya (*Carica papaya* L.). Changing scenario in the production systems of horticultural crops. *S. Indian Hort.* **49**: 71-75.

Jiji, T. and Dale, D. 1999. Vermiculture for sustainable agriculture and waste management. *Proc. of National Seminar on Participatory Approaches for Horticulture Development*, Thiruvanathapuram. Kerala Hort. Devp. Programme, Kerala, pp. 237-240.

Jonathan, E.I., Umamaheswari, R. and Bommaraju, P. 2006. Bioefficacy of native plant growth promoting rhizobacteria against *Meloidogyne incognita* and *Phytophthora capsici. Indian J. Nematol.* **36**: 230-233.

Jonathan, E.I., Raghuchander, T., Vetrivelkalai, P. and Archana, S. 2010. Evaluation of consortial formulation for the management of root-knot and burrowing nematodes in black pepper. *National Conf. on Innovations in Nematological Research for Agricultural Sustainability – Challenges and A Roadmap Ahead.* Tamil Nadu Agri. Univ., Coimbatore, p. 91.

Joseph, K.T. 1971. Nutrient content and nutrient removal in bananas as an initial guide for assessing fertilizer needs. *Planter* **47:** 7-10.

Kadalli, V.G., Bankapur, V.M., Patil, A.P. and Hulmani, N.C. 1988. Studies on growth and yield of green chilli with French bean and onion as companion crops. *S. Indian Hort.* **36**: 148-151.

Kalra, S.K., Sidhiv, P.I.S., Dhaliwal, J.S. and Singh, R. 1994. Effect of different spacing on yield of guava cv. Allahabad Safeda. *Indian J. Hort.* **51**: 272-274.

Kanthaswamy, V. and Selvaraj, P. 2006. Studies on the effect of nitrogen and *Azospirillum* on growth and yield of chillies. *National Symp. on Improving Input Use Efficiency in Hort.*, Indian Inst. of Hort. Res., Bangalore, p. 124.

Kanthaswamy, V. and Yassin, M.G. 2009. Efficacy of specialty fertilizers on yield of okra (Abstr.). *Int. Conf. on Hort.*, Bangalore, pp. 134-135.

Katan, J. 1981. Solar heating (solarization) of soil for control of soil borne pests. *Ann. Rev. Phytopathol.* **19**: 211-236.

Katan, J., Rotem, Y., Finkel, Y. and Daniel, J. 1980. Solar heating of the soil for the control of pink root and other soil borne diseases of onions. *Phytoparasitica* **8**: 39-50.

Katwal, J.R. and Saraf, R.K. 1990. Effect of growth regulator and urea spray on the growth and yield of chillies. *Orissa J. Hort.* **18**: 52-56.

Kavino, M., Kumar, N., Soorianathasundaram, K. and Jeyakumar, P. 2002. Effect of sources of fertilizers for fertigation on yield and quality of banana cv. Robusta. *S. Indian Hort.* **50**: 301-307.

Kavitha, C., Vadivel, E. and Thangamani, C. 2006. Effect of organic manures and inorganic fertilizers on seed yield and L DOPA content of *Mucuna prurensis* (L.) DC. *National Symp. on Improving Input Use Efficiency in Hort.*, Indian Inst. of Hort. Res., Bangalore, p. 159.

Kavitha, M., Kumar, N. and Jeyakumar, P. 2000a. Effect of Zinc and Boron on biochemical and quality characters of papaya cv.CO-5. *S. Indian Hort.* **48**: 1-5.

Kavitha, M., Kumar, N. and Jeyakumar, P. 2000b. Role of zinc and boron on fruit yield and associated characters in papaya cv. CO-5. *S. Indian Hort.* **48**: 6-10.

Keitzar, S and Tiameren A.N. 2003. Application of plastics in agriculture: Some experiences in Nagaland. In: *Plasticulture Intervention for Agriculture Develop-*

*ment in North Eastern Region*. (Satpathy, K. K. and Ashwani Kumar, eds.), pp. 331-335. ICAR Research Complex for NEH Region, Umiam, Meghalaya.

Kempe Gowda, K., Krishna Manohar, R., Kumar, D.P., and Lavanya, M. 2009. Influence of drip irrigation and plant density on growth and yield off garden pea (*Pisum sativum* L.) (Abstr.). *Int. Conf. on Hort.*, Bangalore, p. 154.

Kempe Gowda, K., Kumar, D.P., Krishna Manohar, R. and Ravi, S. 2009. Effect of integrated management on growth and yield in bhendi (*Abelmoschus esculentus* (L.) Moench) cv., Arka Anamika (Abstr.). *Int. Conf. on Hort.*, Bangalore, pp. 136-137.

Keshavan, G. 2004. Effect of Different Sources of Nitrogen on Yield and Quality of Banana cv. Ney Poovan (AB). M.Sc. (Hort.) thesis, Tamil Nadu Agricultural University, Coimbatore.

Khaderkhan, H., Nataraju, M.S. and Nagaraja, G.N. 1998a. Economics of IPM in tomato. In *Advances in IPM for Horticultural Crops* (P. Parvatha Reddy, N.K. Krishna Kumar and A. Verghese, eds.), pp. 151-152. AAPMHE, Bangalore.

Khaderkhan, H., Nataraju, M.S. and Nagaraja, G.N. 1998b. Integrated pest management in cabbage: A cost benefit-constraint analysis. In *Advances in IPM for horticultural crops* (P. Parvatha Reddy, N.K. Krishna Kumar and A. Verghese, eds.), pp. 153-154. AAPMHE, Bangalore.

Khan, A.A., Sharma, S.K. and Singh, J. 1965. Intercropping of potato with wheat. *Indian Fmg.* **14**: 7, 25.

Khan, M.M., Shivashankar, K., Farooki, A.A., Krishna Manohar, R., Kariyanna and Sreerama, R. 2001. *Research Highlights of Studies on Fertigation in Horticultural Crops.* Univ. of Agri. Sci., Bangalore, 28 pp.

Khan, M.R., Bhattacharya, I., Chattopadhyay, S.B. and Ghosh, S. 2009. Integrated management of root-knot (*Meloidogyne incognita*) and other nematodes in pointed gourd (*Trichosanthes dioica*). *Indian J. Nematol.* **39**: 25-28.

Khan, R.M. and Parvatha Reddy, P. 1994. Nematode problems of ornamental crops and their management. In *Floriculture-Technology, Trades and Trends* (J. Prakash and K.R. Bhandary, eds.), pp.468-472. Oxford & IBH Publishing Co. Pvt. Ltd., New Delhi, 667 pp.

Khan, Z.U., McNeil, D.L. and Samad, A. 1998. Root pruning of apple trees grown at ultra high-density affects carbohydrate reserves distribution in vegetative and reproductive growth. *New Zealand J. Crop Hort. Sci.* **26**: 291-197.

Khanduja, S.D. and Garg, Y.K. 1984. Macro-nutrient element composition of leaves from jujube (*Zizyphus mauritiana* Lamk.) tree in north India. *Indian J. Hort.* **41**: 22-24.

Kishore, K., Pathak, K.A., Brajendra and Bharali, R. 2006. Effect of organic sources of nutrients on yield and fruit quality of 'purple' passion fruit (*Passiflora edulis* Sims.). *National Symp. on Improving Input Use Efficiency in Hort.*, Indian Inst. of Hort. Res., Bangalore, p. 122.

Klein, I., Levin, I., Bar, Yosef, B., Assaf, R. and Berkovitz, A. 1989. Drip nitrogen fertigation of 'Starking Delicious' apple trees. *Plant and Soil* **119**: 305-314.

Kling, J., 1996. Could transgenic super crops one day breed super weeds? *Science* **274**: 180-181.

Kotur, S.C., Shilpashree, Y.P., Sheshashayee, M.S. and Ramesh, P.R. 2006. Nitrogen use efficiency and yield of tomato (*Lycopersicon esculentum* L.) and French bean (*Phaseolus vulgaris* L.) as influenced by coating of urea with neem oil and graded levels of nitrogen. *National Symp. on Improving Input Use Efficiency in Hort.*, Indian Inst. of Hort. Res., Bangalore, p. 150.

Krishnamoorthy, P.N. and Krishna Kumar, N.K. 2002. Advances in the use of botanicals for the IPM of major vegetable crops. *Proc. of Int. Conf. on Vegetables,* Bangalore.

Krishnamoorthy, P.N., Krishna Kumar, N.K., Girija, G., Varalakshmi, B. and Prabhakar, M. 2003. *Integrated Pest Management in Cabbage Cultivation*. Extn. Bull. No. 1, Indian Inst. of Hort. Res., Bangalore, 10 pp.

Krishnamoorthy, P.N., Krishna Kumar, N.K., Selvaraj, C. and Daniel, J.S. 1998. Neem seed kernel extract applications for diamondback moth management: Transfer of technology for mechanized farming. *Pest Mangmt. Hort. Ecosystems* **4**: 128-130.

Krishnan, B.M. and Shanmugavelu, K.G. 1979. Studies on water requirements of banana cv. Robusta. Effect on morphological characters, crop duration, yield and quality of fruits. *Mysore J. Agric. Sci.* **13**: 136-137.

Krishnan, B.M. and Shanmugavelu, K.G. 1980. Studies on water requirements of banana cv. Robusta. Total consumptive use and water use efficiency. *Mysore J. Agric. Sci.* **14**: 433-441.

Kumar, A. and Shukla, H.S. 2009. Response of foliar application of zinc sulphate and gibberellic acid on physical quality and yield of guava (*Psidium guajava* L.) (Abstr.). *Int. Conf. on Hort.*, Bangalore, p. 168.

Kumar, D. and Pandey, V. 2006a. Growth and yield of mango, Arka Anmol under drip irrigation. *National Symp. on Improving Input Use Efficiency in Hort.*, Indian Inst. of Hort. Res., Bangalore, p. 96.

Kumar, D. and Pandey, V. 2006b. Mulching and drip irrigation in mango Lat Sundari. *National Symp. on Improving Input Use Efficiency in Hort.*, Indian Inst. of Hort. Res., Bangalore, p. 97.

Kumar, D.P., Thamara, M.P., Krishna Manohar, R., Jayaprasad, K.V. and Raju, B. 2009. Higher income generation from China aster (*Callistephus chinensis* (L.) Nees. cultivation through fertigation (Abstr.). *Int. Conf. on Hort.*, Bangalore, p. 184.

Kumar, J.M. and Patil, C.S. 2006. Bioefficacy of some newer insecticides against leaf miner, *Liriomyza trifolii* (Burgess) and thrips, *Thrips palmi* Karny infesting watermelon through seed treatment. *National Symp. on Improving Input Use Efficiency in Hort.*, Indian Inst. of Hort. Res., Bangalore, pp. 196-197.

Kumar, N. and Nalina, S. 2001. Research on high density planting of banana in Tamil Nadu – A resume. *S. Indian Hort.* **49**: 1-5.

Kumar, R. 1997. Response of planting densities, size and type of planting materials on growth, yield and quality of Kew pineapple. *Indian J. Hort.* **34**: 13.

Kumar, R. and Hooda, M.N. 1977. Crop regulation studies in Allahabad Safeda guava. *Indian J. Hort.* **54**: 40-43.

Kumar, S. 2009. Management of root-knot nematode in papaya by *Pochonia chlamydosporia*. *Indian J. Nematol.* **39**: 105-107.

Kumar, S. and Patel, B.N. 2006a. Evaluation of imidacloprid against hopper and thrips on mango. *National Symp. on Improving Input Use Efficiency in Hort.*, Indian Inst. of Hort. Res., Bangalore, p. 179.

Kumar, S. and Patel, B.N. 2006b. Extent of avoidable incidence and loss due to mango hopper. *National Symp. on Improving Input Use Efficiency in Hort.*, Indian Inst. of Hort. Res., Bangalore, p. 179.

Lal, B. and Mishra, D. 2006. Pruning in mango cv. Chausa to improve production and quality. *National Symp. on Improving Input Use Efficiency in Hort.*, Indian Inst. of Hort. Res., Bangalore, pp. 168-169.

Lal, S., Tewari, J.P. and Mishra, K.K. 1996. Effect of plant spacing and planting density on flowering and fruiting of guava. *Ann. Agron.* **17**: 83-89.

Lee, S.A. 1977. *Malaysian Agri. J.* **51**: 223-227.

Lichou, J., Garien, A. and Auduber, A. 1989. Apricots: Training trees and the prospects for intensive cultivation. Infos Centre Technique. *Interprofessional-des-Fruits-et-Legues*, France **50**: 22-26.

Loganathan, M., Swarnakumari, N., Sivakumar, M., Prakasam, V., Mohan, L., Ramraj, B. and Samiyappan, R. 2001. Biological suppression of fungal nematode complex diseases of major cruciferous vegetables. *S. Indian Hort.* **49**: Special issue.

Mahajan, A.R., Lal, S. and Tiwari, J.P. 2005. Effect of different planting systems on growth and yield of guava. *Prog. Hort.* **37**: 27-31.

Mahalakshmi, M., Kumar, N., Jeyakumar, P and Soorianathasundaram, K. 2001. Fertigation studies in banana under normal system of planting. *S. Indian Hort.* **49**: 86-91.

Mahalakshmi, M., Kumar, N. and Selvaraj, P. 2006. Data development analysis for fertigation studies in banana cv. Robusta (AAA). *National Symp. on Improving Input Use Efficiency in Hort.*, Indian Inst. of Hort. Res., Bangalore, p. 135.

Mahantesh, B., Ravi Prasad Sajjan, M. and Harshavardhan, M. 2009a. Investigations on the impact of biofertilizers with levels of NPK on nutrient uptake and yield of onion (*Allium cepa* L.) cv. Bellary Red in kharif season under irrigated situation in central dry zone of Karnataka (Abstr.). *Int. Conf. on Hort.*, Bangalore, p. 140.

Mahantesh, B., Ravi Prasad Sajjan, M. and Harshavardhan, M. 2009b. Influence of inorganic and biofertilizers on dry matter production and yield of rainy season onion (*Allium cepa* L.) cv. Bellary Red (Abstr.). *Int. Conf. on Hort.*, Bangalore, p. 141.

Mahendran, P.P and Kumar. N. 1996. A note on the effect of biofertilizers in garlic (*Allium sativum* L.). *S. Indian Hort.* **44:** 170-171.

Mahendran, P. P. and Kumar. N. 1998. Effect of biofertilizers on tuber yield and certain quality parameters of potato cv. Kufri Jyoti. *S. Indian Hort.* **46:** 97 - 98.

Majumdar, P.K., Sharma, D.K. and Singh, R.N. 1982. A study on high density orcharding in mango (*Mangifera indica* L.) var. Amrapali. *Punjab Hort. J.* **22**: 123-127.

Mandal, R.C., Singh, K.D., Saraswat, V.N. and Velappan, E. 1972. Possibilities of growing tuber crops in multiple cropping. *Proc. Multiple Cropping*, Haryana Agri. Univ., Hisar, p. 307.

Malik, R.S., Kumar, K. and Bhandari, A.R. 1994. Effect of urea application through drip irrigation system on nitrate distribution in loam sand soils and pea yield. *J. Indian Soc. Soil Sci.* **42**: 6-10.

Malik, T.P., Tehlan, S.K., Nandal, J.K. and Bhatia, A.K. 2009. Effect of rhizobacteria on growth and seed yield of coriander. *Int. Conf. on Hort.*, Bangalore, p. 357.

Maniram, Chatterjee, B.N., Yadav, R.L. and Singh, D.V. 1987. Effect of different nitrogen carriers on growth, accumulation pattern and yield of essential oil in Japanese mint. *Indian J. Agron.* **32**: 417-424.

Manoj Kumar, R., Rao, M.S. and Ramachandran, N. 2010. Biomanagement of disease complex caused by *Meloidogyne incognita* and *Phytophthora parasitica* Dastur in protected cultivation of gerbera using formulation of *Trichoderma harzianum* and *Pseudomonas fluorescens. National Conf. on Innovations in Nematological Research for Agricultural Sustainability – Challenges and A Roadmap Ahead*, Tamil Nadu Agri. Univ., Coimbatore, pp. 97-98.

Mantur, S.M., Patil, H.B. and Patil, D.R. 2006. Influence of nutrition and planting geometry on productivity of capsicum grown under shade house. *National Symp. on Improving Input Use Efficiency in Hort.*, Indian Inst. of Hort. Res., Bangalore, p. 119.

Martin-Prevel, P. 1964. Nutrient elements in the banana plant and fruit. *Fertilizer News* **22**: 3-14.

Martin-Prevel, P. 1992. Banana. In *IFA, World Fertilizer Manual* (W. Wichman, ed.), pp.398-409. International Fertilizer Industry Association, Paris.

Martin-Prevel, P., Lacoevilhe, J.J. and Marshall, J. 1968. Mineral elements in the Gros Michael banana plant in Cameroon. *Potash Rev. Sub.* **27**(3): 65.

Masood Ali. 1986. *Trop. Pest Mangmt.* **34**: 384-387.

Mathew, J. and Mini, C. 2006. Improving resource use efficiency. *National Symp. on Improving Input Use Efficiency in Hort.*, Indian Inst. of Hort. Res., Bangalore, p. 139.

Mathur, N.M., Sharma, G.K. and Qureshi, Q.G. 1998. Fruit borer (*Earias* spp.) management on okra in semi-arid region of Rajasthan. In *Advances in IPM for Horticultural Crops* (P. Parvatha Reddy, N.K. Krishna Kumar and A. Verghese, eds.), pp. 121-123. Association for Advancement of Pest Management in Horticultural Ecosystems, Bangalore.

Mathura Rai. Potentialities of horticultural development in eastern plateau. *National Seminar on Sustainable Horticultural Production in Tribal Regions,* Central Hort. Expt. Stn., Ranchi, pp. 23-30.

McLand, M. and Ystaas, J. 1998. Yield and fruit quality of 'Van' sweet cherry in four high density production systems over seven years. *Acta Hort.* **468**: 425-432.

Meena, S.S., Anwar, M.M., Lal, G. and Kakani, R.K. 2009. Effect of weed management practices on productivity of cumin (*Cuminum cyminum* L.) (Abstr.). *Int. Conf. on Hort.*, Bangalore, p. 199.

Mellor, F.C. and Stace-Smith, R., 1977. Virus-free potatoes by tissue culture. In *Applied and Fundamental Aspects of Plant Cell, Tissue and Organ Cultures* (J. Reinert and Y.P.S. Bajaj, eds.), pp. 616-635. Springer-Verlag, Berlin.

Meng, F.W., Wang, J.B., Cheng, Y.J., Chen, L.Q. and Wn, L.Y. 1997. Summary of early production techniques for high density apple orchard established on river basin soil. *China Fruits* **4**: 32-33.

Mirshekari, B., Frahvash, F., Javanshir, A. and Moghbeli, A.H.H.Z. 2009a. Essential oil of fennel (*Foeniculum vulgare* L.) affected by seed inoculation with *Azospirillum* strains and seed coating with micro elements in semi-arid region (Abstr.). *Int. Conf. on Hort.*, Bangalore, p.202.

Mirshekari, B., Frahvash, F., Javanshir, A., Sadegi, M.N. and Moghbeli, A.H.H.Z. 2009b. Management of irrigation and nitrogen fertilizing of fennel (*Foeniculum vulgare* L.) in semi-arid region (Abstr.). *Int. Conf. on Hort.*, Bangalore, p.203.

Mishra, D., Lal, B. and Verma, A.K. 2006. Canopy management in mango cv. Mallika to enhance solar energy utilization. *National Symp. on Improving Input Use Efficiency in Hort.*, Indian Inst. of Hort. Res., Bangalore, p. 169.

Mistry, P.M. and Patel, B.N. 2006. Improving land use efficiency through high density plantations on yield of mango cv. Kesar. *National Symp. on Improving Input Use Efficiency in Hort.*, Indian Inst. of Hort. Res., Bangalore, p. 122.

Mohammed, S., Wilson, L.A. and Prendergast, N. 1984. Gauva meadow orchard: Effect of ultra-high density plantings and growth regulators on growth, flowering and fruiting. *Trop. Agri.* **61**: 197-301.

Monastra, F., Salvadar, F.R., de Salvador, F.R. and Krayiannis, I. 1999. Comparison between vertical axis and y-shaped apricot training system. *Acta Hort.* **488**: 469-476.

Montagut, G and Martin-Prevel, P. 1965. Essais sol-plate sur bananeirs. 1. Besoins en engrais des bananeraries antillaises. 2. (et Lacoeuilhe, J.J) Nutrition minerale comparee dans six essais. *Fruits* **20:** 265-273, 398-410.

Morel, G. and Martin, C., 1952. Guérison de dahlias atteints d'une maladie à virus. C.R. *Acad. Sci.* **235**: 1324-1325.

Mosella, L.C., Signoret, P.A. and Jonard, R., 1980. Sur la mise au point de techniques de microgreffage d'apex en vue de l'élimination de deux types de particules virales chez le pêcher (*Prunus persica* Batsch). C.R. *Acad. Sci.* **290**: 287-290.

Mukhopadhyaya, M.C. and Dalal, M.R. 1971. Effect of two nematicides on *Tylenchulus semipenetrans* and on sweet lime yield. *Indian J. Nematol.* **1**: 95-97.

Murali Krishnasamy, S., Veerabhadran, V., Kumar, V. and Sakthivel, S. 2006a. Drip irrigation and fertigation in chillies (*Capsicum annuum*). *National Symp. on Improving Input Use Efficiency in Hort.*, Indian Inst. of Hort. Res., Bangalore, pp. 107-108.

Murali Krishnasamy, S., Veerabhadran, V., Kumar, V. and Sakthivel, S. 2006b. Micro sprinkler irrigation and fertigation in onion (*Allium cepa*). *National Symp. on Improving Input Use Efficiency in Hort.*, Indian Inst. of Hort. Res., Bangalore, pp. 108-109.

Mutnal, S.M. 1984. M. Sc. (Agri.) thesis, Univ. of Agri. Sci., Dharwad, Karnataka.

Nagesh, M. and Janakiram, T. 2004. Root-knot nematode problem in polyhouse roses and its management using dazomet, neem cake and *Pochonia chlamydosporia* (*Verticillium chlamydosporium*). *J. Orn. Hort.* **7**: 147-152.

Nagesh, M. and Parvatha Reddy, P. 1995. Comparative efficacy of *Paecilomyces lilacinus* (Thom.) Samson and *Verticillium lecanii* (A. Zimmerman) Viegas in combination with botanicals against *Meloidogyne incognita* (Kofoid and White, 1919) Chitwood, 1949 infecting *Crossandra undulaefolia* L. *J. Biol. Control* **9**: 109-112.

Nagesh, M. and Parvatha Reddy, P. 2005. Management of carnation and gerbera to control the root-knot nematode, *Meloidogyne incognita*, in commercial polyhouses. *Nematol. Mediter.* **33**: 157-162.

Nagesh, M., Parvatha Reddy, P. and Rao, M.S. 1998. Comparative efficacy of VAM fungi in combination with neem cake against *Meloidogyne incognita* on *Crossandra undulaefolia. Mycorrhiza News* **11**: 11-13.

Nagesh, M., Parvatha Reddy, P., Vijaya Kumar, M.V. and Nagaraju, B.M. 1998. Management of *Meloidogyne incognita* on *Polianthes tuberosa* using parasitic fungus, *Paecilimyces lilacinus* in combination with oil cakes. In *Advances in IPM for Horticultural Crops* (P. Parvatha Reddy, N.K. Krishna Kumar and A. Verghese, eds.), pp. 339-344. Association for Advancement of Pest Management in Horticultural Ecosystems, Bangalore.

Nagre, P.K., Gurav, S.B., Katwate, S.M., Raut, V.S. and Nandre, D.R. 2006b. Development of greenhouse production technology for tomato in plains. *National Symp. on Improving Input Use Efficiency in Hort.*, Indian Inst. of Hort. Res., Bangalore, pp. 136-137.

Nagre, P.K., Katwate, S.M., Gurav, S.B., Gonge, V.S. and Bharad, S.G. 2006a. Development of greenhouse production technology for capsicum in plains. *National Symp. on Improving Input Use Efficiency in Hort.*, Indian Inst. of Hort. Res., Bangalore, p. 136.

Naik, D. 2004. Biotechnological Approaches for the Management of Wilt Disease Complex in Capsicum (*Capsicum annum* L.) and Egg plant (Solanum melongena) with Special Emphasis on Biological Control. Ph.D. thesis, Kuvempu Univ., Shimoga.

Naik, D., Rao, M.S., Shylaja, M., Reddy, M.S. and Rahiman, B.A. 2003. Management of wilt disease complex in capsicum using *Bacillus pumilus* and *Paecilomyces lilacinus. Proc. of Int. Plant Growth Promoting Rhizobacteria Workshop*, Calicut, India, pp. 276-281.

Nalina, L. 2002. Standardization of Fertilizer Requirement for Tissue Cultured Banana cv. Robusta. Ph.D. thesis, Tamil Nadu Agricultural University, Coimbatore.

Nambiar, K.K.N. (ed.). 1996. *Basal Stem Rot Disease of Coconut*. Tech. Bull., Central Plantation Crops Res. Inst., Kasargod, Kerala.

Nandi, A., Paul, P.K., Nayak, S.C., Beura, S.K., Prasad, G. and Dash, S.K. 2006. Efficiency of foliar micronutrient sprays in respect of yield and economics of hybrid cabbage (*Brassica oleracea* var. *capitata*). *National Symp. on Improving Input Use Efficiency in Hort.*, Indian Inst. of Hort. Res., Bangalore, p. 131.

Nankare, J.T. 1990. Scope and prospects for intercropping of potato with sugarcane in Maharashtra state. *Field Crop Res.* **25**: 123-132.

Narayanan, K. and Gopalakrishnan, G. 1987. Effect of nuclear polyhedrosis virus for the control of *Heliothis armigera* on tomato. *Int. Heliothis Biol. Control Newsl.* **6**: 10.

Nataraja, A., Kumar, D.P., Vasundhara, M., Srikantaprasad, D. and Nuthan, D. 2009. Studies on effect of moisture regimes, INM and stage of harvest on growth, yield and quality of vetiver (*Chrysopogon zizanioides* (L.) Roberty) (Abstr.). *Int. Conf. on Hort.*, Bangalore, p. 200.

Natarajan, S., Sasikala, S. and Kumaresan, G.R. 2005. Influence of growing media, irrigation regime, nutrient management and mulching on growth, yield and economics of tomato under polyhouse condition. *Int. Conf. on Plasticulture and Precision Farming,* TNAU, Coimbatore.

Navarro, l. and Juarez, J. 1977. Tissue culture techniques used in Spain to recover virus-free citrus plants. *Acta Hort.* **78**: 425-423.

Nayak, M.G., Swamy, K.R.M. and Bhat, M.G. 2006. Efficacy of paclobutrazol in enhancing the space and labour use efficiency in cashew (*Anacardium occidentale* L.). *National Symp. on Improving Input use Efficiency in Hort.*, Indian Inst. of Hort. Res., Bangalore, p. 212.

Nel, Chris. 1993. Transforming the mango industry. *Farmers' Weekly,* September 10, 1993, pp. 14-17.

Nielsen, G.H., Nielsen, D., Herbert, L.C. and Hogue, E.J. 2004. Response of apple to fertigation of N and K under conditions susceptible to the development of K deficiency. *J. Am. Soc. Hort. Sci.* **129**: 26-31.

Niranjana, N.S., Meera Bai, M. And Sivaprasad, P. 1999. Interrelationship between fertilizer application and disease incidence an investigation in *Amaranthus tricolour* L. *Proc. of National Seminar on Participatory Approaches for Horticulture Development*, Thiruvanathapuram. Kerala Hort. Devp. Programme, Kerala, pp. 258-260.

Ogawa, K. and Komada, H. 1984. Biological control of Fusarium wilt of sweet potato with cross protection by non-pathogenic *Fusarium oxysporum*. In *Ecology and Management of Soil-borne Plant Pathogens* (C.A. Parker, A.D. Rovira, K.J. Moore, P.T.W. Wong and J.F. Kollmorgen, eds.), pp. 121-123. The American Phytopathological Society, St. Paul, Minn.

Ogawa, K. and Komada, H. 1986. Induction of systemic resistance against Fusarium wilt of sweet potato by non-pathogenic *Fusarium oxysporum*. *Ann. Phytopath. Soc. Japan* **52**: 15-21.

Ojha, K.L and Yadav, B.P. 2000. Rajendra Agricultural University, Pusa, Bihar. In Progress Report for the Period Sept. 1997 - Aug. 2000 of National Network Project on Phytophthora Diseases of Horticultural Crops (S.R. Sarma, ed.), pp. 79-88. Indian Inst. of Spices Research, Calicut, Kerala.

Padmadevi, K., Jawaharlal, M. and Vijayakumar, M. 2004. Effect of Biofertilizers on floral characters and vase life of Anthurium (*Anthurium andreanum* Lind.) cv. Temptation. *S. Indian Hort.* **52**: 228-231.

Palaniappan, S.P., Jeyabal, A. and Chellaiah, S. 1999. Response of tomato and chilli to foliar application of specialty fertilizers. *Veg. Sci.* **26**: 198-200.

Palwe, C.R., Patil, T.D., Shinde, S.S., Budgujar, C.D. and Deshmukh, S.S. 1999a. Effect of different levels of potassium on yield and uptake of Basrai banana. *Seminar on Technological Advancement in Banana Production, Handling and Processing Management*, Jalgaon, Maharashtra, pp. 65-67.

Palwe, C.R., Patil, T.D., Shinde, S.S., Budgujar, C.D. and Deshmukh, S.S. 1999b. Standardization of organic and inorganic fertilizer requirement in banana cv. Basrai. *Seminar on Technological Advancement in Banana Production, Handling and Processing Management*, Jalgaon, Maharashtra, pp. 68-71.

Pampanna, Y. 1992. *Studies on the Softwood Grafting in Sapota.* M. Sc. (Ag.) thesis, Univ. of Agri. Sci., Dharwad.

Pandey, R. 1994b. Comparative performance of oil seed cakes and pesticides in management of root-knot nematode disease of davana. *Nematol. Medit.* **22**:17-19.

Pandey, R. 2005b. Management of *Meloidogyne incognita* in *Artemisia pallens* with bioagents. *Phytoparasitica* **33**: 304-308.

Pandey, R., Gupta, M.L., Singh, H.B., Singh, H.N. and Kumar, S. 2000. Disease management potentialities of VAM fungi to *Meloidogyne incognita* on *Hyoscyamus muticus. Proc. of Int. Conf. Integrated Plant Dis. Mangmt. for Sustainable Agri.*, Indian Agri. Res. Inst., New Delhi, pp. 1239-1240.

Pandey, R. and Kalra, A. 2005a. Chemical activators: A novel and sustainable management approach for *Meloidogyne incognita* (Kofoid and White) Chitwood in *Chamomile recutita* L. *Archives of Phytopathology and Plant Protection* **38**: 107-111.

Pandey, R. and Kalra, A. 2005b. Root-knot disease of ashwagandha *Withania somnifera* and its eco-friendly cost effective management. *J. Mycol. Plant Pathol.* **33**: 240-245.

Pandey, R., Kalra, A., Gupta, M.L and Sharma, S.C. 2002. *Bacopa monnieri* damaged by root-knot nematode and remedial measures. *J. Trop. Medicinal Plants*

Pandey, R., Kalra, A., Gupta, M.L and Sharma, P. 2003. Phytonematodes: Major pest of MAPs. In *Proceedings of First National Interactive Meet on Medicinal and Aromatic Plants* (A.K. Mathur, S. Dwivedi, D.D. Patra, G.D. Bagchi, N.S. Sangwan, A. Sharma and S.P.S. Khanuja, eds.), pp. 188-197. CIMAP, Lucknow, India.

Pandey, R., Singh, H.B., and Gupta, M.L. 1997. Antagonistic impact of vesicular arbuscular mycorrhizae (VAM) on *Meloidogyne incognita* population development in Japanese mint. *Int. J. Trop. Plant Dis.* **15**: 237-245.

Pandey, R., Singh, H.B., Gupta, M.L., Singh, H.N., Sharma, S. and Kumar, S. 1998. Efficacy of VAM fungi, green manure, organic materials and pesticides in the management of root-knot nematode in menthol mint. *Proc. of National Symp. on Rational Approaches in Nematode Mangmt. for Sustainable Agri* (S.C. Dhawan and K.K. Kaushal, eds.), pp. 47-50. Gujarat Agri. Univ., Anand

Pandey, S.N. and Majumdar, P.K. 1988. Research Report, Fruit Research Workshop on Subtropical and Temperate Fruits, Pusa, Bihar, p. 34.

Papadopoulos, I. 1987. Nitrogen fertigation of greenhouse grown strawberries. *Fertilizer Res.* **13**: 269-276.

Pariari, A. and Khan, S. 2009. Integrated nutrient management of coriander (*Coriandrum sativum*) (Abstr.). *Int. Conf. on Hort.*, Bangalore, p. 192.

Paruthi, I.J., Nandal, S.N. and Verma, K.K. 2010. Field evaluation of mustard and neem cake against rot-knot nematode (*Meloidogyne* spp) in bottle gourd. *National Conf. on Innovations in Nematological Research for Agricultural Sustainability – Challenges and A Roadmap Ahead.* Tamil Nadu Agri. Univ., Coimbatore, p. 84.

Parvatha Reddy, P. and Nagesh, M. 2002. *Avermectins – Isolation, Fermentation, Preliminary Charecterization and Screening for Nematicidal Activity.* Tech. Bull.17, Indian Inst. of Hortil. Research, Bangalore, 28 pp.

Parvatha Reddy, P., Nagesh, M. and Devappa, V. 1997. Effect of integration of *Pasteuria penetrans, Paecilomyces lilacinus* and neem cake for the management of root-knot nematode infecting tomato. *Pest Mangmt. Hortil. Ecosystems* **3**: 100-104.

Parvatha Reddy, P., Nagesh, M. and Devappa, V. 1998a. Management of *Meloidogyne incognita* on tomato by integration of *Trichoderma harzianum* and neem cake. In *Advances in IPM for Horticultural Crops* (P. Parvatha Reddy, N.K. Krishna Kumar and A. Verghese, eds.), pp. 83-86. Association for Advancement of Pest Management in Horticultural Ecosystems, Bangalore.

Parvatha Reddy, P., Nagesh, M. and Devappa, V. 1998b. Management of *Meloidogyne incognita* on tomato by integration of *Trichoderma harzianum, Glomus fasciculatum* and neem cake. In *Advances in IPM for Horticultural Crops* (P. Parvatha Reddy, N.K. Krishna Kumar and A. Verghese, eds.), pp. 349-352. Association for Advancement of Pest Management in Horticultural Ecosystems, Bangalore.

Parvatha Reddy, P., Rao, M.S. and Nagesh, M. 2002. Integrated management of burrowing nematode (*Radopholus similis*) using endomycorrhiza (*Glomus mosseae*) and oil cakes in banana. In *Banana – Improvement, Production and Utilization* (H.P. Singh and K.L. Chadha, eds.), pp. 344-348. AIPUB, Trichy.

Patel, C.B., Patel, A.N. and Patel, A.R. 1999. Identification of critical stages of weeding and their influence on yield of banana cv. Basrai. *Seminar on Technological Advancement in Banana Production, Handling and Processing Management*, Jalgaon, Maharashtra, pp. 87-90.

Patel, D.J., Patel, B.A., Patel, S.K. and Patel, A.D. 2010. Innovations in extension nematology. *National Conf. on Innovations in Nematological Research for Agricultural Sustainability – Challenges and A Roadmap Ahead*, Tamil Nadu Agri. Univ., Coimbatore, pp. 63-64.

Patel, G.J., Vadar, H.R. and Patel, R.A. 2006. Drip irrigation for okra and its economics in middle Gujarat condition. *National Symp. on Improving Input Use Efficiency in Hort.*, Indian Inst. of Hort. Res., Bangalore, p. 95.

Patel, J.A., Patel, G.J., Vadar, H.R. and Patel, R.A. 2006. Effect of drip irrigation on yield of cabbage. *National Symp. on Improving Input Use Efficiency in Hort.*, Indian Inst. of Hort. Res., Bangalore, p. 95.

Pathak, R.K. and Pathak, R.A. 1991. Peaches. In *Temperate Fruits* (S.K. Mitra, T.K. Bose and D.S. Rathore, EDS.), PP. 150-180. Horticulture and Allied Publishers, Kolkata.

Patil, B.C., Naik, H.B., Kulkarni, B.S., Jagadeesh, H.L. and Nalwadi, U.G. 2004. Comparison of effect of split application of nitrogen at different stages on growth and flower yield of galliardia on red and black soils. *J. Orn. Hort.* **7**: 138-143.

Patil, M.B., Kadam, S.B., Patil, R.J. and Shitole, D.S. 2006. Effect of organic and inorganic fertilizers on seed production of onion cv. Phule Suvarna. *National Symp. on Improving Input Use Efficiency in Hort.*, Indian Inst. of Hort. Res., Bangalore, p. 123.

Patil, M.P., Athani, S.I., Gasti, V.D., Mulge, R., Madalgeri, M.B., Ustad, A.I. and Kale, U.B. 2006. *National Symp. on Improving Input Use Efficiency in Hort.*, Indian Inst. of Hort. Res., Bangalore, p. 155.

Patil, N.M., Dharne, P.K. and Deshmukh, S.S. 1999a. Sucker dip treatment in monocrotophos with intercrops for the management of plant parasitic nematodes in banana. *Seminar on Technological Advancement in Banana Production, Handling and Processing Management*, Jalgaon, Maharashtra, pp. 126-128..

Patil, N.M., Dharne, P.K. and Deshmukh, S.S. 1999b. Disinfection of suckers with integration of monocrotophos for the management of plant parasitic nematodes in banana. *Seminar on Technological Advancement in Banana Production, Handling and Processing Management*, Jalgaon, Maharashtra, pp. 129-130.

Patil, S.R., Reddy, B.S. and Prashant, J.M. 2004. Effect of organic, inorganic and *in situ* vermiculture on chlorophyll content and flower yield of *Jasminum sambac* Ait. *J. Orn. Hort.* **7**: 164-167.

Patil, S.R., Reddy, B.S., Prashant, J.M. and Kulkarni, B.S. 2004. Effect of growth substances on growth and yield of golden rod (*Solidago canadensis* L.). *J. Orn. Hort.* **7**: 159-163.

Pattanayak, S.K. and Rao, D.L.N. 2006. Biofertilizers for improved nutrient use efficiency in vegetable crops. *National Symp. on Improving Input Use Efficiency in Hort.*, Indian Inst. of Hort. Res., Bangalore, p. 142.

Peter, K.V. 1999. Spices research and development – An updated review. *National Seminar on Sustainable Horticultural Production in Tribal Regions,* Central Hort. Expt. Stn., Ranchi, pp. 48-54.

Prabhakar, M. and Hebbar, S.S. 1998. Water management in chilli under rainfed conditions. In *Water and Nutrient Management for Sustainable Production and Quality of Spices*, Madikeri, Karnataka (A.K. Sadanandan *et al.*, eds.), pp. 131-135.

Prabhakar, M. and Hebbar, S.S. 2005. Improved technologies for production of vegetables. In *Souvenir of Rashtriya Bagwani Kisan Mela* (R.R. Suresh, D. Sreenivasa Murthy, Leela Sahijram and V. Keshava Rao, eds.), pp. 62-66. Indian Inst. of Hort. Res., Bangalore.

Prabhakar, M., Hebbar, S.S., Anjaneyulu, K. and Gayatri, M. 2006. Nitrogen and water use efficiency and crop performance under fertigation in watermelon (*Citrullus lanatus*). *National Symp. on Improving Input Use Efficiency in Hort.*, Indian Inst. of Hort. Res., Bangalore, p. 146.

Prabhakar, M., Naik, C.L. and Savanur, V. 2001. Fertigation studies in hybrid tomato. *S. Indian Hort.* **49** (Special): 98-100.

Prabhu, M. 2001. *Studies on Nitrogen and Phosphorus Requirement of Brinjal Hybrid COBH 1.* M.Sc. (Hort.) thesis, Tamil Nadu Agricultural University, Coimbatore.

Prasad, K.K., Chowdhary, B.M., Kumar Amrendra and Kumar, A. 1997. Response of tomato to boron application in Chotanagpur region. *J. Res. Birsa Agri. Univ.* **9**: 145-147.

Prasad, V.M., Marabi, J.S., Fatmi, U., Roshan, R.K. and Sharma, M.C. 2009. Effect of biofertilizers and organic manures on growth and yield of isabgol (*Plantago ovata* Forsk) cv. Niharika. *Int. Conf. on Hort.*, Bangalore, p. 358.

Punitha, A. 2007. Studies on the Effect of Pre-harvest Management and Post-harvest Techniques in Carnation (*Dianthus caryophyllus* L.) under Polyhouse Condition. Ph. D. thesis. Tamil Nadu Agri. Univ., Coimbatore.

Puttaraju, T.B., Nache Gowda, V., Krishna, H.C., Lingaiah, H.B. and Suresh Kumar, B. 2009. Effect of weed management on growth and yield of garlic (*Allium sativum* L.) cv. Rajelle under eastern dry zone of Karnataka (Abstr.). *Int. Conf. on Hort.*, Bangalore, p. 142.

Qureshi, Q.G., Mathur, N.M. and Srivastava, R.C. 1998. Management of brinjal shoot and fruit borer, *Leucinodes orbonalis* Guenee, using *Bacillus thuringiensis* var. *kurstaki*. In *Advances in IPM for Horticultural Crops* (P. Parvatha Reddy, N.K. Krishna Kumar and A. Verghese, eds.), pp. 81-83. Association for Advancement of Pest Management in Horticultural Ecosystems, Bangalore.

Raghumoorthi, K.N. and Subba Rao, P.V. 1998. IPM for moringa fruitfly *Gitona distigma* Meigon (Diptera: Drosophilidae). In *Advances in IPM for Horticultural Crops* (P. Parvatha Reddy, N.K. Krishna Kumar and A. Verghese, eds.), pp. 140-144. Association for Advancement of Pest Management in Horticultural Ecosystems, Bangalore.

Raghuramulu, Y. 1996. Water management in coffee. *Proc. of the All India Seminar on Modern Irrigation Techniques*, Bangalore, pp. 129-134.

Raina, J.N., Thakur, B.C. and Bhandari, A.R. 1999. Effect of mulches on soil hydrothermal regime, root growth, yield and quality of strawberry. *64th Conv. Indian Soc. Soil Sci.* **1**: 13.

Raina, J.N., Thakur, B.C. and Verma, M.L. 1999. Effect of drip irrigation and polythene mulch on yield, quality and water use efficiency of tomato. *Indian J. Agric. Sci.* **69**: 430-433.

Raj, H. and Bhardwaj, M.L. 2000. Soil solarization for controlling soil borne pathogens in vegetable crops. *Indian J. Agric. Sci.* **70**: 305-307.

Raja, J., Rajendran, B. and Pappiah, C.M. 1998a. Management of egg plant shoot and fruit borer *Leucinodes orbonalis* Guen. In *Advances in IPM for Horticultural Crops* (P. Parvatha Reddy, N.K. Krishna Kumar and A. Verghese, eds.), pp. 84-86. Association for Advancement of Pest Management in Horticultural Ecosystems, Bangalore.

Raja, J., Rajendran, B. and Pappiah, C.M. 1998b. Management of bhendi fruit borer, *Earias vitella* (F.). In *Advances in IPM for Horticultural Crops* (P. Parvatha Reddy, N.K. Krishna Kumar and A. Verghese, eds.), pp. 118-120. Association for Advancement of Pest Management in Horticultural Ecosystems, Bangalore.

Rajagopal, V., Arulraj, S. and Sairam, C.V. 2005. Role of production technology for sustainable growth in Indian coconut sector. In *Souvenir of Rashtriya Bagwani Kisan Mela* (R.R. Suresh, D. Sreenivasa Murthy, Leela Sahijram and V. Keshava Rao, eds.), pp. 85-91. Indian Inst. of Hort. Res., Bangalore.

Rajalingam, G.V. and Kumar, N. 2001. Effect of digested coir pith compost and biofertilizers on green leaf yield of tea. In *Changing Scenario in the Production of Horticultural Crops*, pp. 142-143.

Rajendran, B. 1998. Effect of neem oil and insecticide on the population and parasitization of eggplant spotted beetle, *Henosepilachna vignitioctopunctata* F. (Coccinellidae: Coleoptera). In *Advances in IPM for Horticultural Crops* (P. Parvatha Reddy, N.K. Krishna Kumar and A. Verghese, eds.), pp. 102-105. Association for Advancement of Pest Management in Horticultural Ecosystems, Bangalore.

Rajesh Kanna, R. 1999. *Studies on Nutrient Requirement of Tomato Hybrid COTH 1.* M.Sc. (Hort.) thesis, Tamil Nadu Agricultural University, Coimbatore.

Rajput, S.G., Joshi, K.I., Kakade, D.K. and Kulkarni, G.U. 2006. Effect of foliar application of Fe and Zn on growth, flowering and yield of China aster (*Calistephus chinensis* L. Nees). *National Symp. on Improving Input Use Efficiency in Hort.*, Indian Inst. of Hort. Res., Bangalore, pp. 161.

Ram, S. 2000. Transforming the mango industry through high density. *J. Appl. Hort.* **2**: 2-5.

Ram, S., Singh, C.P. and Kumar, S. 1997. A success story of high density orcharding in mango. *Acta Hort.* 455: 375-382.

Ram, S., Singh, C.P. and Shukla, P. 2001. Effect of different planting densities on growth and yield of mango. *Indian J. Hort.*

Ramakrishnan, S. 2010. Field evaluation of super Pseudomonas in the management of root-knot nematode Meloidogyne incognita in medicinal crops. National Conf. on Innovations in Nematological Research for Agricultural Sustainability – Challenges and A Roadmap Ahead. Tamil Nadu Agri. Univ., Coimbatore, p. 93.

Raman, S. 1999. Status of research on micro-irrigation for improving water use efficiency in some horticultural crops. *Proc. of National Seminar on Problems and Prospects of Micro-irrigation – A Critical Appraisal*, Bangalore, pp. 31-45.

Ramanujam, B., Prasad, R.D. and Rangeswaran, R. 2003. Achievements in biological control of diseases with antagonistic organisms at Project Directorate of Biological Control, Bangalore. *Proc. of the Group Meeting on Antagonistic Organisms in Plant Disease Management*, Project Directorate of Biological Control, Bangalore, pp. 258-271.

Ramaswamy, N. 1971. *Studies on the Effect of Nitrogen on the Growth and Development of 'Robusta'* (*Musa cavendishi* L.). M. Sc. (Ag.) thesis, Annamalai Univ., Annamalinagar.

Ramesh Kumar, A and Kumar, N. 2007. Sulfate of potash foliar spray effects on yield, quality and post harvest life of banana. *Better crops* **91:** 22-24.

Ranpise, S.A, Bharmal, V.S. and Darwade, R.T. 2004. Effect of different levels of indole butyric acid (IBA) on rooting, growth and flower yield of chrysanthemum cv. Sonali Tara. *J. Orn. Hort.* **7**: 331-337.

Rao, M.M. 1997. Studies on the improvement of finger size in Munavalli (Musa AAB) banana. In *Research and Development in Fruit Crops in North Karnataka* (M.M. Rao and G.S. Sulikeri, eds.), pp. 59-61. Univ. of Agri. Sci., Dharwad, Karnataka.

Rao, M.M., Narasimhan, P., Nagaraja, N. and Anandaswamy, B. 1968. Effect of naphthalene acetic acid and p-chlorophenoxy acetic acid on control of berry drop in Anab-e-Shahi grape. *J. Food Sci. Tech.* **5**: 127-128.

Rao, M.M. and Shankarnarayanan, H.N. 1982. Effect of horticultural practices and growth regulators on colouration and ripening of Gulabi grapes. *Mahrashtra J. Hort. Sci.* **1**: 41-42.

Rao, M.M., Srihari, D., Patil, V.S. and Madalgeri, M.B. 1997. Further studies on chemical induction of flowering on fruited shoots in "off" phase Alphanso mango trees. *Karnataka J. Agri. Sci.* **10**: 598-601.

Rao, M.M. and Sulikeri, G.S. 1997. Technology for mass production of planting material in guava. In *Research and Development in Fruit Crops in North Karnataka* (M.M. Rao and G.S. Sulikeri, eds.), pp. 43-45. Univ. of Agri. Sci., Dharwad, Karnataka.

Rao, M.S., Naik, D. and Shylaja, M. 2003. Management of *Meloidogyne incognita* on egg-plant using a formulation of *Pochonia chlamydosporia* Zare *et al.* (*Verticillium chlamydosporium* Goddard). *Pest Mangmt. Hort. Ecosystems* **9**: 71-76.

Rao, M.S., Naik, D. and Shylaja, M. 2004. Bio-intensive management of root-knot nematodes on bell pepper using *Pochonia chlamydosporia* and *Pseudomonas fluorescens. Nematol. Medit.* **32**: 159-163.

Rao, M.S., Naik, D., Shylaja, M. and Parvatha Reddy, P. 2002. Prospects for the management of the root-knot nematode and bacterial wilt disease complex in capsicum by integration of fungal and bacterial biocontrol agents. *Proc. Int. Conf. Vegetables,* Bangalore, pp. 347-351.

Rao, M.S. and Parvatha Reddy, P. 2001. Control of *Meloidogyne incognita* on egg plant using *Glomus mosseae* integrated with *Paecilomyces lilacinus* and neem cake. *Nematol. Medit.* **29**: 153-157.

Rao, M.S., Parvatha Reddy, P. and Mohan Das, S. 1995. Effect of integration of endomycorrhiza (*Glomus mosseae*) and neem cake on the control of root-knot nematode on tomato. *J. Pl. Dis. Prot.* **102**: 526-529.

Rao, M.S., Parvatha Reddy, P. and Nagesh, M. 1997. Integrated management of *Meloidogyne incognita* on okra by castor cake suspension and *Paecilomyces lilacinus*. *Nematol. Mediter.* **25**: 17-19.

Rao, M.S., Parvatha Reddy, P. and Nagesh, M. 1998. Integrated management of *Meloidogyne incognita* on tomato using *Verticillium chlamydosporium* and *Pasteuria penetrans. Pest Mangmt. Hort. Ecosystems* **4**: 32-35.

Rao, M.S., Ramachandran, N., Sowmya, D.S., Ratnamma, K., Chaya, M.K. and Manoj Kumar, R. 2009. Biological control of nematode induced disease complex in certain vegetable crops. *Int. Conf. on Hort.*, Bangalore.

Rao, M.S. and Shylaja, M. 2003. Management of root-knot nematode (*Meloidogyne incognita*) in carnation using a formulation of *Pochonia chlamydosporia* (*Verticillium chlamydosporium*). *J. Orn. Hort.* **6**: 202-206.

Rao, M.S. and Shylaja, M. 2004. Role of *Pseudomonas fluorescens* (Migula) in induction of systemic resistance (ISR) and managing *Rotylenchulus reniformis* (Linford and Oliveira) on carrot. *Pest Mangmt. Hort. Ecosystems* **10**: 87-93.

Rao, M.S., Shylaja, M. and Naik, D. 2003. Management of nematode induced wilt disease complex in tuberose (*Polianthes tuberosa* L.) cultivar Prajwal using *Pochonia chlamydosporia* (*Verticillium chlamydosporium*) and *Trichoderma harzianum*. *J. Orn. Hort.* **6**: 341-346.

Rao, M.S., Shylaja, M. and Parvatha Reddy, P. 2004. Bio-management of *Meloidogyne incognita* on tuberose using a formulation of *Pochonia chlamydosporia*. *Nematol. Medit.* **32**: 165-167.

Rao, M.S., Sowmya, D.S., Chaya, M.K. and Manoj, R. 2010. Management of nematodes on banana using bio-agents. *National Conf. on Innovations in Nematological Research for Agricultural Sustainability – Challenges and A Roadmap Ahead.* Tamil Nadu Agri. Univ., Coimbatore, pp. 99-100.

Rathore, D.S. 1991. Cherries. In *Temperate Fruits* (S.K. Mitra, T.K. Bose and D.S. Rathore, eds.), pp. 309-343. Horticulture and Allied Publishers, Kolkata.

Rathour, K.S, Sharma, A. and Ganguly, S. 2010. Yield enhancement of brinjal by using entomopathogenic nematodes under field conditions. *National Conf. on Innovations in Nematological Research for Agricultural Sustainability – Challenges and A Roadmap Ahead.* Tamil Nadu Agri. Univ., Coimbatore, pp. 99-134.

Raturi, G.B. and Hiwale, S.S. 1999. Horti-silvi-pastoral system – A boon to the tribal farmers. *National Seminar on Sustainable Horticultural Production in Tribal Regions,* Central Hort. Expt. Stn., Ranchi, pp. 34-39.

Ravi, K., Nanje Gowda, D. and Parvatha Reddy, P. 2001. Integrated management of disease complex in banana involving burrowing nematode and Panama wilt – An ecofriendly approach. In *IPM in Horticultural Crops* (A. Verghese and P. Parvatha Reddy, eds.), p. 154. AAPMHE, Bangalore.

Ravichandra, N.G. and Reddy, B.M.R. 2008. Efficacy of *Pasteuria penetrans* in the management of *Meloidogyne incognita* infecting tomato. *Indian J. Nematol.* **38**: 172-175.

Ravichandran, V., Kumar, N., Jeyakumar, P., Soorianathasundaram, K. and Vijayakumar, R.M. 2002. Influence of planting density and nutrient levels on growth and yield of papaya cv. CO-2. *S. Indian Hort.* **55**: 23-29.

Reader, B.D., Bowen, H.H. and Aldred, W.H. 1980. Peach tree training and spacing. *Hort. Sci.* **15**: 580-581.

Redenbaugh, K., 1993. *Synseeds. Applications of Synthetic Seeds to Crop Improvement*. CRC Press, Boca Raton, Ann Arbor, London, Tokyo, 481 pp.

Redenbaugh, K., Fujii, J., Slade, D., Viss, P., and Kossler, M., 1991. Artificial seeds encapsulated somatic embryos. In *Biotechnology in Agriculture and Forestry*, Vol. 17, *High-Tech and Micropropagation 1* (Y.P.S. Bajaj, ed.), pp. 395-416. Springer-Verlag, Berlin.

Reish, B., 1983. Genetic variability in regenerated plants. In *Handbook of Plant Cell Culture, Vol. 1* (W.R. Sharp, D.A. Evans, P.V. Ammirato, Y. Yamada, eds.), pp 748-769. Macmillan Publishing Company, New York.

Renuka, B. and Ravishankar, C. 2001. Effect of organic manures on growth and yield of tomato. *S. Indian Hort.* **49**: 249-251.

Reuther, W., Batchelor, H.J. and Webber, H.J. 1962. *Citrus Industry, Vol. II.* Univ. of California Press, Berkeley, California.

Rizvi, Zeenat, Sharma, V.K. and Rizvi, Z. 1990. Synergistic effect of cyanobacteria and DAP on tomato yield. *Sci. Cult.* **56**: 129-131.

Roy, S.S. and Hore, J.K. 2009. Influence of biofertilizer along with inorganic nutrition on growth, yield and quality of turmeric grown as intercrop in areca nut plantation (Abstr.). *Int. Conf. on Hort.*, Bangalore, pp. 194-195.

Sabina Ayyanna, B. and Farooki, A.A. 1995. *Studies on Sprinkler Irrigation with Normal and Water soluble NPK Fertilizers in Geranium (Pelargonium graveolens).* M.Sc. (Agri.) thesis, Univ. of Agri. Sci., Bangalore.

Sachan, S.K., Singh, D.V. and Chaudhary, A.S. 2006. IPM approach enhances the yield of capsicum. *National Symp. on Improving Input Use Efficiency in Hort.*, Indian Inst. of Hort. Res., Bangalore, pp. 187-188.

Sadanandan, A.K. 2001. Integrated plant nutrient system for yield and quality of spices. In *Spices Indica – Silver Jubilee Souvenir, Indian Inst. of Spices Research, Calicut-12* (Y.R. Sarma, B. Sasikumar and B. Champakam, eds.), pp. 168-171. Indian Inst. of Spices Research, Calicut, Kerala.

Saikia, M.K. and Borah, A. 2008. Comparative efficacy of fungal and bacterial bioagents with nematicides against *Meloidogyne incognita* on brinjal. *Indian J. Nematol.* **38**: 159-164.

Samra, J.S., Thakur, R.S. and Chadha, K.L. 1978. Comparison of some mango cultivars in terms of their macronutrient status in fruiting and non-fruiting terminals. *Indian J. Hort.* **35**: 144-187.

Sankar, V., Tripathi, P.C., Qureshi, A.A. and Lawande, K.E. 2001. Effect of organic sea weed extract on growth, yield and quality of onion (*Allium cepa* L.) var. N-2-4-1. *S. Indian Hort.* **49**: 247-248.

Sanwal, S.K., Patel, K.K. and Yadav, D.S. 2004. Vegetable production under protected conditions in NEH region: Problems and prospects. *Envis Bull. Himalayan Ecology* **12**(2): 6-12.

Saraswathi, T. 2005. In: *Horticultural Scientist Meet Report.* Tamil Nadu Agricultural University, Coimbatore.

Saravanan, T. 2007. Effect of Bioinoculants in Combination with Organic Manure and Inorganic Fertilizers on Growth, Yield and Quality of Mandarin Orange (*Citrus reticulata* Blanco). M.Sc. (Agri.) thesis, TNAU, Coimbatore.

Sarkar, A. and Ghosh, B. 2009. Growth regulators in controlling fruit drop, augmentation of fruit retention, yield, quality and economics in litchi cv. Bombai (Abstr.). *Int. Conf. on Hort.*, Bangalore, pp. 173-174.

Sarma, Y.R., Anandraj, M. And Venugopal, M.N. 1996. Biological control of diseases of spices. In *Biological Control in Spices* (M. Anandraj and K.V. Peter, eds.), pp. 1-19. Indian Inst. of Spices, Calicut, Kerala.

Sasikala, S., Natarajan, S., Kavitha, M. and Tamil Selvi, C. 2007. Influence of growing media, irrigation regime, integrated nutrient management and mulching and the performance of sweet pepper (*Capsicum annum* L. cv. Grossum) hybrid under polyhouse condition. *Int. J. Agric. Sci.* **3:** 56-59.

Sasser, J.N. and Freckman, D.W. 1987. A world perspective on Nematology: The role of the Society. In *Vistas on Nematology* (J.A. Veech and D.W. Dickson, eds.), pp. 7-14. Society of Nematologists, Hyattsville, Maryland.

Sathiyamurthy, V.A. and Pugalendhi, L. 2009. Studies on the effect of organic manures in okra-tomato-cowpea cropping system (Abstr.). *Int. Conf. on Hort.*, Bangalore, p. 350.

Satyanarayan, M. and Rao, R.R. 1985. Influence of plant density on growth and yield of Poovan banana. *J. Res. APAU* **13**: 35-38.

Sawke, D.P., Mane, M.S., Mahadkar, U.V. and Thorat, T.N. 2006. Effect of fertilizer and irrigation applied through diffuser on cashew nut yield in lateritic soil of Konkan. *National Symp. on Improving Input Use Efficiency in Hort.*, Indian Inst. of Hort. Res., Bangalore, p. 109.

Seenivasan, N and Devarajan, K. 2008. Integrated approach for the management of root-knot nematode, *Meloidogyne incognita* in medicinal coleus. *Indian J. Nematol.* **38**: 154-158.

Seibert, M., 1976. Shoot initiation from carnation shoot apices frozen to -196°C. *Science* **191**: 1178-1179.

Semal, J. and Lepoivre, P. 1992. Biotechnologie et Agriculture: Impact et perspectives. *Cahiers Agriculteurs* **13**: 153-162.

Senin, V.I. and Kovaleva, A.F. 1992. Productivity of apple trees on M-9 rootstock in density planted orchards. *Sadovodstvoi – vinogradarstvo.* **2**: 11-13.

Senthamarai, M., Poornima, K. and Subramanian, S. 2006. Bio-management of root-knot nematode, *Meloidogyne incognita* on *Coleus forskohlii* Briq. *Indian J. Nematol.* **36**: 206-208.

Senthil Kumar, P.S., Aruna Geetha, S., Savithri, P., Mahendran, P., Jagadeeswaran, R. and Ragunath, K.P. 2004. Effect of Zn enriched organic manures and zinc solubilizer application on yield, curcumin content and nutrient status of soil under turmeric cultivation. *J. Appl. Hort.* **6**: 82-86.

Senthil Kumar, T. and Rajendran, G. 2004. Biocontrol agents for the management of disease complex involving root-knot nematode, *Meloidogyne incognita* and *Fusarium moniliforme* on grapevine (*Vitis vinefera*). *Indian J. Nematol.* **34**:49-51.

Seshadri, K.V. and Madhavi, M. 2001. Effect of organic and inorganic manuring on twenty years old seedlings of sweet orange (*Citrus sinensis* ) cv. Sathgudi. *S. Indian Hort.* **49:** 122-125.

Shalini, M. and Patil, V.S. 2004. Effect of integrated weed management practices on vegetative, reproductive and yield parameters in gerbera (*Gerbera jamesonii* H. Bolus). *J. Orn. Hort.* **7**: 144-147.

Shanmugam, K.S. and Velayutham, K.S. 1972. Better manure your bananas. *Agric. Dig.* **3:** 17-29.

Shanthi, A. 2010. Management of root-knot nematode, Meloidogyne incognita in tomato using bioinoculants. National Conf. on Innovations in Nematological Research for Agricultural Sustainability – Challenges and A Roadmap Ahead. Tamil Nadu Agri. Univ., Coimbatore, p. 93.

Shanthi, A., Rajeswari, S. and Sivakumar, C.V. 1998. Soil application of *Pseudomonas fluorescens* (Migula) for the control of root-knot nematode (*Meloidogyne incognita*) on grapevine (*Vitis vinefera* Linn.). In *Nematology – Challenges and Opportunities in 21st Century,* (Usha K. Mehta, ed.), pp. 203-206. Sugarcane Breeding Institute, Coimbatore.

Shanthi, V.P. 2004. Utilization of Cement Kiln Flue Dust (CKFD) as a Source of Potassium for Banana Production. Ph.D. thesis, Tamil Nadu Agricultural University, Coimbatore.

Sharma, B.D., Dhandar, D.G. and Bhargava, R. 2006a. Water use efficiency in pomegranate under micro-irrigation systems. *National Symp. on Improving Input Use Efficiency in Hort.*, Indian Inst. of Hort. Res., Bangalore, p. 100.

Sharma, B.D., Dhandar, D.G. and Bhargava, R. 2006b. Nitrogen use efficiency in ber (*Ziziphus mauritiana* var. *rotundifolia)* under drip irrigation. *National Symp. on Improving Input Use Efficiency in Hort.*, Indian Inst. of Hort. Res., Bangalore, pp. 100-101.

Sharma, H.K., Pankaj and Lal, J. 2010. Management of *Rotylenchulus reniformis* and *Tylenchorhynchus* sp. on cowpea. National Conf. on Innovations in Nematological Research for Agricultural Sustainability – Challenges and A Roadmap Ahead. Tamil Nadu Agri. Univ., Coimbatore, p. 123.

Sharma, H.L., Singh, C.M. and Kapur, B.L. 1986. Effect of neem cake blended urea on germination and yield of potato. *Indian J. Agric. Sci.* **56**: 802-804.

Sharma, J.N., Chandran, G.S., Vij, V.K. and Monga, P.K. 1992. Effect of spacing on growth, yield and quality of kinnow mandarin. *Indian J. Hort.* **49**: 158-164.

Sharma, J.P. and Kumar, S. 2009. Management of *Ralstonia* wilt of tomato through microbes, plant extract and combination of cake and chemicals. *Indian Phytopath.* **62**: 417-423.

Sharma, R.C., Grewal, J.S., Sharma, A.K. and Sharma, H.C. 1980. Relative efficacy of calcium ammonium nitrate, dimethyl urea and urea mixed with neem cake for potato. *Indian J. Agric. Sci.* **50**: 152-157.

Shashidhar, T.R., Dharmatti, P.R., Chavan, M.L. and Mannikeri, I.M. 2006. Influence of different organic manures on growth and yield of garlic (*Allium sativum* L.). *National Symp. on Improving Input Use Efficiency in Hort.*, Indian Inst. of Hort. Res., Bangalore, pp. 161.

Sheela, M.S., Jiji, T. and Nisha, M.S. 2002. Evaluation of different control strategies for the management of nematodes associated with vegetables (brinjal) (Abstr.). *Int. Conf. Vegetables,* Bangalore, p. 268.

Shekhawat, G.S., Gadewar, A.V. and Chakravarti, S.K. 2000. *Potato Bacterial Wilt in India*. Tech. Bull. No. 38 (Revised), Central Potato Res. Inst., Shimla, 56 pp.

Shende, D.G. 1977. Effect of N, P and K on Growth, Yield and Quality of Pomegranate (*Punica granatum* L.). M. Sc. (Ag.) thesis, Mahtma Phule Agri. Univ., Rahuri, Maharashtra.

Shikamany, S.D. 2001. Fertigation techniques and economics in grape cultivation. *S. Indian Hort.* **49**: 63-68.

Shinde, P.P., More, V.G., Ramtek, J.R. and Chavan, S.A. 2002. Response of brinjal to fertigation. *J. Maharashtra Agri. Univ.* **27**: 260-262.

Shinde, S.S., Palwe, C.R., Patil, T.D. and Deshmukh, S.S. 1999. Standardization of optimum dose and time of nitrogen application for Basrai banana. *Seminar on Technological Advancement in Banana Production, Handling and Processing Management*, Jalgaon, Maharashtra, pp. 62-64.

Shiozaki, Y., Kikuchi, T., Kon, M., Fuji, A. and Mikami, M. 1988. Studies on the semi-intensive orcharding system of apple trees of *Malus prunifolia* rootstock. I. Performance of centre-leader pyramid trees as compared with open-centre trees. *Bull. of Faculty of Agri.*, Hirosoki Univ. **49**: 22-23.

Shiragur, M., Shirol, A.M., Gorabal, K., Reddy, B.S. and Kulkarni, B.S. 2004. Evaluation of standard carnation cultivars for their flowering, flower quality and yield parameters under protected conditions. *J. Orn. Hort.* **7**: 206-211.

Shirgure, P.S., Lallan Ram, Marathe, R.A. and Yadav, R.P. 1999. Effect of nitrogen fertigation on vegetative growth and leaf nutrient content of acid lime (*Citrus aurantifolia*) in central India. *Indian J. Soil Conservation* **27**:45-49.

Shirgure, P.S., Srivastava, A.K. and Singh, S. 2001. Growth, yield and quality of Nagpur mandarin (*Citrus reticulata* Blanco) in relation to irrigation and fertigation. *Indian J. Agric. Sci.* **71**: 547-550.

Shivananda, T.N., George, S., Hegde, M.R. and Srinivasappa, K.N. 2009. Micronutrient formulation in French bean var. Arka Anoop. *Int. Conf. on Hort.*, Bangalore, p. 373.

Shivaprasad, M., Mohan Kumar, H.D., Chittapur, B.M., Tatagar, M.H. and Mesta, R.K. 2009. Effect of organic and inorganic sources of nutrients on yield and economics of chilli (cv. Byadgi Dabbi) (Abstr.). *Int. Conf. on Hort.*, Bangalore, p. 132.

Shivaramu, K. and Kulkarni, K.A. 2001. Integrated pest management strategy for of *Helicoverpa armigera* (Hubner) in chilli. In *IPM in Horticultural Crops: Emerging Trends in the New Millennium* (A. Verghese and P. Parvatha Reddy, eds.), p. 64. Association for Advancement of Pest Management in Horticultural Ecosystems, Bangalore.

Shivayya, V. and Ashok Kumar, C.T. 2006. Management of melon fly, *Bactrocera cucurbitae* Coq. (Diptera: Tephritidae) using food attractants. *National Symp. on Improving Input Use Efficiency in Hort.*, Indian Inst. of Hort. Res., Bangalore, pp. 188-189.

Shrivastava, P.K., Parikh, M.M., Sawani, N.G. and Raman, S. 1994. Effect of drip irrigation and mulching on tomato yield. *Agri. Water Mangmt.* **25**: 179-184.

Shukla, A.K. and Singh, B.P. 2006. Effect of plant spacing and level of N on flowering, fruit characters and yield of papaya. *National Symp. on Improving Input Use Efficiency in Hort.*, Indian Inst. of Hort. Res., Bangalore, pp. 160.

Shukla, V. and Prabhakar, B.S. 1988. Effect of mixatoll on production potential of brinjal. *Indian J. Hort.* **43**: 403-407.

Sibi, M., 1990. Genetic basis of variation from *in vitro* tissue culture. In *Somaclonal Variation in Crop Improvement* (Y.P.S. Bajaj, ED.), pp. 112-133. Springer Verlag, Berlin.

Singh, A., Kumar, P., Yadav, A.C., Sharma, N.K. and Dhankar, B.S. 2006. Effect of irrigation intensities and dates of sowing on water use efficiency and production of radish. *National Symp. on Improving Input Use Efficiency in Hort.*, Indian Inst. of Hort. Res., Bangalore, pp. 102.

Singh, B. and Dhaulakhandi, A.B. 2002. Scope of protected cultivation of vegetables in high altitude areas. In *Hi-tech Horticulture* (K.L. Chadha, M.L. Choudhary and K.V. Prasad, eds.), pp. 181-192. Horticultural Soc. of India, New Delhi.

Singh, B., Dwivedi, S.K., Singh, N. and Paljor, E. 1998. Sustainable horticulture practices for cold arid areas. *Int. Symp. on sustainable Agri. in Hill Areas*, Himachal Pradesh Agri. Univ., Palampur, pp. 35-36.

Singh, C. 1974. Intercropping of potato in sugarcane. *Coop. Sugar* **5**: 119-120.

Singh, D.B. and Kingsly, A.R.P. 2006. Efficiency of boron application on control of fruit cracking and improvement in yield and quality of pomegranate fruits. *National Symp. on Improving Input Use Efficiency in Hort.*, Indian Inst. of Hort. Res., Bangalore, pp. 104.

Singh, D.K., Dubey, B.K. and Bhonde, S.R. 2006. Fertigation studies in onion. *National Symp. on Improving Input Use Efficiency in Hort.*, Indian Inst. of Hort. Res., Bangalore, p. 153.

Singh, G. 2006. Land-use-efficiency through meadow orcharding in guava. *National Symp. on Improving Input Use Efficiency in Hort.*, Indian Inst. of Hort. Res., Bangalore, pp. 165.

Singh, G., Kirad, K.S., Fatmi, U. and Pebam, N. 2009. Response of ashwagandha [*Withania somnifera* (L.) Dunal var. *poshita*] to organic manures and biofertilizers under Allahabad conditions (Abstr.). *Int. Conf. on Hort.*, Bangalore, p. 196.

Singh, H.P. 1985. Studies on Growth and Productivity of Acid Lime (*Citrus aurantifoloia* Swingle) cv. Kagzi Lime as Affected by Moisture Regimes. Ph. D. thesis, Univ. of Agri. Sci., Bangalore.

Singh, H.P. and Samuel, J.C. 2000. Sustainable development of horticulture with efficient use of natural resources. In *Advances in Land Resource Management* (S.P. Gawane, J.S. Bali, D.C. Das and D.K. Das, eds.). Soil Conservation Soc. of India, New Delhi, 350 pp.

Singh, H.P., Samuel, J.C. and Kumar, A. 2000. Micro-irrigation in horticultural crops. *Indian Hort.* **45**(3): 37-43.

Singh, Jarnail, Gorcha, M.S., Khanna, A.S and Singh, J. 1992. Red ripe fruit yield of hot chillies (*Capsicum annuum* L.) under the influence of etherel application. *Indian J. Plant Physiol.* **35**: 285-287.

Singh, M. and Mathur, V.C. 2008. Structural changes in horticulture sector in India. *Indian J. Agri. Econ.* **63**: 332-348.

Singh, R., Jangir, R.P. and Poonia, B.L. 1995. Evaluation of herbicides for control of weeds in chilli. *Indian. J. Agric. Sci.* **65**: 723-726.

Singh, R.K., Sharma, J. and Singh, S. 2006. Effect of irrigation and nitrogen on potato growth cracks and tuber yield. *National Symp. on Improving Input Use Efficiency in Hort.*, Indian Inst. of Hort. Res., Bangalore, p. 113.

Singh, R.V. and Gill, J.S. 1998. Important nematode problems on vegetables and their management. In *Pathological Problems of Economic Crop Plants and Their Management* (S.M. Paul Khurana, ed.), pp. 557-570. Scientific Publishers (India), Jodhpur, India.

Singh, R.V., Mathura Rai and Reddy, N.N. 1999. Effect of kharif intercropping on growth of mango cv. Maldah and additional productivity of orchard (Abstr.). *National Seminar on Sustainable Horticultural Production in Tribal Regions*, Central Hort. Expt. Stn., Ranchi, p. 81.

Singh, S.K. and Lal, S.S. 2006. Effect of integrated nutrient management in potato (*Solanum tuberosum* L.) under irrigated conditions of south Bihar. *National Symp. on Improving Input Use Efficiency in Hort.*, Indian Inst. of Hort. Res., Bangalore, p. 121.

Singh, S.P. 1995. *Commercial Fruits.* Kalyani Publishers, Ludhiana, 293 pp.

Singh, S.S., Kumar, S. and Singh, S.P. 2002. Effect of organic and inorganic fertilizers on production of tomato. In *Impact of Vegetable Research in India* (Sant Kumar, P.K. Joshi and Suresh Pal, eds.), p. 212. National Centre for Agricultural Economics and Policy Research, New Delhi.

Singh, V.K., Bhriguvanshi, S.R., Garg, N., Singh, G. and Singh, D.K. 2006. Increasing nutrient uptake efficiency, root growth and yield of mango (*Mangifera indica* L.) with mulching. *National Symp. on Improving Input Use Efficiency in Hort.*, Indian Inst. of Hort. Res., Bangalore, pp. 164.

Sinsinwar, P.S., Mathur, I.N., Singh, P. and Gupta, P.K. 2006. Effect of irrigation schedules and sulphur application on the bulb yield of garlic in Chambal command of Rajasthan. *National Symp. on Improving Input Use Efficiency in Hort.*, Indian Inst. of Hort. Res., Bangalore, p. 101.

Sivakumar, M. and Vadivelu, S. 1999. Management of *Meloidogyne incognita* on grapevine using biocontrol agents, botanicals and biofertilizers. *Pest Mangmt. Hortil. Ecosystems* **5**: 127-131.

Smith, D.R., 1997. The role of *in vitro* methods in pine plantation establishment: the lesson from New Zealand. *Plant Tissue Culture and Biotechnology* **32**: 63-73.

Soman, S. 2002. Microirrigation in onion. In *Onion and Garlic: Production – Utilization* (H.P. Singh, J.S. Mann, U.B. Pandey, Lallan Singh and S.R. Bonde, eds.), pp. 36-45. National Horticultural Research and Development Foundation, Nashik, Maharashtra.

Somasekhara, Y.M. and Ravichandra, N.G. 2010. Evaluation of bio-pesticides for the management of Meloidogyne incognita in carrot. National Conf. on Innovations in Nematological Research for Agricultural Sustainability – Challenges and A Roadmap Ahead. Tamil Nadu Agri. Univ., Coimbatore, p. 73.

Sonwalkar, N.H. and Barad, A.V. 2009. Effect of foliar application of growth regulators on rose (*Rosa hybrida* Linn.) cv. "Gladiator" under protected conditions (Abstr.). *Int. Conf. on Hort.*, Bangalore, p. 168.

Sowmya, D.S. and Rao, M.S. 2010. Investigations on the synergistic effect of Pseudomonas fluorescens strains against Meloidogyne incognita on gladiolus. National Conf. on Innovations in Nematological Research for Agricultural Sustainability – Challenges and A Roadmap Ahead. Tamil Nadu Agri. Univ., Coimbatore, pp. 99-100.

Sowmya, D.S., Rao, M.S., Gopalakrishnan, C. and Ramachandran, N. 2010. Biomanagement of *Meloidogyne incognita* and *Erwinia carotovora* s.sp. carotovora infecting carrot (*Dacus carota* L.). National Conf. on Innovations in Nematological

Research for Agricultural Sustainability – Challenges and A Roadmap Ahead, Tamil Nadu Agri. Univ., Coimbatore, p. 97.

Srihari, D. and Rao, M.M. 1996. Induction of flowering in "off" phase mango trees by soil application of paclobutrazol. *Vth Int. Mango Symp.*, Tel Aviv, Israel.

Srinivas, K. 2006. Fertigation studies in high density mango. *National Symp. on Improving Input Use Efficiency in Hort.*, Indian Inst. of Hort. Res., Bangalore, pp. 150-151.

Srinivasan, K and Krishna Moorthy, P.N. 1991. Indian mustard as a trap crop for management of major lepidopterous pests on cabbage. *Trop. Pest Mangmt.* **37**: 26-32.

Srinivasan, K and Krishna Moorthy, P.N. 1993. Evaluation of neem products and other standard chemicals for the management of major pest complex on cabbage: Comparison between standard spray regime and IPM involving mustard as a trap crop. In *Neem and Environment, Vol. I* (R.P. Singh, M.S. Chari, A.K. Raheja and W. Kraus, eds.), pp. 447-458. Oxford and IBH Publishing Co. (Pvt.) Ltd., New Delhi.

Stone, O.M., 1963. Factors effecting the growth of carnation plants from shoot apices. *Ann. Appl. Biol.* **52**: 199-209.

Subramanian, S. and Sivakumar, M. 2010. Management of Heterodera cajani in pigeon pea by non-chemical approaches. National Conf. on Innovations in Nematological Research for Agricultural Sustainability – Challenges and A Roadmap Ahead. Tamil Nadu Agri. Univ., Coimbatore, p. 93.

Suganthi, L. 2002. Fertigation Management Studies in Banana cv. Red Banana (AAA) Under Different Planting Densities. M. Sc. thesis, Tamil Nadu Agricultural University, Coimbatore-3.

Sujatha, K., Gowda, J.V.N. and Khan, M.M. 2002. Effects of different fertigation levels on gerbera under low cost greenhouse. *J. Orn. Hort.* **5**: 54-59.

Sunaina, V., Kishore, V. and Shekhawat, G.S. 2001. Impact of growth promoting *Bacillus* bioagents on potato yields in field plot experiments. *J. Indian Potato Association* **28**: 107-108.

Sundarababu, R. 1992. Nematodes in jasmine. *TNAU Newsl.* **21**(11): 2.

Sundaram, V., Kanthaswamy, V. and Selvaraj, P. 2006. Influence of water soluble fertilizers on growth and yield of bhendi. *National Symp. on Improving Input Use Efficiency in Hort.*, Indian Inst. of Hort. Res., Bangalore, p. 125.

Suta, A., Cotorobai, M. and Stan, S. 1986. The yield potential of some apple in high density planting systems. *Acta Hort.* **160**: 305-314.

Swaminathan, M.S. 2008. Marking globalization work for the poor: Technology and trade.

Swamy, K.R.M. and Sadashiva, A.T. (eds.). 2004. *Vegetable Varieties/Hybrids of IIHR*. Tech. Bull. 2, Indian Inst. of Hort. Res., Banaglore, 30 pp.

Tanksley, S.D., Medina-Filho and Rick, C.M. 1981. The effect of isozyme selection on metric characters in an interspecific backcross of tomato-basis of an early screening procedure. *Theor. Appl. Genet.* **60**: 291-296.

Thind, T.S., Kaur, S. and Thind, S.K. 2000. Punjab Agricultural University, Ludhiana, Punjab. In Progress Report for the Period Sept. 1997 - Aug. 2000 of

National Network Project on Phytophthora Diseases of Horticultural Crops (S.R. Sarma, ed.), pp. 63-78. Indian Inst. of Spices Research, Calicut, Kerala.

Thiyagarajan, G., Ranghaswami, M.V., Uma Devi, R., Rajakumar, D., Arunadevi, K. and Muralidharan, J. 2006. Water-yield relationship of tomato (*Lycopersicon esculentum* M.) under drip irrigation system. *National Symp. on Improving Input Use Efficiency in Hort.*, Indian Inst. of Hort. Res., Bangalore, p. 96.

Tirakannanavar, S., Shekharagouda, M., Merwade, M.N., Prabhuling, G. and Jagadeesha, C.C. 2006. Effect of fertilizers on sex expression and seed yield in bitter gourd (*Memordica charantia* L.). *National Symp. on Improving Input Use Efficiency in Hort.*, Indian Inst. of Hort. Res., Bangalore, p. 157.

Tribulato, E., Continella, G. and La Rosa, G. 1994. Research on higher density planting for orange and lemon. In *VII Int. Citrus Cong.*, Aureale, Italy, pp. 702-704.

Tumbare, A.D. and Bhoite, S.U. 2002. Effect of solid soluble fertilizer applied through fertigation on growth and yield of chilli. *Indian J. Agric. Sci.* **72**: 109-111.

Tumbare, A.D., Shinde, B.N. and Bhoite, S.U. 1999. Effect of liquid fertilizer through drip irrigation on growth and yield of okra. *Indian J. Agron.* **44**: 176-178.

Uma Jyothi, K., Anka Rao, A. and Vishnu Sankara Rao, D. 2004. Integrated nutrient management strategies on yield and economics of sapota. *First Indian Hort. Cong.*, p. 257.

Umamaheswari, R., Somasekhar, N. and Joseph, T.A. 2010a. Effect of organic amendments on potato cyst nematodes, *Globodera rostochiensis* and *G. pallida* infesting potato. *National Conf. on Innovations in Nematological Research for Agricultural Sustainability – Challenges and A Roadmap Ahead.* Tamil Nadu Agri. Univ., Coimbatore, p. 85.

Umamaheswari, R., Somasekhar, N. and Joseph, T.A. 2010b. Evaluation of biocontrol agents on potato cyst nematodes, *Globodera rostochiensis* and *G. pallida* infesting potato. *National Conf. on Innovations in Nematological Research for Agricultural Sustainability – Challenges and A Roadmap Ahead.* Tamil Nadu Agri. Univ., Coimbatore, p. 86.

Uthaiah, B.C., Bhandary, K.R. and Rao, M.M. 1976. Effect of pre-conditioning treatments on the rooting of air layers of sapota. *S. Indian Hort.* **24**: 135-139.

Veerannah, L., Selvaraj, P. and Alagiamanavalan, R.S. 1976. Studies on the nutrient uptake in Robusta and Poovan. *Indian J. Hort.* **33:** 203-208.

Veeraraghavan, P.G. 1972. Manuring cum liming experiment on Nendran banana. *Agric. Res. J. Kerala* **10**: 116-118.

Venkataraman, D., Raghuramulu, Y. and Shivashankar, K. 1998. Feasibilities of fertigation in coffee. In *Fertigation Studies with Water Soluble NPK Fertilizers in Crop Production* (K. Shivasankhar, M.M. Khan and A.A. Farooki, eds.), p. 13. Kremia Oy. Finland and Micro-Irrigation Soc. of India.

Venkitaswamy, R and Hameed Khan, H. 2004. Response of coconut to drip irrigation in the southern tropical low rainfall zone of India. In: *Proceedings I. Review Papers and Guest Presentations.* 75th Anniversary of Coconut Research Institute of Sri Lanka, Colombo, Sri Lanka.

Verma, A.C., Singh, H.K and Nehal Khan, M. 2005. Management of root-knot nematode, *Meloidogyne incognita* through antagonistic approaches in pointed gourd. *Indian J. Nematol.* **35**: 78-79.

Verma, L.R. and Arha. 2004. Regulation of flowering in African marigold (*Tagetes erecta* L.) by the application of $GA_3$, etherel and MH. *J. Orn. Hort.* **7**: 168-170.

Verma, L.R. and Gupta, R.P. 2009. Effect of fertigation in garlic (Abstr.). *Int. Conf. on Hort.*, Bangalore, pp. 143-144.

Verma, R.A., Singh, R.V. and Dwivedi, K. 2010. Effect of organic amendments on management of Meloidogyne incognita in sponge gourd (*Luffa cylindrica*). National Conf. on Innovations in Nematological Research for Agricultural Sustainability – Challenges and A Roadmap Ahead. Tamil Nadu Agri. Univ., Coimbatore, p. 75.

Vyas, R.V., Patel, B.A., Patel, B.N. and Patel, J.G. 2009. Integrated management of root-knot nematode in brinjal under field conditions. *Indian J. Nematol.* **39**: 35-37.

Wahid, P.A., Nambiar, P.K.N., Jose, A.I. and Rajaram, K.P. 1988. Soils and nutrition. In: *Six Decades of Coconut Research; Kerala Agricultural University*, Vellanikkara, Trichur, pp. 46-80.

Wang, P.J. 1977. Regeneration of virus-free potato from tissue culture. In *Plant Tissue Culture and its Biotechnological Application* (W. Barz, E. Reinhard and M.H. Zenk, eds.), pp. 386-391. Springer Verlag, Berlin, Hedelberg, New York.

Wang, P.J. and Charles, A. 1991. Micropropagation through meristem culture. In *Biotechnology in Agriculture and Forestry, 17, High Tech and Micropropagation I.* (Y.P.S. Bajaj, ed.), pp. 32-52. Springer-Verlag, Berlin.

Wavhal, K.N., Rane, D.A., Kumble, K.B. and Choudhary, S.M. 1985. Nutritional studies in pomegranate. *Third National Workshop on Arid Zone Fruit Research*, Mahtma Phule Agri. Univ., Rahuri, Maharashtra.

Wheatan, T.A., Castle, W.S., Whitney, J.D. and Tucker, D.H.P. 1991. Performance of citrus scion cultivars and rootstocks in high density planting. *Hort. Sci.* **26**: 837-840.

Wustenberghs and Keulemons, J. 1997. Comparison of three plum tree training systems during the first five years of growth. *Acta Hort.* **451**: 625-631.

Yadav, A.C., Singh, A., Dhankar, S.K., Kamboj, A.P. and Mehta, C.P. 2006. Yield performance of okra varieties under various plant spacings and fertility levels. *National Symp. on Improving Input Use Efficiency in Hort.*, Indian Inst. of Hort. Res., Bangalore, p. 132.

Yadav, S.S. and Yadav, N. 2006. Comparative study of drip and check basin irrigation in tomato. *National Symp. on Improving Input Use Efficiency in Hort.*, Indian Inst. of Hort. Res., Bangalore, pp. 106.

Yassin, G.M. and Kanthaswamy, V. 2009. Effect of water soluble fertilizers on growth and yield of brinjal (Abstr.). *Int. Conf. on Hort.*, Bangalore, p. 131.

Yeledahalli, N.A., Prakash, S.S. and Nagaraja, H. 2006. Effect of soil waste amendification on growth and yield of okra [*Abelmoschus esculentus* (L.) Moench] and some soil properties. *National Symp. on Improving Input Use Efficiency in Hort.*, Indian Inst. of Hort. Res., Bangalore, p. 130.

Ziv, M., 1990. Morphogenesis of gladiolus buds in bioreactors. Implication for scaled-up propagation of geophytes. In *Progress in Plant Cellular and Molecular Biology*. (H.J.J. Nijkamp, L.H.W. Van der Plas and J. Van Aartrijk, eds.), pp. 119-124. Kluwer Academic Publishers, Dordrecht.

**APPENDIX**

# ACRONYMS

| Name of University/ Institute | Expansion |
|---|---|
| AAU | Assam Agricultural University |
| AC&RI | Agricultural College & Research Institute |
| AES | Agricultural Experiment Station |
| AICRP | All India Coordinated Research Project |
| AINRP on M&AP | All India Network Research Project on Medicinal & Aromatic Plants |
| ANGRAU | Acharya NG Ranga Agricultural University |
| ARS | Agricultural Research Station |
| ARU | Agricultural Research Unit |
| BAC | Bihar Agricultural College |
| BAU | Birsa Agricultural University |
| BCKV | Bidhan Chandra Krishi Viswa Vidhyalaya |
| CARI | Central Agricultural Research Institute |
| CCSHAU | Chaudhary Charan Singh Haryana Agricultural University |
| CHES | Central Horticultural Experiment Station |
| CIAH | Central Institute for Arid Horticulture |
| CISH | Central Institute for Subtropical Horticulture |
| CPCRI | Central Plantation Crops Research Institute |
| CPRI | Central Potato Research Institute |
| CRC | Cardamom Research Centre |
| CRS | Coconut Research Station |
| CSAUAT | Chandra Sekhar Azad University of Agriculture & Technology |
| CTCRI | Central Tuber Crops Research Institute |
| DARL | Defence Agricultural Research Laboratory |
| DBSKKV | Dr. Balasaheb Sawant Konkan Krishi Vidhyapeeth |

| | |
|---|---|
| FRS | Fruit Research Station |
| GAU | Gujarat Agricultural University |
| GBPUAT | Govind Bhallabh Pant University of Agriculture & Technology |
| HARP | Horticulture & Agroforestry Research Programme |
| HAU | Haryana Agricultural University |
| HC&RI | Horticulture College & Research Institute |
| HPAU | Himachal Pradesh Agricultural University |
| HPKV | Himachal Pradesh Krishi Viswa Vidhyalaya |
| HRS | Horticultural Research Station |
| IARI | Indian Agricultural Research Institute |
| ICAR RC NEH Region | ICAR Research Complex for NEH Region |
| ICRI | Indian Cardamom Research Institute |
| IGAU | Indira Gandhi Agricultural University |
| IGKV | Indira Gandhi Krishi Viswa Vidhyalaya |
| IIHR | Indian Institute of Horticultural Research |
| IISR | Indian Institute of Spices Research |
| IIVR | Indian Institute of Vegetable Research |
| JNKVV | Jawaharlal Nehru Krishi Viswa Vidhyalaya |
| KAU | Kerala Agricultural University |
| KKV | Konkan Krishi Vidhyapeeth |
| MPKV | Mahtma Phule Krishi Vidhyapeeth |
| NAU | Navsari Agricultural University |
| NBPGR | National Bureau of Plant Genetic Resources |
| NDUAT | Narendra Deva University of Agriculture & Technology |
| NHRDF | National Horticultural Research & Development Foundation |
| NRC | National Research Centre |
| NRCOG | National Research Centre for Onion & Garlic |
| OUAT | Orissa University of Agriculture & Technology |
| PAU | Punjab Agricultural University |
| PDKV | Dr. Panjabrao Deshmukh Krishi Vidhyapeeth |
| RARS | Regional Agricultural Research Station |
| RAU | Rajendra Agricultural University |
| RS | Regional Station |
| SKN | Shri Karan Narendra College of Agriculture |
| SKUA&T | Sher-e-Kashnir University of Agriculture & Technology |
| SKUAS&T | Sher-e-Kashnir University of Agricultural Sciences & Technology |
| SKUAST(K) | Sher-e-Kashnir University of Agricultural Sciences & Technology, Kashmir |

| | |
|---|---|
| SVBPUA&T | Sardar Vallabha Bhai Patel University of Agriculture & Technology |
| TNAU | Tamil Nadu Agricultural University |
| TNAU RS | Tamil Nadu Agricultural University Research Station |
| UAS | University of Agricultural Sciences |
| UBKVV | Uttar Banga Krishi Viswa Vidhyalaya |
| VPKAS | Vivekananda Parvatiya Krishi Anusandhan Sansthan |
| VRS | Vegetable Research Station |
| YSPUHF | Dr. Y.S. Parmar University of Horticulture & Forestry |

# SUBJECT INDEX